BioMEMS and Biomedical Nanotechnology

Volume IV
Biomolecular Sensing, Processing and Analysis

BioMEMS and Biomedical Nanotechnology

Mauro Ferrari, Ph.D., Editor-in-Chief
Professor, Brown Institute of Molecular Medicine Chairman
Department of Biomedical Engineering
University of Texas Health Science Center, Houston, TX

Professor of Experimental Therapeutics
University of Texas M.D. Anderson Cancer Center, Houston, TX

Professor of Bioengineering
Rice University, Houston, TX

Professor of Biochemistry and Molecular Biology
University of Texas Medical Branch, Galveston, TX

President, the Texas Alliance for NanoHealth
Houston, TX

Volume IV
Biomolecular Sensing, Processing and Analysis

Edited by

Rashid Bashir and Steve Wereley
Purdue University, West Lafayette, IN

Rashid Bashir
Purdue University
West Lafayette, Indiana

Steve Wereley
Purdue University
West Lafayette, Indiana

Mauro Ferrari
Ohio State University
Columbus, Ohio

Library of Congress Cataloging-in-Publication Data

Volume IV
ISBN-10: 0-387-25566-4   e-ISBN 10: 0-387-25845-0   Printed on acid-free paper.
ISBN-13: 978-0387-25566-8   e-ISBN-13: 978-0387-25845-4
Set
ISBN-10: 0-387-25661-3   e-ISBN:10: 0-387-25749-7
ISBN-13: 978-0387-25561-3   e-ISBN:13: 978-0387-25749-5

© 2006 Springer Science+Business Media LLC
All rights reserved. This work may not be translated or copied in whole or in part without the written permission of the publisher (Springer Science+Business Media LLC, 233 Spring Street, New York, NY 10013, USA), except for brief excerpts in connection with reviews or scholarly analysis. Use in connection with any form of information storage and retrieval, electronic adaptation, computer software, or by similar or dissimilar methodology now known or hereafter developed is forbidden.
The use in this publication of trade names, trademarks, service marks and similar terms, even if they are not identified as such, is not to be taken as an expression of opinion as to whether or not they are subject to proprietary rights.

9 8 7 6 5 4 3 2 1   SPIN 11408253

springer.com

Dedicated to Richard Smalley (1943–2005), in Memoriam

To Rick,

father founder of nanotechnology
prime inspiration for its applications to medicine
gracious mentor to its researchers
our light—forever in the trenches with us

(Rick Smalley received the 1996 Chemistry Nobel Prize
for the co-discovery of carbon-60 buckeyballs)

Contents

List of Contributors... xv

Foreword... xix

Preface.. xxi

I. Micro and Nanoscale Biosensors and Materials............................. 1

1. Biosensors and Biochips.. 3
Tuan Vo-Dinh
1.1 Introduction 3
1.2 Biosensors... 5
    1.2.1 Different Types of Bioreceptors................................... 5
    1.2.2 Types of Transducers.. 11
1.3 Biochips... 14
    1.3.1 Microarray Systems.. 14
    1.3.2 Integrated Biochip Systems.. 16
1.4 Conclusion... 18
    Acknowledgements.. 18
    References.. 19

2. Cantilever Arrays... 21
Min Yue, Arun Majumdar, and Thomas Thundat
2.1 Introduction 21
2.2 Theory... 22
2.3 Readout Techniques... 24
    2.3.1 Optical Beam Deflection of 1D Cantilever Array.................... 24
    2.3.2 Optical Beam Deflection of 2D Array............................... 24
    2.3.3 Piezoresistive Cantilever Array................................... 27
2.4 Microfluidics.. 27
2.5 Biomolecular Reaction Assays... 28
    2.5.1 Detection of DNA.. 29
    2.5.2 Detection of PSA.. 30
2.6 Conclusions.. 31
    References.. 32

3. An On-Chip Artificial Pore for Molecular Sensing ... 35
O. A. Saleh and L. L. Sohn
- 3.1 Introduction ... 35
- 3.2 The Basic Device: Fabrication and Measurement ... 35
  - 3.2.1 Fabrication of the Pore ... 36
  - 3.2.2 Pore Measurement ... 37
  - 3.2.3 PDMS-Based Pore ... 40
- 3.3 Applications ... 44
  - 3.3.1 An All-Electronic Immunoassay ... 44
  - 3.3.2 Summary ... 50
  - 3.3.3 Single Molecule Detection ... 51
- 3.4 Conclusions ... 52
- References ... 52

4. Cell Based Sensing Technologies ... 55
Cengiz S. Ozkan, Mihri Ozkan, Mo Yang, Xuan Zhang, Shalini Prasad, and Andre Morgan
- 4.1 Overview ... 55
- 4.2 Cell-Based Biosensors ... 56
  - 4.2.1 Cellular Microorganism Based Biosensors ... 57
  - 4.2.2 Fluorescence Based Cellular Biosensors ... 58
  - 4.2.3 Impedance Based Cellular Biosensors ... 58
  - 4.2.4 Intracellular Potential Based Biosensors ... 59
  - 4.2.5 Extracellular Potential Based Biosensors ... 61
- 4.3 Design and Methods ... 63
  - 4.3.1 Requirements for Cell Based Sensors ... 63
  - 4.3.2 Cell Manipulation Techniques ... 64
  - 4.3.3 Principles of Dielectrophoresis (DEP) ... 64
  - 4.3.4 Cell Manipulation Using Dielectrophoresis (DEP) ... 66
  - 4.3.5 Cell Types and Parameters for Dielectrophoretic Patterning ... 67
  - 4.3.6 Biosensing System ... 67
  - 4.3.7 Cell Culture ... 68
  - 4.3.8 Experimental Measurement System ... 69
- 4.4 Measurements ... 69
  - 4.4.1 Long Term Signal Recording *in vivo* ... 69
  - 4.4.2 Interpretation of Bioelectric Noise ... 73
  - 4.4.3 Influence of Geometry and Environmental Factors on the Noise Spectrum ... 74
  - 4.4.4 Signal Processing ... 76
  - 4.4.5 Selection of Chemical Agents ... 76
  - 4.4.6 Control Experiment ... 79
  - 4.4.7 Chemical Agent Sensing ... 79
- 4.5 Discussion and Conclusion ... 87
- References ... 89

5. Fabrication Issues of Biomedical Micro Devices ... 93
Nam-Trung Nguyen
- 5.1 Introduction ... 93
- 5.2 Materials for Biomedical Micro Devices ... 94
  - 5.2.1 Silicon and Glass ... 94
  - 5.2.2 Polymers ... 95
- 5.3 Polymeric Micromachining Technologies ... 97
  - 5.3.1 Lithography ... 97
  - 5.3.2 Polymeric Surface Micromachining ... 100
  - 5.3.3 Replication Technologies ... 104
  - 5.3.4 Laser Machining ... 106
- 5.4 Packaging of Biomedical Micro Devices ... 109
  - 5.4.1 Thermal Direct Bonding ... 109
  - 5.4.2 Adhesive Bonding ... 110
  - 5.4.3 Interconnects ... 110
- 5.5 Biocompatibility of Materials and Processes ... 113
  - 5.5.1 Material Response ... 113
  - 5.5.2 Tissue and Cellular Response ... 113
  - 5.5.3 Biocompatibility Tests ... 113
- 5.6 Conclusions ... 114
- References ... 114

6. Intelligent Polymeric Networks in Biomolecular Sensing ... 117
Nicholas A. Peppas and J. Zachary Hilt
- 6.1 Intelligent Polymer Networks ... 118
  - 6.1.1 Hydrogels ... 119
  - 6.1.2 Environmentally Responsive Hydrogels ... 121
  - 6.1.3 Temperature-Sensitive Hydrogels ... 121
  - 6.1.4 pH-Responsive Hydrogels ... 122
  - 6.1.5 Biohybrid Hydrogels ... 123
  - 6.1.6 Biomolecular Imprinted Polymers ... 124
  - 6.1.7 Star Polymer Hydrogels ... 125
- 6.2 Applications of Intelligent Polymer Networks as Recognition Elements ... 126
  - 6.2.1 Sensor Applications: Intelligent Polymer Networks as Recognition Matrices ... 127
  - 6.2.2 Sensor Applications: Intelligent Polymer Networks as Actuation Elements ... 128
- 6.3 Conclusions ... 129
- References ... 129

II. Processing and Integrated Systems ... 133

7. A Multi-Functional Micro Total Analysis System (µTAS) Platform ... 135
Abraham P. Lee, John Collins, and Asuncion V. Lemoff
- 7.1 Introduction ... 135

7.2 MHD Micropump for Sample Transport Using Microchannel
    Parallel Electrodes.. 136
    7.2.1 Principle of Operation 136
    7.2.2 Fabrication of Silicon MHD Microfluidic Pumps 138
    7.2.3 Measurement Setup and Results 139
    7.2.4 MHD Microfluidic Switch 142
    7.2.5 Other MHD Micropumps and Future Work 144
7.3 Microchannel Parallel Electrodes for Sensing Biological Fluids 145
    7.3.1 MHD Based Flow Sensing 145
    7.3.2 MHD Based Viscosity Meter 146
    7.3.3 Impedance Sensors with MicroChannel Parallel Electrodes 146
7.4 Parallel Microchannel Electrodes for Sample Preparation................. 153
    7.4.1 A Microfluidic Electrostatic DNA Extractor 153
    7.4.2 Channel Electrodes for Isoelectric Focusing Combined with
          Field Flow Fractionation... 155
7.5 Summary... 156
    References... 157

8. Dielectrophoretic Traps for Cell Manipulation 159
Joel Voldman
8.1 Introduction 159
8.2 Trapping Physics 160
    8.2.1 Fundamentals of Trap Design 160
    8.2.2 Dielectrophoresis 162
    8.2.3 Other Forces.. 169
8.3 Design for Use with Cells... 172
8.4 Trap Geometries 175
    8.4.1 n-DEP Trap Geometries 175
    8.4.2 p-DEP Trap Geometries 179
    8.4.3 Lessons for DEP Trap Design 179
8.5 Quantitating Trap Characteristics 180
8.6 Conclusions 182
8.7 Acknowledgements... 182
    References... 183

9. BioMEMS for Cellular Manipulation and Analysis 187
Haibo Li, Rafael Gómez-Sjöberg, and Rashid Bashir
9.1 Introduction 187
9.2 BioMEMS for Cellular Manipulation and Separation.................... 188
    9.2.1 Electrophoresis 189
    9.2.2 Dielectrophoresis 190
9.3 BioMEMS for Cellular Detection 193
    9.3.1 Optical Detection 194
    9.3.2 Mechanical Detection 196
    9.3.3 Electrical Detection 197

CONTENTS

    9.4 Conclusions and Future Directions 200
          Acknowledgements 201
          References 201

10. Implantable Wireless Microsystems 205
Babak Ziaie

   10.1 Introduction 205
   10.2 Microsystem Components 206
         10.2.1 Transducers 206
         10.2.2 Interface Electronics 207
         10.2.3 Wireless Communication 207
         10.2.4 Power Source 208
         10.2.5 Packaging and Encapsulation 209
   10.3 Diagnostic Microsystems 210
   10.4 Therapeutic Microsystems 213
   10.5 Rehabilitative Microsystems 216
   10.6 Conclusions and Future Directions 219
         References 220

11. Microfluidic Tectonics 223
J. Aura Gimm and David J. Beebe

   11.1 Introduction 223
   11.2 Traditional Manufacturing Methods 224
         11.2.1 Micromachining 224
         11.2.2 Micromolding 224
   11.3 Polymeric μFluidic Manufacturing Methods 225
         11.3.1 Soft Lithography 225
         11.3.2 Other Methods 226
         11.3.3 Liquid Phase Photopolymerization—Microfluidic
                   Tectonics (μFT) 226
         11.3.4 Systems Design 227
         11.3.5 Valves 228
         11.3.6 Pumps 229
         11.3.7 Filters 230
         11.3.8 Compartmentalization: "Virtual Walls" 231
         11.3.9 Mixers 232
         11.3.10 Hydrogel as Sensors 234
         11.3.11 Sensors That Change Shape 234
         11.3.12 Sensors That Change Color 235
         11.3.13 Cell-gel Sensors 236
         11.3.14 Liposome Sensor 237
         11.3.15 E-gel 237
   11.4 Concluding Remarks 240
         References 240

12. AC Electrokinetic Stirring and Focusing of Nanoparticles 243
Marin Sigurdson, Dong-Eui Chang, Idan Tuval, Igor Mezic, and Carl Meinhart
- 12.1 Introduction 243
- 12.2 AC Electrokinetic Phenomena 244
- 12.3 DEP: A System Theory Approach 244
- 12.4 Non-Local DEP Trapping 247
- 12.5 Electrothermal Stirring 249
- 12.6 Enhancement of Heterogeneous Reactions 251
- 12.7 Conclusions 253
  - Acknowledgements 254
  - References 254

III. Micro-fluidics and Characterization 257

13. Particle Dynamics in a Dielectrophoretic Microdevice 259
S.T. Wereley and I. Whitacre
- 13.1 Introduction and Set up 259
  - 13.1.1 DEP Device 259
  - 13.1.2 Dielectrophoresis Background 260
  - 13.1.3 Micro Particle Image Velocimetry 261
- 13.2 Modeling/Theory 262
  - 13.2.1 Deconvolution Method 262
  - 13.2.2 Synthetic Image Method 263
  - 13.2.3 Comparison of Techniques 264
- 13.3 Experimental Results 266
- 13.4 Conclusions 275
  - Acknowledgements 275
  - References 275

14. Microscale Flow and Transport Simulation for Electrokinetic and Lab-on-Chip Applications 277
David Erickson and Dongqing Li
- 14.1 Introduction 277
- 14.2 Microscale Flow and Transport Simulation 278
  - 14.2.1 Microscale Flow Analysis 278
  - 14.2.2 Electrical Double Layer (EDL) 280
  - 14.2.3 Applied Electrical Field 282
  - 14.2.4 Electrokinetic Microtransport Analysis 284
- 14.3 Numerical Challenges Due to Length Scales and Resulting Simplification 285
- 14.4 Case Study I: Enhanced Species Mixing Using Heterogeneous Patches 286
  - 14.4.1 Flow Simulation 288
  - 14.4.2 Mixing Simulation 289
- 14.5 Case Study II: AC Electroosmotic Flows in a Rectangular Microchannel 290
  - 14.5.1 Flow Simulation 291
- 14.6 Case Study III: Pressure Driven Flow over Heterogeneous Surfaces for Electrokinetic Characterization 293

	14.6.1 System Geometry, Basic Assumptions and Modeling Details	293
	14.6.2 Flow Simulation	295
	14.6.3 Double Layer Simulation	295
14.7	Summary and Outlook	298
	References	298

15. Modeling Electroosmotic Flow in Nanochannels ... 301
A. T. Conlisk and Sherwin Singer

15.1	Introduction	301
15.2	Background	304
	15.2.1 Micro/Nanochannel Systems	304
	15.2.2 Previous Work on Electroosmotic Flow	304
	15.2.3 Structure of the Electric Double Layer	306
15.3	Governing Equations for Electrokinetic Flow	308
15.4	Fully Developed Electroosmotic Channel Flow	314
	15.4.1 Asymptotic and Numerical Solutions for Arbitrary Electric Double Layer Thickness	314
	15.4.2 Equilibrium Considerations	318
15.5	Comparison of Continuum Models with Experiment	320
15.6	Molecular Dynamics Simulations	324
	15.6.1 Introduction	324
	15.6.2 Statics: the Charge Distribution in a Nanochannel	325
	15.6.3 Fluid Dynamics in Nanochannels	326
15.7	Summary	328
	Acknowledgements	329
	References	329

16. Nano-Particle Image Velocimetry ... 331
Minami Yoda

16.1	Introduction	331
16.2	Diagnostic Techniques in Microfluidics	332
16.3	Nano-Particle Image Velocimetry Background	334
	16.3.1 Theory of Evanescent Waves	334
	16.3.2 Generation of Evanescent Waves	337
	16.3.3 Brownian Diffusion Effects in nPIV	338
16.4	Nano-PIV Results in Electroosmotic Flow	342
	16.4.1 Experimental Details	342
	16.4.2 Results and Discussion	343
16.5	Summary	346
	Acknowledgements	347
	References	348

17. Optical MEMS-Based Sensor Development with Applications to Microfluidics ... 349
D. Fourguette, E. Arik, and D. Wilson

17.1	Introduction	349
17.2	Challenges Associated with Optical Diagnostics in Microfluidics	350

17.3 Enabling Technology for Microsensors: Computer Generated
     Hologram Diffractive Optical Elements 350
17.4 The Miniature Laser Doppler Velocimeter 352
     17.4.1 Principle of Measurement 353
     17.4.2 Signal Processing 354
     17.4.3 A Laser Doppler Velocimeter for Microfluidics 355
17.5 Micro Sensor Development 356
     17.5.1 Micro Velocimeter 356
     17.5.2 Micro Sensors for Flow Shear Stress Measurements 358
17.6 Application to Microfluidics: Velocity Measurements in Microchannels . 363
     17.6.1 Velocity Measurements in Polymer Microchannels 363
     17.6.2 Test on a Caliper Life Science Microfluidic Chip using
            the MicroV 365
17.7 Conclusions 369
     Bibliography 369

18. Vascular Cell Responses to Fluid Shear Stress 371
Jennifer A. McCann, Thomas J. Webster, and Karen M. Haberstroh
     Abstract 371
18.1 Introduction 371
     18.1.1 Vessel Physiology 372
     18.1.2 Vessel Pathology 374
18.2 Hemodynamics of Blood Flow 375
18.3 Techniques for Studying the Effects of Shear Stress on Cell Cultures 378
     18.3.1 Cone and Plate Viscometer 378
     18.3.2 Parallel Plate Flow Chamber 379
18.4 Modifications to Traditional Flow Chambers 382
18.5 Nontraditional Flow Devices 382
18.6 Laminar Shear Stress Effects on Endothelial Cells 383
18.7 Endothelial Cell Response to Altered Flows 385
18.8 Laminar Shear Stress Effects on Vascular Smooth Muscle Cells 386
18.9 Mechanotransduction 388
     18.9.1 Shear Stress Receptors 388
18.10 Transduction Pathways 389
     18.10.1 Ras-MAPK Pathways 389
     18.10.2 IKK-NF·κ B Pathway 391
18.11 Applications to Clinical Treatment 391
18.12 Summary 391
      References 392

About the Editors 395
Index 397

List of Contributors

VOLUME IV

E. Arik, VioSense Corporation, Pasadena, California USA

Rashid Bashir, Birck Nanotechnology Center and Bindley Biosciences Center, Discovery Park, School of Electrical and Computer Engineering, Weldon School of Biomedical Engineering, Purdue University, West Lafayette, Indiana USA

David J. Beebe, Dept. of Biomedical Engineering, University of Wisconsin-Madison, Madison, Wisconsin USA

Dong-Eui Chang, Dept. of Mechanical Engineering, University of California, Santa Barbara, Santa Barbara, California USA

John Collins, Dept. of Biomedical Engineering, University of California, Irvine, Irvine, California USA

A.T. Conlisk, Dept. of Mechanical Engineering, The Ohio State University, Columbus, Ohio USA

David Erickson, Sibley School of Mechanical and Aerospace Engineering, Cornell University Ithaca, NY

Mauro Ferrari, Ph.D., Professor, Brown Institute of Molecular Medicine Chairman, Department of Biomedical Engineering, University of Texas Health Science Center, Houston, TX; Professor of Experimental Therapeutics, University of Texas M.D. Anderson Cancer Center, Houston, TX; Professor of Bioengineering, Rice University, Houston, TX; Professor of Biochemistry and Molecular Biology, University of Texas Medical Branch, Galveston, TX; President, the Texas Alliance for NanoHealth, Houston, TX

D. Fourguette, VioSense Corporation, Pasadena, California USA

J. Aura Gimm, Dept. of Biomedical Engineering, University of Wisconsin-Madison, Madison, Wisconsin USA

Rafael Gómez-Sjöberg, Birck Nanotechnology Center and Bindley Biosciences Center, Discovery Park, School of Electrical and Computer Engineering, Weldon School of Biomedical Engineering, Purdue University, West Lafayette, Indiana USA

Karen M. Haberstroh, Weldon School of Biomedical Engineering, Purdue University, West Lafayette, Indiana USA

J. Zachary Hilt, Dept. of Chemical and Materials Engineering, The University of Kentucky, Lexington, Kentucky USA

Abraham P. Lee, Dept. of Biomedical Engineering, Mechanical & Aerospace Engineering, University of California, Irvine, Irvine, California USA

Asuncion V. Lemoff, Biotechnology Consultant, Union City, California USA

Dongqing Li, Department of Mechanical Engineering, Vanderbilt University Nashville, TN

Haibo Li, Birck Nanotechnology Center and Bindley Biosciences Center, Discovery Park, School of Electrical and Computer Engineering, Weldon School of Biomedical Engineering, Purdue University, West Lafayette, Indiana USA

Arun Majumdar, Dept. of Mechanical Engineering, University of California, Berkeley, Berkeley, California USA

Jennifer A. McCann, Weldon School of Biomedical Engineering, Purdue University, West Lafayette, Indiana USA

Carl Meinhart, Dept. of Mechanical Engineering, University of California, Santa Barbara, Santa Barbara, California USA

Igor Mezic, Dept. of Mechanical Engineering, University of California, Santa Barbara, Santa Barbara, California USA

Andre Morgan, Mechanical Engineering Dept., University of California, Riverside, California USA

Nam-Trung Nguyen, School of Mechanical and Production Engineering, Nanyang Technological University, Singapore, China

Cengiz S. Ozkan, Mechanical Engineering Dept., University of California, Riverside, California USA

Mihri Ozkan, Mechanical Engineering Dept., University of California, Riverside, California USA

Nicholas A. Peppas, Dept. of Chemical Engineering, Biomedical Engineering, Pharmaceutics, The University of Texas, Austin, Texas USA

LIST OF CONTRIBUTORS

Shalini Prasad, Mechanical Engineering Dept., University of California, Riverside, California USA

O.A. Saleh, Laboratoire de Physique Statistique, Ecole Normale Supérieure, Paris, France

Marin Sigurdson, Dept. of Mechanical Engineering, University of California, Santa Barbara, Santa Barbara, California USA

Sherwin Singer, Chemistry Department, The Ohio State University, Columbus, Ohio USA

L.L. Sohn, Dept. of Mechanical Engineering, University of California, Berkeley, Berkeley, California USA

Thomas Thundat, Life Sciences Division, Oak Ridge National Laboratory, Oak Ridge, Tennessee USA

Idan Tuval, Dept. of Mechanical Engineering, University of California, Santa Barbara, Santa Barbara, California USA

Tuan Vo-Dinh, Center for Advanced Biomedical Photonics, Oak Ridge National Laboratory, Oak Ridge, Tennessee USA

Joel Voldman, Dept. of Electrical Engineering, Massachusetts Institute of Technology, Cambridge, Massachusetts USA

Thomas J. Webster, Weldon School of Biomedical Engineering, Purdue University, West Lafayette, Indiana USA

S.T. Wereley, School of Mechanical Engineering, Purdue University, West Lafayette, Indiana USA

I. Whitacre, School of Mechanical Engineering, Purdue University, West Lafayette, Indiana USA

D. Wilson, Jet Propulsion Laboratory, Pasadena, California USA

Mo Yang, Mechanical Engineering Dept., University of California, Riverside, California USA

Minami Yoda, G.W. Woodruff School of Mechanical Engineering, Georgia Institute of Technology, Atlanta, Georgia USA

Min Yue, Dept. of Mechanical Engineering, University of California, Berkeley, Berkeley, California USA

Xuan Zhang, Mechanical Engineering Dept., University of California, Riverside, California USA

Babak Ziaie, School of Electrical and Computer Engineering, Purdue University, West Lafayette, Indiana USA

Foreword

Less than twenty years ago photolithography and medicine were total strangers to one another. They had not yet met, and not even looking each other up in the classifieds. And then, nucleic acid chips, microfluidics and microarrays entered the scene, and rapidly these strangers became indispensable partners in biomedicine.

As recently as ten years ago the notion of applying nanotechnology to the fight against disease was dominantly the province of the fiction writers. Thoughts of nanoparticle-vehicled delivery of therapeuticals to diseased sites were an exercise in scientific solitude, and grounds for questioning one's ability to think "like an established scientist". And today we have nanoparticulate paclitaxel as the prime option against metastatic breast cancer, proteomic profiling diagnostic tools based on target surface nanotexturing, nanoparticle contrast agents for all radiological modalities, nanotechnologies embedded in high-distribution laboratory equipment, and no less than 152 novel nanomedical entities in the regulatory pipeline in the US alone.

This is a transforming impact, by any measure, with clear evidence of further acceleration, supported by very vigorous investments by the public and private sectors throughout the world. Even joining the dots in a most conservative, linear fashion, it is easy to envision scenarios of personalized medicine such as the following:

- patient-specific prevention supplanting gross, faceless intervention strategies;
- early detection protocols identifying signs of developing disease at the time when the disease is most easily subdued;
- personally tailored intervention strategies that are so routinely and inexpensively realized, that access to them can be secured by everyone;
- technologies allowing for long lives in the company of disease, as good neighbors, without impairment of the quality of life itself.

These visions will become reality. The contributions from the worlds of small-scale technologies are required to realize them. Invaluable progress towards them was recorded by the very scientists that have joined forces to accomplish the effort presented in this 4-volume collection. It has been a great privilege for me to be at their service, and at the service of the readership, in aiding with its assembly. May I take this opportunity to express my gratitude to all of the contributing Chapter Authors, for their inspired and thorough work. For many of them, writing about the history of their specialty fields of *BioMEMS and Biomedical Nanotechnology* has really been reporting about their personal, individual adventures through scientific discovery and innovation—a sort of family album, with equations, diagrams, bibliographies and charts replacing Holiday pictures....

It has been a particular privilege to work with our Volume Editors: Sangeeta Bhatia, Rashid Bashir, Tejal Desai, Michael Heller, Abraham Lee, Jim Lee, Mihri Ozkan, and Steve Werely. They have been nothing short of outstanding in their dedication, scientific vision, and generosity. My gratitude goes to our Publisher, and in particular to Greg Franklin for his constant support and leadership, and to Angela De Pina for her assistance.

Most importantly, I wish to express my public gratitude in these pages to Paola, for her leadership, professional assistance throughout this effort, her support and her patience. To her, and our children Giacomo, Chiara, Kim, Ilaria and Federica, I dedicate my contribution to BioMEMS and Biomedical Nanotechnology.

With my very best wishes

Mauro Ferrari, Ph.D.
Professor, Brown Institute of Molecular Medicine Chairman
Department of Biomedical Engineering
University of Texas Health Science Center, Houston, TX

Professor of Experimental Therapeutics
University of Texas M.D. Anderson Cancer Center, Houston, TX

Professor of Bioengineering
Rice University, Houston, TX

Professor of Biochemistry and Molecular Biology
University of Texas Medical Branch, Galveston, TX

President, the Texas Alliance for NanoHealth
Houston, TX

Preface

BioMEMS and its extensions into biomedical nanotechnology have tremendous potential both from a research and applications point of view. Exciting strides are being made at intersection of disciplines and BioMEMS and biomedical nanotechnology is certainly one of these very interdisciplinary fields, providing many opportunities of contribution from researchers from many disciplines. In the specific areas of bimolecular sensing, processing and analysis, BioMEMS can play a critical role to provide the various technology platforms for detection of cells, microorganisms, viruses, proteins, DNA, small molecules, etc. and the means to interface the macroscale realm to the nanoscale realm.

We are very pleased to present volume 4 in the Handbook of BioMEMS and Biomedical Nanotechnology, published by Kluwer Academic Press. This volume contains 18 chapters focused on 'Biomolecular Sensing, Processing and Analysis', written by experts in the field of BioMEMS and biomedical nanotechnology. The chapters are groups into three broad categories of Sensors and Materials, Processing and Integrated Systems, and Microfluidics.

Prof. Taun Vo-Dinh from Oakridge National Labs begins the Sensors and Materials section by providing a review of biosensors and biochips. This review is followed by an example of mechanical cantilever sensor work described by Prof. Arun Majumdar's group at UC Berkeley and Prof. Tom Thundat at Oakridge National Laboratory. An example of a nano-scale sensor electrical sensor, an artificial pore, integrated in a microscale device is presented next by Prof. Lydia Sohn's group at UC Berkeley. Cell based sensors are an important class of electrical sensors and Profs. Cengiz Ozkan and Mihri Ozkan at UC Riverside present a review of their work in this area. These chapters on sensors are followed by a review chapter on silicon and glass BioMEMS processing by Prof. Nam Trung Nguyen at Nanyang Technological University. Polymers and hydrogels are an important class of bioMEMS materials and Profs. Nicholas Peppas at UT Austin and Zach Hilt at University of Kentucky provide a review chapter in this area to close off this section.

The Processing and Integrated Systems section is focused on means to manipulate biological and fluidic samples in BioMEMS device and examples of integrated BioMEMS systems. Prof. Abe Lee from UC Irvine presents a review of magnetohydrodynamic methods and their utility in BioMEMS and micro-total-analysis (μTAS) systems. Dielectrophoresis (DEP) is being increasing used at the microscale and in BioMEMS applications and Prof. Joel Voldman from MIT provides a review of DEP and applications, especially for cellular analysis and manipulation. Prof. Rashid Bashir and his group from Purdue present an overview of BioMEMS sensors and devices for cellular sensing, detection and manipulation. Microsystems and BioMEMS integrated with wireless and RF devices for in-vivo applications is a growing field and Prof. Babak Zaiae, previously of University of Minnesota, and now at Purdue, presents an overview of this area. As reviewed in the first section, polymers and hydrogels are a very important class of BioMEMS materials and Prof. David Beebe from University of Wisconsin presents an overview of the work in his group on polymer based self-sensing and actuating microfluidic systems. Lastly, mixing and stirring of fluids is an important problem to be addressed at the microscale due to the fact that Reynold's numbers are small, flows are laminar, and it is challenge to create mixing. Prof. Meinhart and colleagues at UC Santa Barbara present the use of AC electrokinetic methods, including DEP, for mixing of fluids in BioMEMS devices.

The Microfluidics section describes work in a very important supporting field for BioMEMS—microfluidics. Since nearly all life processes occur in or with the help of water, microfluidics is a key technology necessary in miniaturizing biological sensing and processing applications. This section starts off with a contribution by Prof. Steve Wereley's group at Purdue University quantitatively exploring how DEP influences particle motion and proposing a new experimental technique for measuring this influence. Prof. David Erickson from Cornell University and Prof. Dongqing Li from Vanderbilt University have contributed an article reviewing emerging computational methods for simulating flows in microdevices. Prof. Terry Conlisk and Prof. Sherwin Singer's (both of Ohio State University) contribution focused exclusively on modeling electroosmotic flow in nanochannels—a challenging domain where Debye length is comparable to channel dimension. This is followed by a contribution from Prof. Minami Yoda at Georgia Tech describing a new version of the versatile micro-Particle Image Velocimetry technique demonstrating spatial resolutions smaller than 1 micron, a requirement for making measurements in nanochannels. Viosense Corporation, led by Dominique Fourguette, has contributed an article on the development of optical MEMS-based sensors, an area of distinct important to BioMEMS. The last contribution to this section is certainly the most biological. Jennifer McCann together with Profs Thomas Webster and Karen Haberstroh (all of Purdue) have contributed a study of how flow stresses influence vascular cell behavior.

Our sincere thanks to the authors for providing the very informative chapters and to Prof. Mauro Ferrari and Kluwer Academic Press for initiating this project. We hope the text will serve as an excellent reference for a wide ranging audience, from higher level undergraduates and beginning graduate students, to industrial researchers, and faculty members.

With best regards
Rashid Bashir (bashir@purdue.edu)
Steve Wereley (wereley@purdue.edu)
Purdue University, West Lafayette, IN

Mauro Ferrari
*Professor, Brown Institute of Molecular Medicine Chairman
Department of Biomedical Engineering
University of Texas Health Science Center, Houston, TX*

*Professor of Experimental Therapeutics
University of Texas M.D. Anderson Cancer Center, Houston, TX*

Professor of Bioengineering, Rice University, Houston, TX

*Professor of Biochemistry and Molecular Biology
University of Texas Medical Branch, Galveston, TX*

President, the Texas Alliance for NanoHealth, Houston, TX

I

Micro and Nanoscale Biosensors and Materials

1

Biosensors and Biochips

Tuan Vo-Dinh

Center for Advanced Biomedical Photonics, Oak Ridge National Laboratory, Bethel Valley Road;
MS-6101, P.O. Box 2008, Oak Ridge, TN 37831-6101, U.S.A.

This chapter provides an overview of the various types of biosensors and biochips that have been developed for biological and medical applications, along with significant advances and over the last several years in these technologies. Various classification schemes that can be used for categorizing the different biosensor and biochip systems are also discussed.

1.1. INTRODUCTION

A biosensor can be generally defined as a device that consists of a biological recognition system, often called a bioreceptor, and a transducer. In general, a biochip consists of an array of individual biosensors that can be individually monitored and generally are used for the analysis of multiple analytes. The interaction of the analyte with the bioreceptor is designed to produce an effect measured by the transducer, which converts the information into a measurable effect, such as an electrical signal. Figure 1.1 illustrates the conceptual principle of the biosensing process. Biosensors that include transducers based on integrated circuit microchips are often referred to as biochips.

There are several classification schemes possible. Biosensors and biochips can be classified either by their bioreceptor or their transducer type (see Figure 1.2). A bioreceptor is a biological molecular species (e.g., an antibody, an enzyme, a protein, or a nucleic acid) or a living biological system (e.g., cells, tissue, or whole organisms) that utilizes a biochemical mechanism for recognition. The sampling component of a biosensor contains a bio-sensitive layer. The layer can either contain bioreceptors or be made of bioreceptors covalently attached to the transducer. The most common forms of bioreceptors used in biosensing are based on 1) antibody/antigen interactions, 2) nucleic acid interactions,

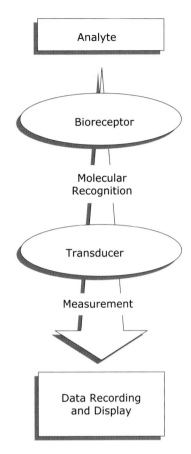

FIGURE 1.1. Conceptual diagram of the biosensing principle.

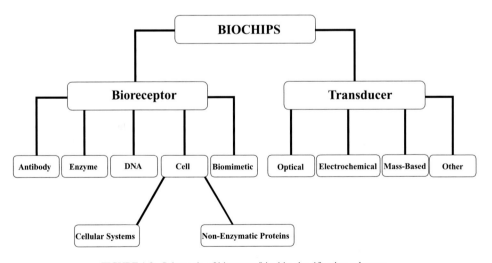

FIGURE 1.2. Schematic of biosensor/biochip classification schemes.

3) enzymatic interactions, 4) cellular interactions (i.e. microorganisms, proteins) and 5) interactions using biomimetic materials (i.e., synthetic bioreceptors). For transducer classification, conventional techniques include: 1) optical measurements (i.e. luminescence, absorption, surface plasmon resonance, etc.) 2) electrochemical and 3) mass-sensitive measurements (i.e. surface acoustic wave, microbalance, etc.).

The development of biosensors was first reported in the early 1960s [6]. Biosensors have now seen an explosive growth and seen a wide variety of applications primarily in two major areas, biological monitoring and environmental sensing applications.

1.2. BIOSENSORS

1.2.1. Different Types of Bioreceptors

The key to specificity for biosensor technologies involves bioreceptors. They are responsible for binding the analyte of interest to the sensor for the measurement. These bioreceptors can take many forms and the different bioreceptors that have been used are as numerous as the different analytes that have been monitored using biosensors. However, bioreceptors can generally be classified into five different major categories. These categories include: 1) antibody/antigen, 2) enzymes, 3) nucleic acids/DNA, 4) cellular structures/cells and 5) biomimetic. Figure 1.3 shows a schematic diagram of two types of bioreceptors: the structure of an immunoglobulin G (IgG) antibody molecule (Fig. 1.3A), and DNA and the principle of base pairing in hybridization (Fig. 1.3B).

1.2.1.1. Antibody Bioreceptors An antibody is a complex biomolecule, made up of hundreds of individual amino acids arranged in a highly ordered sequence. Antibodies are biological molecules that exhibit very specific binding capabilities for specific structures. For an immune response to be produced against a particular molecule, a certain molecular size and complexity are necessary: proteins with molecular weights greater then 5000 Da are generally immunogenic. The way in which an antigen and its antigen-specific antibody interact may be understood as analogous to a lock and key fit, by which specific geometrical configurations of a unique key enables it to open a lock. In the same way, an antigen-specific antibody "fits" its unique antigen in a highly specific manner. This unique property of antibodies is the key to their usefulness in immunosensors where only the specific analyte of interest, the antigen, fits into the antibody binding site.

Radioimmunoassay (RIA) utilizing radioactive labels have been applied to a number of fields including pharmacology, clinical chemistry, forensic science, environmental monitoring, molecular epidemiology and agricultural science. The usefulness of RIA, however, is limited by several shortcomings, including the cost of instrumentation, the limited shelf life of radioisotopes, and the potential deleterious biological effects inherent to radioactive materials. For these reasons, there are extensive research efforts aimed at developing simpler, more practical immunochemical techniques and instrumentation, which offer comparable sensitivity and selectivity to RIA. In the 1980s, advances in spectrochemical instrumentation, laser miniaturization, biotechnology and fiberoptic research have provided opportunities for novel approaches to the development of sensors

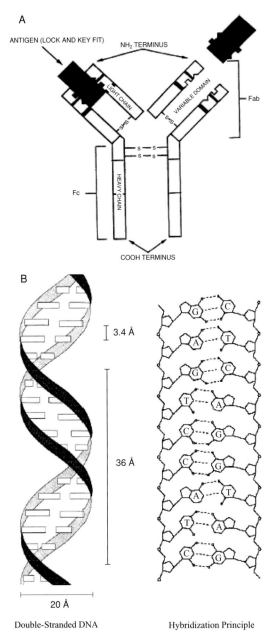

FIGURE 1.3. Schematic diagrams of two types of bioreceptors: **A)** IgG antibody. **B)** DNA and the hybridization principle.

for the detection of chemicals and biological materials of environmental and biomedical interest.

The first fiberoptic immunosensor was developed for *in situ* detection of the chemical carcinogen benzo[a]pyrene [52]. Nowadays, antibodies are often used in biosensors today. Biomolecular interactions can be classified in two categories, according to the test format performed (i.e., direct and indirect). In a direct format the immobilized target molecule interacts with a ligand molecule or the immobilized ligand interacts with a target molecule directly. For immunosensors, the simplest situation involves *in situ* incubation followed by direct measurement of a naturally fluorescent analyte [52]. For nonfluorescent analyte systems, *in situ* incubation is followed by development of a fluorophor-labeled second antibody. The resulting antibody sandwich produces a fluorescence signal that is directly proportional to the amount of bound antigen. The sensitivity obtained when using these techniques increases with increasing amounts of immobilized receptor. The indirect format involves competition between fluorophor-labeled and unlabeled antigens [42]. In this case, the unlabeled analyte competes with the labeled analyte for a limited number of receptor binding sites. Assay sensitivity therefore increases with decreasing amounts of immobilized reagent.

Antibody-based biosensors have been developed for use in an electrochemical immunoassay for whole blood [4]. The assay is performed on a conducting redox hydrogel on a carbon electrode on which avidin and choline oxidase have been co-immobilized. Biotinylated antibody was then bound to the gel. When the antigen binds to the sensor, another solution of complementary horseradish peroxidase labeled antibody is bound to the antigen, thus creating an electrical contact between the redox hydrogel and the peroxidase. The hydrogel then acts as an electrocatalyst for the reduction of hydrogen peroxide water.

Binding of the bioreceptor to the measurement support or the transducer is an important aspect of biosensor fabrication. A method for the immobilization of histidine-tagged antibodies onto a gold surface for surface plasmon resonance measurements was reported [15]. A synthetic thioalkane chelator is self-assembled on a gold surface. Reversible binding of an anti-lysozyme F-ab fragment with a hexahistidine modified extension on the C terminal end is then performed. Infrared spectroscopy was used to determine that the secondary structure of the protein was unaffected by the immobilization process. Retention of antibody functionality upon immobilization was also demonstrated. Due to the reversible binding of such a technique, this could prove a valuable method for regeneration of biosensors for various applications. Enzyme immunoassays can further increase the sensitivity of detection of antigen-antibody interactions by the chemical amplification process, whereby one measures the accumulated products after the enzyme has been allowed to react with excess substrate for a period of time [51].

With the use of nanotechnology, submicron fiberoptic antibody-based biosensors have been developed by Vo-Dinh and coworkers for the measurements of biochemicals inside a single cell [1, 8, 50]. Nanometer scale fiberoptic biosensors were used for monitoring biomarkers related to human health effects that are associated with exposure to polycyclic aromatic hydrocarbons (PAHs). These sensors use a monoclonal antibody for benzo[a]pyrene tetrol (BPT), a metabolite of the carcinogen benzo[a]pyrene, as the bioreceptor. Excitation light is launched into the fiber and the resulting evanescent field at the tip of the fiber is used to excite any of the BPT molecules that have bound to the antibody. The fluorescent light is then collected via a microscope. Using these antibody-based nanosensors, absolute detection limits for BPT of ca. 300 zeptomol (10^{-21} moles) have been reported [1]. These nanosensors allow the probing of cellular and subcellular environments in single cells [8, 50] as well as monitoring signaling processes in single cells [18, 46].

1.2.1.2. Enzyme Bioreptors Another type of commonly used bioreceptors involves enzymes, which are often chosen as bioreceptors based on their specific binding capabilities as well as their catalytic activity. In biocatalytic recognition mechanisms, the detection is amplified by a reaction catalyzed by macromolecules called biocatalysts. With the exception of a small group of catalytic ribonucleic acid molecules, all enzymes are proteins. Some enzymes require no chemical groups other than their amino acid residues for activity. Others require an additional chemical component called a cofactor, which may be either one or more inorganic ions, such as Fe^{2+}, Mg^{2+}, Mn^{2+}, or Zn^{2+}, or a more complex organic or metalloorganic molecule called a coenzyme. The catalytic activity provided by enzymes allows for much lower limits of detection than would be obtained with common binding techniques. The catalytic activity of enzymes depends upon the integrity of their native protein conformation. If an enzyme is denatured, dissociated into its subunits, or broken down into its component amino acids, its catalytic activity is destroyed. Enzyme-coupled receptors can also be used to modify the recognition mechanisms. For instance, the activity of an enzyme can be modulated when a ligand binds at the receptor. This enzymatic activity is often greatly enhanced by an enzyme cascade, which leads to complex reactions in the cell [9].

Multiple enzymes have been immobilized onto an array of optical fibers for use in the simultaneous detection of penicillin and ampicillin [29]. These biosensors provide an indirect technique for measuring penicillin and ampicillin based on pH changes during their hydrolysis by penicillinase. Immobilized onto the fibers with the penicillinase is a pH indicator, phenol red. As the enzyme hydrolyzes the two substrates, shifts in the reflectance spectrum of the pH indicator are measured. Various types of data analysis of the spectral information were evaluated using a multivariate calibration method for the sensor array containing biosensors of different compositions.

The development and use of a micrometer-sized fiber-optic biosensor were reported for the detection of glucose [30]. These biosensors are 100 times smaller than existing glucose optodes and represent the beginning of a new trend in nanosensor technology [2]. These sensors are based on the enzymatic reaction of glucose oxidase that catalyses the oxidation of glucose and oxygen into gluconic acid and hydrogen peroxide. To monitor the reaction, an oxygen indicator, tris(1,10-phenanthroline)ruthenium chloride, is immobilized into an acrylamide polymer with the glucose oxidase, and this polymer is attached to the fiber-optic via photopolymerization. A comparison of the response of glucose sensors created on different size fibers was made, and it was found that the micrometer size sensors have response times at least 25 times faster (only 2 s) than the larger fibers. In addition, these sensors are reported to have absolute detection limits of ca. 10^{-15} mol and an absolute sensitivity 5–6 orders of magnitude greater than current glucose optodes [30].

1.2.1.3. Nucleic Acid Bioreceptors Nucleic acids have received increasing interest as bioreceptors for biosensor and biochip technologies. The complementarity of adenine:thymine (A:T) and cytosine:guanosine (C:G) pairing in DNA (Fig. 1.2b) forms the basis for the specificity of biorecognition in DNA biosensors, often referred to as genosensors.

If the sequence of bases composing a certain part of the DNA molecule is known, then the complementary sequence, often called a probe, can be synthesized and labeled with an optically detectable compound (e.g., a fluorescent label). By unwinding the double-stranded DNA into single strands, adding the probe, and then annealing the strands, the labeled probe will hybridize to its complementary sequence on the target molecule.

DNA biosensors have been developed for the monitoring of DNA-ligand interactions [26]. Surface plasmon resonance was used to monitor real-time binding of low molecular weight ligands to DNA fragments that were irreversibly bound to the sensor surface via Coulombic interactions. The DNA layer remained stable over a period of several days and was confirmed using ellipsometry. The sensor was capable of detecting binding effects between 10 and 400 pg/mm^2. Binding rates and equilibrium coverages were determined for various ligands by changing the ligand concentration. In addition, affinity constants, association rates and dissociation rates were also determined for these various ligands.

Another type of biosensor uses a peptide nucleic acid as the biorecognition element [33]. The peptide nucleic acid is an artificial oligo amide that is capable of binding very strongly to complimentary oligonucleotide sequences. Using a surface plasmon resonance sensor, the direct detection of double stranded DNA that had been amplified by a polymerase chain reaction (PCR) has been demonstrated.

Vo-Dinh and coworkers have developed a new type of DNA gene probe based on surface-enhanced Raman scattering (SERS) detection [16, 49]. The SERS probes do not require the use of radioactive labels and have great potential to provide both sensitivity and selectivity via label multiplexing due to the intrinsically narrow bandwiths of Raman peaks. The effectiveness of the new detection scheme is demonstrated using the *gag* gene sequence of the human immunodefficiency (HIV) virus [16]. The development of a biosensor for DNA diagnostics using visible and near infrared (NIR) dyes has been reported [48]. The system employed a two-dimensional charge-coupled device and was used to detect the cancer suppressor *p53* gene.

1.2.1.4. Cellular Bioreceptors Cellular structures and cells have been used in the development of biosensors and biochips [12]. These bioreceptors are either based on biorecognition by an entire cell/microorganism or a specific cellular component that is capable of specific binding to certain species. There are presently three major subclasses of this category: 1)cellular systems, 2) enzymes and 3) non-enzymatic proteins. Due to the importance and large number of biosensors based on enzymes, these have been given their own classification and were previously discussed. One of the major benefits associated with using this class of bioreceptors is that often the detection limits can be very low because of signal amplification. Many biosensors developed with these types of bioreceptors rely on their catalytic or pseudocatalytic properties.

Microorganisms offer a form of bioreceptor that often allows a whole class of compounds to be monitored. Generally these microorganism biosensors rely on the uptake of certain chemicals into the microorganism for digestion. Often, a class of chemicals is ingested by a microorganism, therefore allowing a class-specific biosensor to be created. Microorganisms such as bacteria and fungi have been used as indicators of toxicity or for the measurement of specific substances. For example, cell metabolism (e.g., growth inhibition, cell viability, substrate uptake), cell respiration or bacterial bioluminescence have been used to evaluate the effects of toxic heavy metals. Many cell organelles can be isolated and used as bioreceptors. Since cell organelles are essentially closed systems, they can be used over long periods of time. Whole mammalian tissue slices or *in vitro* cultured mammalian cells are used as biosensing elements in bioreceptors. Plant tissues are also used in plant-based biosensors because they are effective catalysts as a result of the enzymatic pathways they possess [9].

A microbial biosensor has been developed for the monitoring of short-chain fatty acids in milk [34]. Arthrobacter nicotianae microorganisms were immobilized in a calcium-alginate gel on an electrode surface. To this gel was added 0.5 mM $CaCl_2$ to help stabilize it. By monitoring the oxygen consumption of the anthrobacter nicotianae electrochemically, its respiratory activity could be monitored, thereby providing an indirect means of monitoring fatty acid consumption. Detection of short-chain fatty acids, ranging from 4 to 12 carbons in length, in milk was accomplished with butyric acid being the major substrate. A linear dynamic range from 9.5–165.5 µM is reported with a response time of 3 min. Methods for shortening the response time and recovery time of microbial sensors are also discussed.

Many proteins often serve the purpose of bioreception for intracellular reactions that will take place later or in another part of the cell. These proteins could simply be used for transport of a chemical from one place to another, such as a carrier protein or channel protein on a cellular surface. In any case, these proteins provide a means of molecular recognition through one or another type of mechanism (i.e. active site or potential sensitive site). By attaching these proteins to various types of transducers, many researchers have constructed biosensors based on non-enzymatic protein biorecognition.

Detection of endotoxin using a protein bioreceptor based biosensor has been reported [17]. The liposaccharide endotoxin is a causative agent in the clinical syndrome known as sepis, which causes more than 100,000 deaths annually. This work describes an evanescent wave fiber optic biosensor that makes use of a covalently immobilized protein, polymyxin B, as the biorecognition element. The sensor is based on a competitive assay with fluorescently tagged lipopolysaccharide. When this sensor was applied to the detection of lipopolysaccharides in *E. coli*, detection of concentrations of 10 ng/mL in 30 s was reported.

Lipopeptides have been used as bioreceptors for biosensors [3]. A lipopeptide containing an antigenic peptide segment of VP1, a capsid protein of the picornavirus that causes foot-and-mouth diseases in cattle, was evaluated as a technique for monitoring antigen antibody interactions. The protein was characterized via circular dichroism and infrared spectroscopy to verify that upon self-assembly onto a solid surface it retained the same structure as in its free form. Based on surface plasmon resonance measurements, it was found that the protein was still fully accessible for antibody binding. This technique could provide an effective means of developing biomimetic ligands for binding to cell surfaces.

1.2.1.5. Biomimetic Receptors An artificial (man-made) receptor that is fabricated and designed to mimic a bioreceptor is often termed a biomimetic receptor. Several different methods have been developed over the years for the construction of biomimetic receptors. These methods include: genetically engineered molecules, artificial membrane fabrication and molecular imprinting. The molecular imprinting technique, which has recently received great interest, consists of mixing analyte molecules with monomers and a large amount of crosslinkers. Following polymerization, the hard polymer is ground into a powder and the analyte molecules are extracted with organic solvents to remove them from the polymer network. As a result the polymer has molecular holes or binding sites that are complementary to the selected analyte.

Recombinant techniques, which allow for the synthesis or modification of a wide variety of binding sites using chemical means, have provided powerful tools for designing synthetic bioreceptors with desired properties. Development of a genetically engineered single-chain antibody fragment for the monitoring of phosphorylcholine has been reported [27]. In this work, protein engineering techniques are used to fuse a peptide sequence that mimics the binding properties of biotin to the carboxyterminus of the phosphorylcholine-binding fragment of IgA. This genetically engineered molecule was capable of being attached to a streptavidin monolayer and total internal reflection fluorescence was used to monitor the binding of a fluorescently labeled phosphorylcholine analog.

Bioreceptor systems also used artificial membranes for many different applications. Stevens and coworkers have developed an artificial membrane by incorporating gangliosides into a matrix of diacetylenic lipids (5–10% of which were derivatized with sialic acid) [5]. The lipids were allowed to self-assemble into Langmuir-Blodgett layers and were then photopolymerized via ultraviolet irradiation into polydiacetylene membranes. When cholera toxins bind to the membrane, its natural blue color changes to red and absorption measurements were used to monitor the toxin concentration. Using these polydiacetylenic lipid membranes coupled with absorption measurements, concentrations of cholera toxin as low as 20 µg/mL were capable of being monitored.

Bioreceptors based on molecular imprinting have been used for the construction of a biosensor based on electrochemical detection of morphine [20]. A molecularly imprinted polymer for the detection of morphine was fabricated on a platinum wire using agarose and a crosslinking process. The resulting imprinted polymer was used to specifically bind morphine to the electrode. Following morphine binding, an electroinactive competitor, codeine, was used to wash the electrode and thus release some of the bound morphine. One of the major advantages of the molecular imprinting technique is the rugged nature of a polymer relative to a biological sample. The molecularly imprinted polymer can withstand harsh environments such as those experienced in an autoclave or chemicals that would denature a protein. On the other hand, due to their rigid structures, molecular imprint probes do not have the same flexibility and selectivity as compared to actual bioreceptors.

1.2.2. Types of Transducers

Transduction can be accomplished via a great variety of methods. Biosensors can also be classified based upon the transduction methods they employ. Most forms of transduction can be categorized in one of three main classes. These classes are: 1) optical detection methods, 2) electrochemical detection methods and 3) mass detection methods. However, new types of transducers are constantly being developed for use in biosensors. Each of these three main classes contains many different subclasses, creating a nearly infinite number of possible transduction methods or combination of methods.

1.2.2.1. Optical Techniques Optical biosensors can use many different types of spectroscopy (e.g., absorption, fluorescence, phosphorescence, Raman, SERS, refraction, dispersion spectrometry, etc.) with different spectrochemical properties recorded. For this reason, optical transduction, which offers the largest number of possible subcategories, have been developed in our laboratory over the last two decades [1, 2, 8, 9, 16, 29, 30, 42, 48–52]. These properties include: amplitude, energy, polarization, decay time and/or phase.

Amplitude is the most commonly measured parameter of the electromagnetic spectrum, as it can generally be correlated with the concentration of the analyte of interest. The energy of the electromagnetic radiation measured can often provide information about changes in the local environment surrounding the analyte, its intramolecular atomic vibrations (i.e. Raman or infrared absorption spectroscopies) or the formation of new energy levels. Measurement of the interaction of a free molecule with a fixed surface can often be investigated based on polarization measurements. Polarization of emitted light is often random when emitted from a free molecule in solution, however, when a molecule becomes bound to a fixed surface, the emitted light often remains polarized. The decay time of a specific emission signal (i.e. fluorescence or phosphorescence) can also be used to gain information about molecular interactions since these decay times are very dependent upon the excited state of the molecules and their local molecular environment. Vo-Dinh and coworkers reported the development of a phase-resolved fiberoptic fluoroimmunosensor (PR-FIS), which can differentiate the carcinogen benzo[a]pyrene and its metabolite benzopyrene tetrol based on the difference of their fluorescence lifetimes [19]. Another property that can be measured is the phase of the emitted radiation. When electromagnetic radiation interacts with a surface, the speed or phase of that radiation is altered, based on the refractive index of the medium (i.e. analyte). When the medium changes, via binding of an analyte, the refractive index may change, thus changing the phase of the impinging radiation.

Absorption measurements of a pH sensitive dye are used to quantify the amount of urea present [23]. A lipophilic carboxylated polyvinyl chloride membrane containing a pH sensitive dye was used as the sensor transducer. Urease was covalently bound to this membrane, forming a very thin layer. As various concentrations of urea were tested using the sensor, the effective pH change caused a shift in the absorbance profile of the dye that was measured. This sensor allowed for the rapid determination of urea over the concentration range 0.3–100 mM.

A fiber-optic evanescent wave immunosensor for the detection of lactate dehydrogenase has been developed [28]. Two different assay methods, a one-step and a two-step assay process, using the sensor based on polyclonal antibody recognition were described. The response of this evanescent wave immunosensor was then compared to a commercially available surface plasmon resonance based biosensor for lactate dehydrogenase detection using similar assay techniques and similar results were obtained. It was also demonstrated that although the same polyclonal antibody can be used for both the one- and two-step assay techniques, the two-step technique is significantly better when the antigen is large.

1.2.2.2. Electrochemical Techniques Electrochemical detection is another possible means of transduction that has been used in biosensors [11, 31, 41]. This technique is very complementary to optical detection methods such as fluorescence, the most sensitive of the optical techniques. Since many analytes of interest are not strongly fluorescent and tagging a molecule with a fluorescent label is often labor intensive, electrochemical transduction can be very useful. By combining the sensitivity of electrochemical measurements with the selectivity provided by bioreception, detection limits comparable to fluorescence biosensors are often achievable. Electrochemical flow-through enzyme-based biosensors for the detection of glucose and lactate have been developed by Cammann and coworkers [32]. Glucose oxidase and lactate oxidase were immobilized in conducting polymers generated from pyrrole, N-methylpyrrole, aniline and o-phenylenediamine on platinum surfaces. These various sensor matrices were compared based on amperometric measurements of glucose and lactate and it was found that the o-phenylenediamine polymer was the most sensitive. This polymer matrix was also deposited on a piece of graphite felt and used as an enzyme reactor as well as a working electrode in an electrochemical detection system. Using this system, a linear dynamic range of 500 μM − 10 mM glucose was determined with a limit of detection of <500 μM. For lactate, the linear dynamic range covered concentrations from 50 μM − 1 mM with a detection limit of <50 μM.

A biosensor for protein and amino acid estimation is reported [14]. A screen-printed biosensor based on a rhodinized carbon paste working electrode was used in the three electrode configuration for a two-step detection method. Electrolysis of an acidic potassium bromide electrolyte at the working electrode produced bromine which was consumed by the proteins and amino acids. The bromine production occurred at one potential while monitoring of the bromine consumption was performed using a lower potential. The method proved very sensitive to almost all of the amino acids, as well as some common proteins and was even capable of measuring L- and D- praline, which give no response to enzyme based biosensors. This sensor has been tested by measuring proteins and amino acids in fruit juice, milk and urine and consumes approximately 10 μL of sample for direct detection.

An electrochemical biosensor has been developed for the indirect detection of L-phenylalanine via NADH [25]. This sensor is based on a three-step multi-enzymatic/electrochemical reaction. Three enzymes, L-phenylalanine dehydrogenase, salicylate hydroxylase and tyrosinase, are immobilized in a carbon paste electrode. The principle behind this reaction/detection scheme is as follows. First, the L-phenylalanine dehydrogenase upon binding and reacting with L-phenylalanine produces NADH. The second enzyme, salicylate hydroxylase, then converts salicylate to catechol in the presence of oxygen and NADH. The tyrosinase then oxidizes the catechol to o-quinone which is electrochemically detected and reduced back to catechol with an electrode potential of −50 mV vs. a Ag/AgCl reference electrode. This reduction step results in an amplification of signal due to the recycling of catechol from o-quinone. Prior to the addition of the L-phenylalanine dehydrogenase to the electrode, it was tested for its sensitivity to NADH, its pH dependence and its response to possible interferents, urea and ascorbic acid. From these measurements, it was found that the sensor sensitivity for NADH increased 33 fold by introducing the recycling step over just the salicylate hydroxylase system alone.

1.2.2.3. Mass-sensitive Techniques Measurement of small changes in mass is another form of transduction that has been used for biosensors [24, 40]. The principle means of mass analysis relies on the use of piezoelectric crystals. These crystals can be made to vibrate at a specific frequency with the application of an electrical signal of a specific frequency. The frequency of oscillation is therefore dependent on the electrical frequency applied to the crystal as well as the crystal's mass. Therefore, when the mass increases due to binding of chemicals, the oscillation frequency of the crystal changes and the resulting change can be measured electrically and be used to determine the additional mass of the crystal.

A quartz crystal microbalance biosensor has been developed for the detection of Listeria monocytogenes [43]. Several different approaches were tested for immobilization of Listeria onto the quartz crystal through a gold film on the surface. Once bound, the microbalance was then placed in a liquid flow cell where the antibody and antigen were allowed to complex, and measurements were obtained. Calibration of the sensor was accomplished using a displacement assay and was found to have a response range from $2.5 \times 10^5 - 2.5 \times 10^7$ cells/crystal. More recently, Guilbault and coworkers have developed a method for covalently binding antibodies to the surface of piezoelectric crystals via sulfur based self-assembled monolayers [53]. Prior to antibody binding, the monolayers are activated with 1-ethyl-3-[3-(dimethylamino)propyl] carbodiimide hydrochloride and N-hydroxysulfosuccinimide. Using this binding technique, a real time capture assay based on mouse IgG was performed and results were reported.

A horizontally polarized surface acoustic wave biosensor has been reported [10]. This sensor has a dual path configuration, with one path acting as an analyte sensitive path and the other path acting as a reference path [10]. Antibodies were immobilized onto the sensor via protein A, with a mass density of 0.4 ng/mm^2. A theoretical detection limit of 33 pg was calculated based on these experiments, and a sensitivity of 100 kHz/(ng/mm^2) is reported. In addition, a means of inductively coupling a surface acoustic wave biosensor to its RF generating circuitry has been reported recently [21]. This technique could greatly reduce wire bonding associated problems for measurements made in liquids, since the electrodes are coated with a layer of SiO_2.

1.3. BIOCHIPS

1.3.1. Microarray Systems

Within the last couple of decades, the development of integrated biosensors for the detection of multiple biologically relevant species has begun to take place. These integrated biosensor arrays that use the same excitation source for all of the elements and the same measurement process have been termed many things; gene chips, DNA-chips, etc. Most of the different array chips have been based on the use of nucleic acids (i.e. DNA) as the bioreceptors. Figure 1.4 illustrates an example of DNA microarray system with its associated detection system. Other types of bioreceptors such as antibodies, enzymes and cellular components can also be used. It is noteworthy that substrates having microarrays of bioreceptors are often referred to as biochips although most of these systems do not have integrated microsensor detection systems. A few of the more recent applications and advances in biochip technology will be discussed in this review.

A microarray of electrochemical biosensors has been developed for the detection of glucose and lactate on line [54]. This array of electrochemical biosensors was prepared using photolithographic techniques, using glucose oxidase and lactate oxidase as the bioreceptors. The glucose oxidase or lactate oxidase at each of the different sites in the array produces hydrogen peroxide when its appropriate substrate, glucose or lactate, is present. The hydrogen peroxide produced was measured at each element amperometrically.

An optical microarray system using a charge-coupled device (CCD) detector and DNA probes has been developed by Vo-Dinh and coworkers [48]. The evaluation of various system components developed for the DNA multi-array biosensor was discussed. The DNA probes labeled with visible and near infrared (NIR) dyes are evaluated. Examples of application of gene probes in DNA hybridization experiments and in biomedical diagnosis (detection of the *p53* cancer suppressor gene) illustrated the usefulness and potential of the DNA

BIOSENSORS AND BIOCHIPS 15

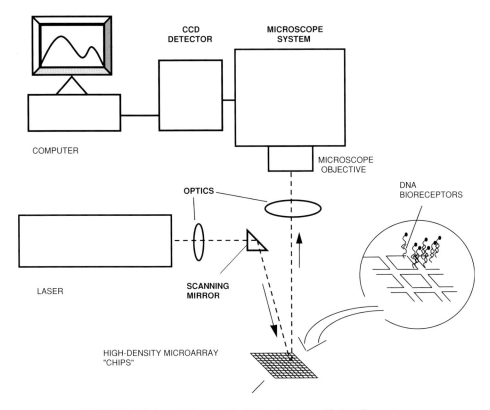

FIGURE 1.4. Schematic diagram of a DNA microarray with detection system.

multiarray device. An optical microarray for the detection of toxic agents using a planar array of antibody probes was described by Ligler and coworkers [13]. Their system was composed of a CCD for detection, an excitation source and a microscope slide with a photoactivated optical adhesive. Antibodies against three different toxins, staphylococcal enterotoxin B (SEB), ricin, and Yersinia pestis, were covalently attached to small wells in the slide formed by the optical adhesive. The microscope slide was then mounted over the CCD with a gradient refractive index (GRIN) lens array used to focus the wells onto the CCD. Toxins were then introduced to the slide followed by Cy5-labeled antibodies. The bound antibodies were then excited and the resulting fluorescence from all of the sensor locations were monitored simultaneously. Concentrations ranging from 5–25 ng/mL were capable of being measured for the different toxins.

High-density oligonucleotide arrays, consisting of greater than 96 000 oligonucleotides have been designed by Hacia *et al.* for the screening of the entire 5.53 kb coding region of the hereditary breast and ovarian cancer *BRCA1* gene for all possible variations in the homozygous and heterozygous states [35]. Single stranded RNA targets were created by PCR amplification followed by *in vitro* transcription and partial fragmentation. These targets were then tested and fluorescence responses from targets containing the four natural bases to greater than 5 592 different fully complimentary 25 mer oligonucleotide probes were found.

To examine the effect of uridine and adenosine on the hybridization specificity, 33 200 probes containing centrally localized base pair mismatches were constructed and tested. Targets that contained modified 5-methyluridine showed a localized enhancement in fluorescence hybridization signals. In general, oligonucleotide microarrays, often referred to as "DNA chips", are generally made by a light-directed chemical reaction that uses photographic masks for each chip [35]. A maskless fabrication method of light-directed oligonucleotide microarrays using a digital microarray has been reported [47]. In this method, a maskless array synthesizer replaces the chrome mask with virtual masks generated on a computer, which are relayed to a digital microarray.

1.3.2. Integrated Biochip Systems

The development of a truly integrated biochip having a phototransistor integrated circuit (IC) microchip has been reported by Vo-Dinh and coworkers [47, 48]. This work involves the integration of a 4 × 4 and 10 × 10 optical biosensor array onto an integrated circuit (Figure 1.5). Most optical biochip technologies are very large when the excitation source and detector are considered, making them impractical for anything but laboratory usage. In this biochip the sensors, amplifiers, discriminators and logic circuitry are all built onto the chip. In one biochip system, each of the sensing elements is composed of 220 individual phototransistor cells connected in parallel to improve the sensitivity of the instrument. The

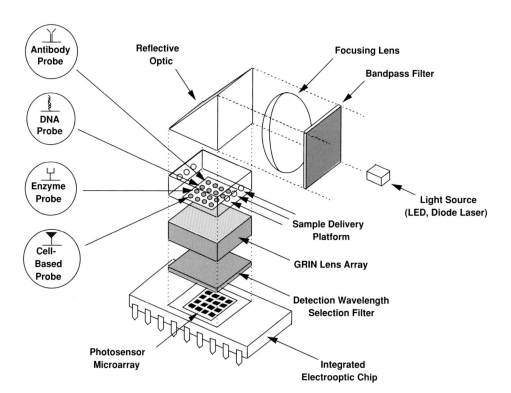

FIGURE 1.5. Schematic diagram of an integrated biochip system with microchip sensor.

ability to integrate light emitting diodes (LEDs) as the excitation sources into the system is also discussed. An important element in the development of the multifunctional biochip (MFB) involves the design and development of an IC electro-optic system for the microchip detection elements using the complementary metal oxide silicon (CMOS) technology. With this technology, highly integrated biochips are made possible partly through the capability of fabricating multiple optical sensing elements and microelectronics on a single system. Applications of the biochip are illustrated by measurements of the HIV1 sequence-specific probes using the DNA biochip device for the detection of a gene segment of the AIDS virus [47]. Recently, a MFB which allows simultaneous detection of several disease endpoints using different bioreceptors, such as DNA, antibodies, enzymes, cellular probes, on a single biochip system was developed [22]. The MFB device was a self-contained system based on an integrated circuit including photodiode sensor arrays, electronics, amplifiers, discriminators and logic circuitry. The multi-functional capability of the MFB biochip device is illustrated by measurements of different types of bioreceptors using DNA probes specific to gene fragments of the *Mycobacterium Tuberculosis* (TB) system, and antibody probes targeted to the cancer related tumor suppressor gene *p53*.

A biochip equipped with a microfluidics sample/reagent delivery system for on-chip monitoring of bioassayshas been developed for *E. coli* detection [39]. The microfluidics system includes a reaction chamber which houses a sampling platform that selectively captures detection probes from a sample through the use of immobilized bioreceptors. The independently operating photodiodes allow simultaneous monitoring of multiple samples. In this study the sampling platform is a cellulosic membrane that is exposed to *E. coli* organisms and subsequently analyzed using a sandwich immunoassay involving a Cy5-labeled antibody probe. Studies show that the biochip has a linear dynamic range of three orders of magnitude observed for conventional assays, and can detect 20 *E. coli* organisms. Selective detection of *E. coli* in a complex medium, milk diluent, is also reported for both off-chip and on-chip assays.

A CMOS biochip coupled to multiplex capillary electrophoresis (CE) system has been developed [36, 37]. This combination of multiplex capillary gel electrophoresis and the IC microchip technology represents a novel approach to DNA analysis on the microchip platform. Separation of DNA ladders using a multiplex CE microsystem of four capillaries was monitored simultaneously using the IC microchip system. The IC microchip-CE system has advantages such as low cost, rapid analysis, compactness, and multiplex capability, and has great potential as an alternative system to conventional capillary array gel electrophoresis systems based on charge-coupled device (CCD) detection.

Antibody-immobilized capillary reactors coupled to biochip detection have been developed for *E. coli* O157:H7 detection using enzyme-linked immunosorbent assay (ELISA), and a biochip system [38]. ISA is very sensitive and selective immunological method to detect pathogenic bacteria. ELISA is also directly adaptable to a miniature biochip system that utilizes conventional sample platforms such as polymer membranes and glass. The antibody immobilized capillary reactor is a very attractive sample platform for ELISA because of its low cost, compactness, reuse, and ease of regeneration. Moreover, an array of capillary reactors can provide high-throughput ELISA. In this report, we describe the use of an array of antibody-immobilized capillary reactors for multiplex detection of *E. coli* O157:H7 in our miniature biochip system. Side-entry laser beam irradiation to an array of capillary reactors contributes significantly to miniaturized optical configuration for this biochip system.

The detection limits of *E. coli* O157:H7 using ELISA and Cy5 label-based immunoassays were determined to be 3 cells and 230 cells, respectively. This system shows capability to simultaneously monitor multifunctional immunoassay and high sensitive detection of *E. coli* O157:H7.

The application of a biochip using the molecular beacon (MB) detection scheme has been reported [Culha et al, 2004]. The medical application of this biochip novel MB detection system for the analysis of the breast cancer gene BRCA1 was illustrated. The MB is designed for the BRCA1 gene and a miniature biochip system is used for detection. The detection of BRCA1 gene is successfully demonstrated in solution and the limit of detection (LOD) is estimated as 70 nM.

1.4. CONCLUSION

For practical medical diagnostic applications, there is currently a strong need for a truly integrated biochip system that comprises probes, samplers, detector as well as amplifier and logic circuitry. Such a system will be useful in physician's offices and could be used by relatively unskilled personnel. Most DNA biosensors previously reported are based on fiberoptic probes or glass and silica plates used as the probe substrates which are externally connected to a photosensing system generally consisting of a conventional detection device, such as a photomultiplier, or a charge-coupled device (CCD). Although the probes on the sampling platform are small (often referred to as a "DNA chip" or "gene chip"), the entire device containing excitation laser sources and detection systems (often a confocal microscope system) is relatively large, e.g., table-top size systems. While these systems have demonstrated their usefulness in gene discovery and genomics research, they are laboratory-oriented and involve relatively expensive equipment.

Biochip technologies could offer a unique combination of performance capabilities and analytical features of merit not available in any other bioanalytical system currently available. With its multichannel capability, biochip technology allows simultaneous detection of multiple biotargets. Biochip systems have great promise to offer several advantages in size, performance, fabrication, analysis and production cost due to their integrated optical sensing microchip. The small sizes of the probes (microliter to nanoliter) minimize sample requirement and reduce reagent and waste requirement. Highly integrated systems lead to a reduction in noise and an increase in signal due to the improved efficiency of sample collection and the reduction of interfaces. The capability of large-scale production using low-cost integrated circuit (IC) technology is an important advantage. The assembly process of various components is made simple by integration of several elements on a single chip. For medical applications, this cost advantage will allow the development of extremely low cost, disposable biochips that can be used for in-home medical diagnostics of diseases without the need of sending samples to a laboratory for analysis.

ACKNOWLEDGEMENTS

This work was sponsored by the Laboratory Directed Research and Development Program (Advanced Nanosystems Project), Oak Ridge National Laboratory, and by the U.S.

Department of Energy, Office of Biological and Environmental Research, under contract DE-AC05-96OR22464 with Lockheed Martin Energy Research Corporation.

REFERENCES

[1] J.P. Alarie and T.Vo-Dinh. *Polycyc. Aromat. Compd.*, 8:45, 1996.
[2] S.L.R. Barker, R. Kopelman, T.E. Meyer, and M.A. Cusanovich. *Anal. Chem.*, 70:971, 1998.
[3] M. Boncheva, C. Duschl, W. Beck, G. Jung, and H. Vogel. *Langmuir*, 12:5636, 1996.
[4] C.N. Campbell, T. de Lumley-Woodyear, and A. Heller. *Fresen. J. Anal. Chem.*, 364:165, 1993.
[5] D. Charych, Q. Cheng, A. Reichert, G. Kuziemko, M. Stroh, J.O. Nagy, W. Spevak, and R.-C. Stevens. *Chem. Biol.*, 3:113, 1996.
[6] L.C. Clark, Jr. and C. Lions. *Ann. Acad. Sci.*, 102:29, 1962.
[7] M. Culha, D.L. Stokes, G.D. Griffin, and T. Vo-Dinh. *Biosens. Bioelectron.*, 19:1007, 2004.
[8] B.M. Cullum, G.D. Griffin, and T. Vo-Dinh. *Anal. Biochem.*, 1999.
[9] D. Diamond. (ed.). *Chemical and Biological Sensors*. Wiley, New York, 1998.
[10] J. Freudenberg, S. Schelle, K. Beck, M. vonSchickfus, and S. Hunklinger. *Biosens. Bioelectron.*, 14:423, 1999.
[11] C. Galan-Vidal, J. Munoz, C. Dominguez, and S. Alegret. *Sensor. Actuat. B-Chem.*, 53:257, 1998.
[12] J.J. Gooding and D.B. Hibbert. *Trac-Trend Anal. Chem.*, 18:525, 1999.
[13] J.G. Hacia, S.A.Woski, J. Fidanza, K. Edgemon, G. McGall, S.P.A. Fodor, and F.S. Collins. *Nucleic Acids Res.*, 26:4975, 1998.
[14] T. Huang, A. Warsinke, T. Kuwana, and F. W. Scheller. *Anal. Chem.*, 70:991–997, 1998.
[15] D. Kroger, M. Liley, W. Schiweck, A. Skerra, and H. Vogel. *Biosens. Bioelectron.*, 14:155, 1999.
[16] N. Isola, D.L. Stokes, and T. Vo-Dinh. *Anal. Chem.*, 70:1352, 1998.
[17] E.A. James, K. Schmeltzer, and F.S. Ligler. *Appl. Biochem. Biotechnol.*, 60:189, 1996.
[18] P.M. Kasili, J.M. Song, and T. Vo-Dinh. *J. Am. Chem. Soc.*, 126:2799–2806, 2004.
[19] R. Koncki, G.J. Mohr, and O.S. Wolfbeis. *Biosens. Bioelectron.*, 10:653, 1995.
[20] D. Kriz and K. Mosbach. *Anal. Chim. Acta*, 300:71, 1995.
[21] I.V. Lamont, R.I. McConnell, and S.P. Fitzgerald. *Clin. Chem.*, 45:A102, 1999.
[22] M. Malmquist. *Biochem. Soc.*, T 27:335, 1999.
[23] T. McCormack, G.O'Keeffe, B.D. MacCraith, and R.O'Kennedy. *Sensor. Actuat. B-Chem.*, 41:89, 1997.
[24] M. Minunni, M. Mascini, R.M. Carter, M.B. Jacobs, G.J. Lubrano, and G.C. Guilbault. *Anal. Chim. Acta*, 325:169, 1996.
[25] Z.H. Mo, X.H. Long, and W.L. Fu. *Anal. Commun.*, 36:281–283, 1999.
[26] J. Piehler, A. Brecht, G. Gauglitz, M. Zerlin, C. Maul, R. Thiericke, and S. Grabley. *Anal. Biochem.*, 249:94, 1997.
[27] R.T. Piervincenzi, W.M. Reichert, and H.W. Hellinga. *Biosens. Bioelectron.*, 13:305, 1998.
[28] T.E. Plowman, W.M. Reichert, C.R. Peters, H.K. Wang, D.A. Christensen, and J.N. Herron. *Biosens. Bioelectron.*, 11:149, 1996.
[29] J. Polster, G. Prestel, M. Wollenweber, G. Kraus, and G. Gauglitz. *Talanta*, 42:2065, 1995.
[30] Z. Rosenzweig and R. Kopelman. *Anal. Chem.*, 68:1408–1413, 1996.
[31] U. Rudel, O. Geschke, and K. Cammann. *Electroanalysis*, 8:1135, 1996.
[32] P. Sarkar and A.P.F. Turner. *Fresen. J. Anal. Chem.*, 364:154, 1999.
[33] S. Sawata, E. Kai, K. Ikebukuro, T. Iida, T. Honda, and I. Karube. *Biosens. Bioelectron.*, 14:397, 1999.
[34] A. Schmidt, C. StandfussGabisch, and U. Bilitewski. *Biosens. Bioelectron.*, 11:1139, 1996.
[35] S. Singh-Gasson, R.D. Green, Y, Yue, C. Nelson, F. Blattner, R. Sussman, and F. Cerrina. *Nat. Biotechnol.*, 10:974, 1999.
[36] J.M. Song and T. Vo-Dinh. *Anal. Bioanal. Chem.*, 373:399, 2002.
[37] J.M. Song, J. Mobley, and T. Vo-Dinh. *J. Chromatogra. B*, 783:501, 2003.
[38] J.M. Song and T. Vo-Dinh. *Anal. Chimi. Acta*, 507:115, 2004.
[39] D.L. Stokes, G.D. Griffin, and T. Vo-Dinh. *Fresen. J. Anal. Chem.*, 369, 2001.
[40] T. Thundat, P.I. Oden, and R.J. Warmack. *Microscale Thermophys. Eng.*, 1:185, 1997.
[41] F. Tobalina, F. Pariente, L. Hernandez, H.D. Abruna, and E. Lorenzo. *Anal. Chim. Acta*, 395:17, 1999.

[42] B.G. Tromberg, M.J. Sepaniak, T. Vo-Dinh, and G.D. Griffin. *Anal. Chem.*, 59:1226, 1987.
[43] R.D. Vaughan, C.K. Sullivan, and G.C. Guilbault. *Fresen. J. Anal. Chem.*, 364:54, 1999.
[44] T. Vo-Dinh. *Sensor. Actuat B-Chem.*, 51:52, 1998.
[45] T. Vo-Dinh. *Proceedings of the 6th Annual Biochip Technologies Conference: Chips for Hits '99*. Berkeley, California, Nov 2–5 1999.
[46] T. Vo-Dinh. *J. Cell. Biochem.*, 39(Suppl.):154, 2002.
[47] T. Vo-Dinh, J.P. Alarie, N. Isola, D. Landis, A.L. Wintenberg, and M.N. Ericson. *Anal. Chem.*, 71:358–363, 1999.
[48] T. Vo-Dinh, N. Isola, J.P. Alarie, D. Landis, G.D. Griffin, and S. Allison. *Instrum. Sci. Technol.*, 26:503, 1998.
[49] T. Vo-Dinh, K. Houck, and D.L. Stokes. *Anal. Chem.*, 33:3379, 1994.
[50] T. Vo-Dinh, B.M. Cullum, J.P. Alarie, and G.D. Griffin. *J. Nanopart. Res.*, (in press).
[51] T. Vo-Dinh, G.D. Griffin, and K.R. Ambrose. *Appl. Spectrosc.*, 40:696, 1986.
[52] T. Vo-Dinh, B.G. Tromberg, G.D. Griffin, K.R. Ambrose, M.J. Sepaniak, and E.M. Gardenhire. *Suppl. Spectrosc.*, 41:735, 1987.
[53] W. Welsch, C. Klein, M. vonSchickfus, and S. Hunklinger. *Anal. Chem.*, 68:2000, 1996.
[54] R.M. Wadkins, J.P. Golden, L. M. Pritsiolas, and F.S. Ligler. *Biosens. Bioelectron.*, 13:407, 1998.

2

Cantilever Arrays: A Universal Platform for Multiplexed Label-Free Bioassays

Min Yue[1], Arun Majumdar[1], and Thomas Thundat[2]

[1] *Department of Mechanical Engineering, University of California, Berkeley, CA 94720*
[2] *Life Sciences Division, Oak Ridge National Laboratory, Oak Ridge, TN*

2.1. INTRODUCTION

Microcantilevers have caught great attention as label-free and ultra-sensitive biological sensors. When molecular adsorption occurs on only one surface of cantilever, the resulting differential surface stress leads to cantilever bending, thus providing a method of detecting molecular adsorption (Figure 2.1). How does the transduction from molecular adsorption to surface stress change occur? Though the underlying science of the transduction is yet to be completely understood, the thermodynamic argument suggests that the reaction-induced free energy reduction on one cantilever surface is balanced by the strain energy increase due to bending, such that at equilibrium the free energy of the whole system reaches the minimum [11]. In other words, the penalty of increasing the strain energy must be compensated by a larger reduction in free energy due to reaction, reflecting the interplay between mechanics and chemistry. Hence, the cantilever bending can be construed as a measure of free energy reduction due to the chemical reaction on one surface. What is worth noting is that because free energy reduction is common for all reactions, the cantilever-based sensing is universal platform for studying reactions. The ability of analyzing molecules without the use of optical or radioactive labels makes this approach rather attractive for biology and medicine.

As DNA microarrays successfully provide means to study genomics in a high-throughput manner, various protein microarrays have been under development to enable quantitative and rapid protein analysis. Researchers have developed protein microarrays

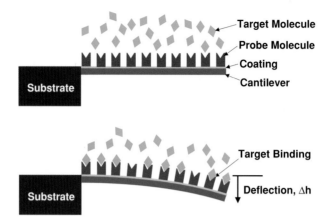

FIGURE 2.1. Specific biomolecular interactions between target and probe molecules alter intermolecular interactions within a self-assembled monolayer on one side of a cantilever beam. This can produce a sufficiently large surface stress to bend cantilever beam and generate motion.

similar to DNA microarrays based on fluorescence technology [2, 9, 10, 17, 22], chemiluminescence detection [2, 9, 13, 15, 16, 33] or radioisotope labeling [8, 39]. Despite the popularity of these protein microarrays, they share a major disadvantage in that labeling is necessary for detections. While the labeling process is itself costly and time-consuming, the issue is more acute since proteins are very labile and their activities can be readily affected by labeling [10]. Surface plasmon resonance (SPR) and the cantilever platform thus are very attractive as both of them are label-free technologies and are capable of yielding kinetic as well as equilibrium information of the biomolecuar interactions. However, multiplexing SPR remains challenging because binding in SPR is measured as an integrated effect over a relatively large area of the surface. Measuring binding of small molecules is another limitation of SPR. For example, it is difficult for SPR to detect phosphorylation of proteins, but it is possible to do so with cantilevers. Hence, the microcantilever platform offers an unparalleled opportunity for the development and mass production of extremely sensitive, low-cost sensors for real-time sensing of many chemical and biochemical species. A major advantage of using cantilevers is that it is a common platform for all reactions because it measures free energy change.

Microcantilevers have been demonstrated as chemical and biological sensors in past decade [4, 5, 6, 7, 12, 21, 28, 29, 32, 34, 35]. While all earlier work focused on studying individual reactions using single cantilever system similar to the one in Figure 2.2, a truly high-throughput technique was missing. Microcantilevers are readily adapted for fabricating multi-element sensor arrays, thus allowing high-throughput multi-analyte detection. In this chapter, we will discuss the principles of operations and the emergence of functional microcantilever arrays.

2.2. THEORY

As molecular reactions on a surface is ultimately driven by free energy reduction of the surface, the free energy reduction leads to a change in surface tension or surface stress.

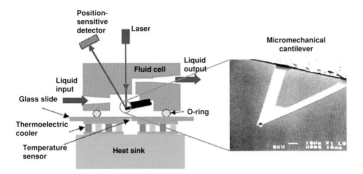

FIGURE 2.2. Schematic diagram of the experimental setup used by Wu et al. [35]. A cantilever was mounted in a fluid cell which allows liquid exchange through I/O ports. A laser was reflected off the cantilever and focused onto a PSD.

While this produces no observable macroscopic change on the surface of a bulk solid, the adsorption-induced surface stresses are sufficient to bend a cantilever if the adsorption is confined to one surface of the beam. However, adsorption-induced forces should not be confused with bending due to dimensional changes such as swelling of thicker polymer films on cantilevers. The sensitivity of adsorption-induced stress sensors can be three orders of magnitude higher than those of frequency variation mass sensors (for resonance frequencies in the range of tens of kHz) [30]. Moreover, the static cantilever bending measurement is ideal for liquid-based applications where frequency-based cantilever sensors suffer from huge viscous damping.

Using Stoney's formula, the deflection at the end of a cantilever, z, can be related to the differential surface stress, σ, as [20, 26]

$$z = \frac{3\sigma(1-v)}{E}\left(\frac{L}{d}\right)^2 \qquad (2.1)$$

where d and L are the cantilever beam thickness and length, respectively; E and v are the elastic modulus and the Poisson ratio of the cantilever material, respectively. Equation 2.1 shows a linear relation between cantilever bending and differential surface stress. For a silicon nitride cantilever of 200 μm long and 0.5 μm thick, with $E = 8.5 \times 10^{10}$ N/m^2 and $v = 0.27$ [36], a surface stress of 0.2 mJ/m^2 will result in a deflection of 1 nm at the end. Because a cantilever's deflection strongly depends on geometry, the surface stress change, which is directly related to biomolecular reactions on the cantilever surface, is a more convenient quantity of the reactions for comparison of various measurements. Changes in free energy density in biomolecular reactions are usually in the range of 1 to 50 mJ/m^2, or as high as 900 mJ/m^2.

The ultimate noise of a cantilever sensor is the thermal vibrational motion of the cantilever. It can be shown from statistical physics [23] that for off-resonance frequencies, thermal vibrations produce a white noise spectrum such that the root-mean-square vibrational noise, h_n, can be expressed as

$$h_n = \sqrt{\frac{2k_B TB}{k\pi f_0 Q}} \qquad (2.2)$$

Here, k_B is the Boltzmann constant (1.38×10^{-23} J/K), T is the absolute temperature (300 K at room temperature), B is the bandwidth of measurement (typically about 1000 Hz for dc measurement), f_0 is the resonance frequency of the cantilever, k is the cantilever stiffness, and Q is the quality factor of the resonance with is related to damping (in liquid $Q \sim 1$). Thermal vibration noise of silicon or silicon nitride cantilevers generally falls in the sub-nm range, which is negligible compared to noise from the detection system. The practical sensing limitation by noise will be discussed in the following for each detection technique.

2.3. READOUT TECHNIQUES

The most common readout technique for cantilever deflection is the optical beam deflection technique. Deflection of a cantilever is transduced into the change in the direction of a light beam reflected off the cantilever. Interferometry optics has also been adapted to read out cantilever motion, in which the deflection is detected by its relative movement to a reference cantilever or substrate. Another attractive readout technique is based on piezoresistivity, whereby the bulk electrical resistivity varies with applied stress. Here, we describe a few examples on how these techniques have been applied to develop cantilever microarrays.

2.3.1. Optical Beam Deflection of 1D Cantilever Array

In an optical beam deflection readout system, a light beam from a laser is focused at the end of the cantilever and reflected to a position sensitive detector (PSD) [24]. The bending of the cantilever results in a large change in the direction of the reflected beam, which can be detected by the PSD signal. Direct multiplexing of such readout of N cantilevers requires the same number of light sources and detectors. Though it could be realized for a few cantilevers, it is far from practical to develop high-throughput cantilever arrays. Using a better approach of multiplexing, Lang *et al.* [14] demonstrated sequential position readout from an array of eight cantilevers for gas sensing (Figure 2.3). Light from eight individual light sources was coupled into an array of multimode fibers and guided onto the sensor array. Upon reflection, the light was collected by a PSD. The eight light sources were switched on and off individually and sequentially at 1.3Hz. Using a time-multiplexed vertical cavity surface-emitting laser (VCSEL) array and a single linear PSD, the same group further developed this technique for sensing of DNA hybridization and protein interactions [3, 7, 19]. The cantilever deflection is calculated with an accuracy of 0.1 nm [20]. However, scaling-up of this readout technique to thousands of cantilevers is very challenging. As one light source is used for each cantilever, it will be difficult to implement for arrays of thousands of cantilevers. On the other hand, linear PSD will not be suitable for detecting the deflections of cantilevers in 2D format.

2.3.2. Optical Beam Deflection of 2D Array

Yue *et al.* [36, 37] developed an innovative whole-field optical readout system for 2D cantilever array based on optical beam deflection technique. The cantilevers are specially

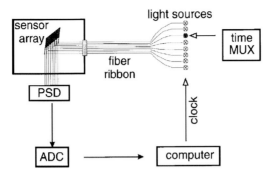

FIGURE 2.3. Schematic setup of optical beam deflection readout system for 1D cantilever array. Quasi-simultaneous readout of eight sensors is achieved by time-multiplexing (MUX) eight light sources which are guided by an optical fiber-ribbon onto the sensor array located in the analysis chamber. The reflected light from the sensors' surface is collected by a PSD, then digitized by an analog-to-digital converter (ADC) and stored in a computer memory for further analysis. The computer also generates the clock pulse for time-multiplexing [14].

designed to enable multiplexed optical readout, as shown in Figure 2.4. Similar to traditional cantilevers, low-pressure chemical vapor deposited (LPCVD) low-stress silicon nitride (SiN_x) was used as the structural material of the cantilevers. A thin gold film was deposited and patterned on one side of the cantilevers to allow immobilization of biomolecules through gold-thiol (Au-S) bonds, as well as to cause enough initial bending in the cantilever beam. However, the rigid paddle at the end of the cantilever, which could act as a flat mirror, made the cantilever different from other traditional ones. The high rigidity of the paddle, or its flatness, was achieved through a close square ridge structure on the paddle that produced a high moment of inertia in that region. The thin arm of the cantilever was usually curved due to the residue stress in and between the gold and silicon nitride layers. When a collimated light beam illuminated the whole area of a cantilever array, the initial curvature of each cantilever diverged the reflection from the thin beam so that only

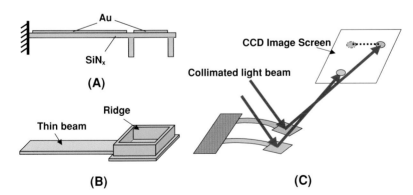

FIGURE 2.4. Innovative cantilever design for optical readout of 2D array. (A) Side view of a cantilever made of silican nitride. Top surface is coated with gold. (B) 3D illustration of the cantilever and the ridge structure on the paddle. (C) A collimated light beam illumination two cantilevers simultaneously. Only reflection from the paddles can be collected on an image screen.

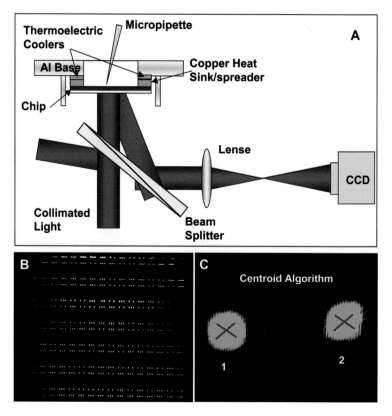

FIGURE 2.5. (A) Whole-field optical readout system; (B) A CCD snap shot of about 500 spots, each spot corresponding to the reflection of the laser beam from the paddle of a cantilever; (C) Individual spot tracking using a centroid algorithm in Matlab.

reflection from the flat paddles could be collected. The initial curvature of cantilevers also enabled the reflections from the flat paddles at the end of the cantilevers form collimated beams in a particular direction, which could then be separated from spurious reflections and directed towards a charged couple device (CCD) camera for imaging (Figure 2.5A). If a cantilever bends, the angle of its paddle and thereby the direction of the reflected light will change, causing the spot to move on the CCD screen. Figure 2.5B shows a CCD image of an entire cantilever array chip, where each spot corresponds to the reflection from the paddle of an individual cantilever. Any motion of a cantilever leads to corresponding motion of the CCD spot, which can be quantified using ray optics. Following acquisition by the CCD camera, images were transferred to a Matlab script, which tracks each cantilever paddle image "spot", by calculating the intensity centroid of each spot (Figure 2.5C).

The thermal bimorph effect of the SiN_x-Au cantilevers results in cantilever motion when temperature fluctuates. A 10 mK change in the temperature can cause ∼2.5 nm end deflection of a silicon nitride cantilever (200 nm long, 0.5 µm thick) with 25-nm thick gold coating. Another major source of the measurement noise is the photon shot noise of the CCD camera. With a typical optical setup, the measurement noise caused by CCD shot noise equals about 3-nm deflection of the same kind of cantilever [38]. The total system noise is the superposition of major noises and can be equivalent to change in surface stress of 0.5 mJ/m^2.

2.3.3. Piezoresistive Cantilever Array

Besides optical approaches, cantilever motion can also be read out electronically. Doped silicon exhibits a strong piezoresistive effect [31]. The resistance of a doped region on a cantilever can change reliably when the cantilever is stressed with deflection. Thaysen et al. [27] developed piezoresistive cantilever sensors with integrated differential readout. Each cantilever had a thin fully encapsulated resistor made of doped Si fabricated on top, of which the resistance would change due to any load on the cantilever. Each sensor was comprised of a measurement cantilever and a built-in reference cantilever, which enabled differential signal readout. The two cantilevers were connected in a Wheatstone bridge and the surface-stress change on the measurement cantilever was detected as the output voltage from the Wheatstone bridge. The researchers later applied the sensor for DNA sensing [18]. The typical signal-to-noise ratio of the resistance measurement was 26 during the experiments. For cantilevers which were 150 μm long, 40 μm wide and 1.3 μm thick, the surface stress sensitivity was $\Delta R/R\sigma^{-1} = 0.44$ mJ/m^2, where $\Delta R/R$ was the relative change in the resistance of integrated piezoresistor. The sensor was determined to have a minimum detectable surface-stress change of approximately 5 mJ/m^2.

Electro-readout technique has several advantages over optical leverage. As electronics for detection is integrated, the sensors can be operated in any solutions, even non-transparent liquids, since the refractive indices of the liquids do not influence the detection. Because no external optics components are required, the sensors integrated with readout electronics can be made very portable suitable for field detections. It is also easy to realize an array of cantilevers with integrated readout, as both cantilevers and readout circuits can be fabricated simultaneously. However, the piezoresistive cantilever sensors developed so far are one order of magnitude less sensitive than those using optical readout techniques. One has to overcome the challenges of improving the sensitivity to develop piezoresistive cantilever arrays for high-throughput biomolecular sensing.

2.4. MICROFLUIDICS

One of the challenges in multiplexing is how to functionalize individual cantilevers. Some researchers have achieved this by inserting cantilevers to microcapillary arrays separately [3, 19]. While this is acceptable for 1D arrays, such an approach is difficult to implement for 2D arrays. Integrating microfluidic chambers with cantilevers provides physical separation for cantilevers thus a direct means for multiplexed experiments. Figure 2.6 illustrates a microfluidic reaction chamber comprised cantilevers, silicon substrate and glass cap [37]. Each such reaction well contains a large fluidic inlet (called big I/O) and two small fluidic outlets (called small I/O). The small I/O is designed to prevent vapor bubbles to be trapped, such that when a fluid sample is injected into the big I/O the gas was ejected through the small I/O's. To effectively combat the inaccuracies arising from sensor drift and fabrication variations between sensors, the design also includes multiple cantilevers per reaction chamber, each of which received the same analytes at all stages of an experiment. The response from all the sensors in a given reaction well

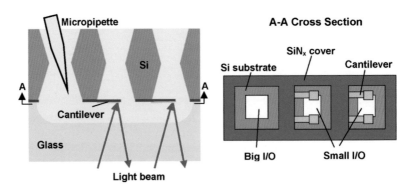

FIGURE 2.6. Schematic diagrams of fluidic design (side view of a bonded reaction well and top view of the Si chip). A single reaction well containing fluidic inlets and outlets in the silicon chip, multiple cantilevers, and the transparent glass/PDMS cover for the laser beam to be used for measuring cantilever deflection.

could then be used to obtain a more statistically relevant response for each well. In order to use the cantilever array chip as a multiplexed sensor array, each reaction well must be physically separated from the neighboring wells. This is achieved using a pyrex substrate that is patterned and etched to produce the reaction well, and bonded to the silicon chip. The bonding is accomplished using an adhesive stamping technique [25]. The fabrication process for the cantilever array chip utilizes conventional microelectromechanical systems (MEMS) fabrication including bulk and surface micromachining, which are described in detail by Yue et al. [36, 37]. The yield (percent of cantilevers on each chip surviving the fabrication process) achieved from this fabrication process ranged from 95–98%. Figure 2.7 shows optical and electron micrographs of the cantilever array chip.

2.5. BIOMOLECULAR REACTION ASSAYS

The microcantilever arrays enable multiplexed label-free analysis for various biomolecular reactions. In this section, we describe recent work on detecting specific biomolecular interactions such as DNA hybridization and antibody-antigen bindings [7, 36, 37].

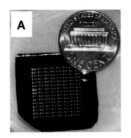 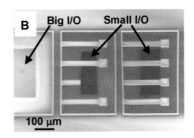

FIGURE 2.7. (A) A cantilever array chip containing a 2-D array of reaction wells, each well containing multiple cantilevers. The array is roughly the size of a penny; (B) Electron micrograph of a single reaction well showing 7 cantilever beams, a big inlet/outlet (I/O) port and two small I/O ports.

CANTILEVER ARRAYS

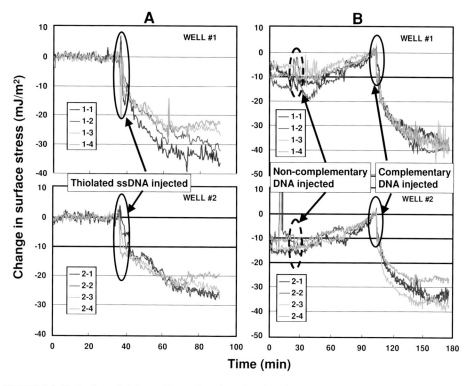

FIGURE 2.8. Deflections of eight cantilevers plotted as a function of time for: (A) DNA immobilization in wells 1 and 2, each well containing four cantilevers; (B) DNA hybridization in the two wells. Dashed circles represent the injection of non-complementary DNA. Solid circles represent the injection of complementary DNA.

2.5.1. Detection of DNA

Single-stranded DNA (ssDNA) can be immobilized using gold-thiol strong bonding on one side of a cantilever by coating that side with gold and using a thiol linker at one end of ssDNA. Single-stranded DNA bound to the cantilever acts as the probe (or receptor) molecule for the target complementary strands. Figure 2.8A shows surface stress change in the cantilevers as a function of time when ssDNA was bound to the cantilevers, *a.k.a.* probe immobilization. In this case, thiolated ssDNA (25-mer oligonucleotide) was injected into two different wells, each of which contained 4 cantilevers respectively. The motion of the cantilevers in multiple wells was monitored simultaneously. Such immobilization resulted in a surface stress change of approximately 25 ± 5 mJ/m^2. The cantilevers were washed several times and re-equilibrated in phosphate buffer after the immobilization was complete. Afterwards, 8μM non-complementary DNA was first injected into the wells in which the cantilevers were functionalized with the thiolated-ssDNA. Only marginal deflection was observed for the non-specific binding (Figure 2.8B). The 5μM complementary DNA was injected to these wells after an hour or so. The specific binding between the DNA strands caused significant deflection of all the cantilevers, corresponding to the surface stress change of 35 ± 5 mJ/m^2. These experiments clearly demonstrated that ssDNA immobilization and DNA hybridization on the gold surface induced significant cantilever deflection while

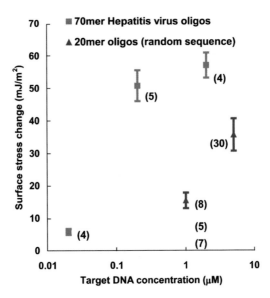

FIGURE 2.9. Summary of quantitative cantilever response to DNA hybridization plotted as a function of target DNA concentration. The numbers in the parenthesis denotes the number of reaction used for statistical analysis.

the deflection from non-specific binding was almost negligible. As evident in Figure 2.8, both reaction steps produced repeatable deflections from the four cantilevers within the same well. Furthermore, the cantilevers in different wells also showed the same degree of deflections, indicating the well-to-well consistency. Figure 2.9 summarizes the quantitative experimental results obtained for DNA hybridization, with DNA of different length and for different target concentrations. Each point represents the average value of the hybridization signals obtained from multiple cantilevers and the error bar is the standard deviation of the signals. The number in the parenthesis next to each point is the number of the cantilevers from which the signals were obtained. It is very clear that the hybridization at lower target DNA concentration caused smaller deflection of the cantilevers, which indicates the equilibrium of the DNA hybridization reaction depended on the DNA concentrations. Figure 2.9 also shows that the hybridization between longer DNA single-strands resulted in larger deflection of the cantilevers, which suggests the total free energy reduction in longer DNA's hybridization is more than that of the shorter ones.

These experiments clearly demonstrate the capability of the multiplexed cantilever chip to quantitatively detect DNA immobilization and hybridization. The platform allows one to rapidly search the parameter space of DNA hybridization, and thus help to understand the origin of nanomechanical forces that lead to cantilever deflection, as well as the dependence of such deflection on the identity and concentration of the target molecules.

2.5.2. Detection of PSA

Antibody-antigen interactions are a class of highly specific protein-protein bindings that play a critical role in molecular biology. When antibody molecules were immobilized to one surface of a cantilever, specific binding between antigens and antibodies produced surface

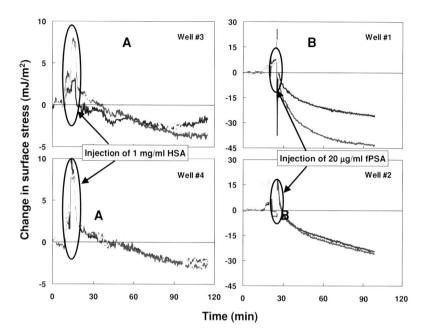

FIGURE 2.10. Deflections of 12 cantilevers in 4 wells plotted as a function of time for protein binding. (A) Non-specific binding of HSA to the MAH-PSA on cantilever surface; (B) Specific binding between injected PSA and the MAH-PSA on cantilever surface.

stress change in the cantilever. Figure 2.10 shows surface stress change in cantilevers as a function of time for quantitative detection of prostate-specific antigen (PSA) using human serum albumin (HSA) as controls. The antibody specific to PSA, mouse anti-human antibody (MAH-PSA), was immobilized to the gold surface of cantilevers using a cross-linker, 3,3′-Dithiobis [sulfo-succinimidylpropionate] (DTSSP). 2-[Methoxy(polyethylenoxy)propyl] trimethoxysilane (PEG-silane) was immobilized on the nitride surface of the cantilevers to block the non-specific absorption of proteins [37]. As HSA only non-specifically binds to MAH-PSA, the cantilevers deflected negligibly upon injection of HSA to the chambers which contained cantilevers functionalized with MAH-PSA. Injection of PSA to other two similar wells resulted in surface stress change of 30 ± 10 mJ/m^2 due to the specific binding between the antigens and antibodies on the cantilever surface. These experiments clearly demonstrated multiplexed protein interaction assay using the microcantilever array.

2.6. CONCLUSIONS

Experimental and theoretical research has shown that when reactions on one surface a cantilever beam, they induce cantilever bending. This occurs increase the free energy reduction of the reaction even at the cost of increase free energy of the cantilever by bending it. Since the cantilever strain energy can be easily calculated, its value provides a quantitative measure of the free energy density of a surface reaction. Since free energy reduction is the common driving force for all reactions in nature, cantilevers form a universal platform for studying all reactions. Furthermore, it is a label-free approach, which can be easily multiplexed to study thousands of reactions simultaneously. In this chapter, we discuss the latest developments in the technology of cantilever arrays and how it can be used to study specific biomolecular reactions involving nucleic acids and proteins.

REFERENCES

[1] P. Angenendt, J. Glökler, Z. Konthur et al. 3D protein microarrays: performing multiplex immunoassays on a single chip. *Anal. Chem.*, 75:4368–4372, 2003.
[2] P. Arenkov, A. Kukhtin, A. Gemmell, S. Voloshchuk, V. Chupeeva, and A. Mirzabekov. Protein microchips: use for immunoassay and enzymatic reactions. *Anal. Biochem.*, 278:123, 2000.
[3] Y. Arntz, J.D. Seelig, H.P. Lang, J. Zhang, P. Hunziker, J.P. Ramseyer, E. Meyer, M. Hegner, and C. Gerber. Label-free protein assay based on a nanomechanical cantilever array. *Nanotechnology*, 14:86, 2003.
[4] R. Berger, E. Delamarche, H. Lang, C. Gerber, J. K. Gimzewski, E. Meyer, and H. J. Güntherodt,. Surface stress in the self-assembly of alkanethiols on gold. *Science*, 276:2021, 1997.
[5] H.J. Butt. A sensitive method to measure changes in surface stress of solids. *J. Coll. Inter. Sci.*, 180:251, 1996.
[6] G.Y. Chen, T. Thundat, E. A. Wachter, and R.J. Warmack. Adsorption-induced surface stress and its effects on resonance frequency of microcantilevers. *J. Appl. Phys.*, 77(8):3618, 1995.
[7] J. Fritz, M.K. Baller, H.P. Lang, H. Rothuizen, P. Vettiger, E. Meyer, H.J. Güntherodt, C. Gerber, and J.K. Gimzewski. Translating biomolecular recognition into nanomechanics. *Science*, 288:316, 2001.
[8] H. Ge. UPA, a universal protein array system for quantitative detection of protein-protein, protein-DNA, protein-RNA and protein-ligand interactions. *Nucleic Acids Res.*, 28:e3, 2000.
[9] D. Guschin, G. Yershov, A. Zaslavsky et al. Manual manufacturing of oligonucleotide, DNA, and protein chips. *Anal. Biochem.*, 250:203, 1997.
[10] B.B. Haab, M.J. Dunham, and P.O. Brown. Protein microarrays for highly parallel detection and quantitation of specific proteins and antibodies in complex solutions. *Genome Biol.*, 2:R4, 2001.
[11] M.F. Hagan, A. Majumdar, and A.K. Chakraborty. Nanomechanical forces generated by surface grafted DNA *J. Phys. Chem. B*, 106:10163, 2002.
[12] K. Hansen, H. Ji, G. Wu, R. Datar, R. Cote, and A. Majumdar. Cantilever-based optical deflection assay for discrimination of DNA single-nucleotide mismatches. *Anal. Chem.*, 73:1567, 2001.
[13] R.-P. Huang, R. Huang, Y. Fan, and Y. Lin. Simultaneous detection of cytokines from conditioned media and patient's sera by an antibody-based protein array system. *Anal. Biochem.*, 294:55, 2001.
[14] H.P. Lang, R. Berger, C. Andreoli, J. Brugger, M. Despont, P. Vettiger, Ch. Gerber, and J.K. Gimzewski. Sequential position readout from arrays of micromechanical cantilver sensors. *Appl. Phys. Lett.*, 72:383, 1998.
[15] T.O. Joos, M. Schrenk, P. Höpfl, K. Kroger, U. Chowdhury, D. Stoll, D. Schörner, M. Dürr, K. Herick, S. Rupp, K. Sohn, and H. Hämmerle. A microarray enzyme-linked immunosorbent assay for autoimmune diagnostics. *Electrophoresis*, 21:2641, 2000.
[16] V. Knezevic, C. Leethanakul, V.E. Bichsel, J.M. Worth, V.V. Prabhu, J.S. Gutkind, L.A. Liotta, P.J. Munson, E.F. Petricoin, and D.B. Krizman. Proteomic profiling of the cancer microenvironment by antibody arrays. *Proteomics*, 1:1271, 2001.
[17] G. MacBeath and S.L. Schreiber. Printing proteins as microarrays for high-throughput function determination. *Science*, 289:1760, 2000.
[18] R. Marie, H. Jensenius, J. Thaysen, C.B. Christensen, and A. Boisen. Adsorption kinetics and mechanical properties of thiol-modified DNA-oligos on gold investigated by microcantilever sensors. *Ultramicroscopy*, 91:29, 2002.
[19] R. McKendry, J. Zhang, Y. Arntz, T. Strunz, M. Hegner, H.P. Lang, M.K. Baller, U. Certa, E. Meyer, H.-J. Guntherodt, and C. Gerber. Multiple label-free biodetection and quantitative DNA-binding assays on a nanomechanical cantilever array. *Proc. Natl. Acad. Sci.*, 99:9783, 2002.
[20] T. Miyatani and M. Fujihira. Calibration of surface stress measurements with atomic force microscopy. *J. Appl. Phys.*, 81:7099, 1997.
[21] R. Raiteri, G. Nelles, H.-J. Butt, W. Knoll, and P. Skládal. Sensing biological substances based on the bending of microfabricated cantilevers. *Sens. Actu. B*, 61:213, 1999.

[22] C.A. Rowe, S.B. Scruggs, M.J. Feldstein, and J.P. Golden, Ligler. An array immunosensor for simultaneous detection of clinical analytes. *Anal. Chem.*, 71:433, 1999.
[23] M.V. Salapaka, S. Bergh, J. Lai, A. Majumdar, and E. McFarland. Multimode noise analysis for scanning probemicroscopy. *J. Appl. Phys.*, 81:2480, 1997.
[24] D. Sarid. *Scanning Force Microscopy.* Oxford University Press, New York, 1994.
[25] S. Satyanarayana, R. Karnik, and A. Majumdar. Stamp and stick room temperature bonding technique for microdevices. *J. MEMS*, (in press), 2004.
[26] G.G. Stoney. The tension of metallic films deposited by electrolysis. *Proc. Roy. Soc. Lond. A*, 82:172, 1909.
[27] J. Thaysen, A. Boisen, O. Hansen, and S. Bouwstra. Atomic force microscopy probe with piezoresistive read-out and a highly symmetrical Wheatstone bridge arrangement. *Sens. Actuators A*, 83:47, 2000.
[28] T. Thundat, R.J. Warmack, G.Y. Chen, and D.P. Allison. Thermal and ambient-induced deflections of scanning force microscope cantilevers. *Appl. Phys. Letts.*, 64:2894, 1994.
[29] T. Thundat, P.I. Oden, and R.J. Warmack. Microcantilever sensors. *Microscale Thermophys. Eng.*, 1:185, 1997.
[30] T. Thundat and A. Majumdar. Microcantilevers for physical, chemical, and biological sensing. In F.G. Barth, J.A.C. Humphrey, T.W. Seecomb, (eds.) *Sensors and Sensing in Biology and Engineering.* Springer-Verlag, New York, 2003.
[31] O.N. Tufte and E.L. Stelzer. Piezoresistive properties of silicon diffused layers. *J. Appl. Phys.*, 34:323, 1993.
[32] E.A. Wachter and T. Thundat. Micromechanical sensor for chemical and physical measurements. *Rev. Sci. Instrum.*, 66:3662, 1995.
[33] R. Wiese, Y. Belosludtsev, T. Powdrill et al. Simultaneous multianalyte ELISA performed on a microarray platform. *Clini. Chem.*, 47(8):1451, 2001.
[34] G. Wu, H. Ji, K. Hansen, T. Thundat, R. Datar, R. Cote, M.F. Hagan, A.K. Chakraborty, and A Majumdar. Origin of nanomechanical cantilever motion generated from biomolecular interactions. *Proc. Natl. Acad. Sci.*, 98:1560, 2001a.
[35] G. Wu, R. Datar, K. Hansen, T. Thundat, R. Cote, and A. Majumdar. Bioassay of prostate specific antigen (PSA) using microcantilevers. *Nat. Biotechnol.*, 19:856, 2001b.
[36] M. Yue, H. Lin, D.E. Dedrick, S. Satyanarayana, A. Majumdar, A.S. Bedekar, J.W. Jenkins, and S. Sundaram. A 2-D microcantilever array for multiplexed biomolecular analysis. *J. MEMS*, 13:290, 2004a.
[37] M. Yue, J.C. Stachowiak, and A. Majumdar. Cantilever arrays for multiplexed mechanical analysis for biomolecular reactions. *MCB*, (in press), 2004b.
[38] M. Yue. High-Throughput Bioassays Based on Nanomechanics and Nanofluidics. Ph.D. Dissertation, University of California, Berkeley, in preparation, 2004.
[39] H. Zhu, J.F. Klemic, S. Chang, P. Bertone, A. Casamayor, K.G. Klemic, D. Smith, M. Gerstein, M.A. Reed, and M. Snyder. Analysis of yeast protein kinase using protein chips. *Nat. Gene.*, 26:283, 2000.

3

An On-Chip Artificial Pore for Molecular Sensing

O. A. Saleh[1] and L. L. Sohn[2]

[1] *Laboratoire de Physique Statistique, Ecole Normale Supérieure, Paris, France*
[2] *Dept. of Mechanical Engineering, University of California, Berkeley, CA. USA*

3.1. INTRODUCTION

Currently, a variety of strategies for developing nanopores for molecular sensing exist—from engineering transmembrane protein pores so that they can detect sequence-specific DNA strands with single-base resolution [1, 2] to "drilling" molecular-scaled holes into silicon nitride membranes to detect the presence of single molecules of DNA [3, 4] to employing gold [5] or carbon [6] nanotubes as the ultimate artificial pores. While all of these strategies have shown early success in molecular sensing, there are major technological hurdles one must overcome: reproducibly creating an effective pore, maintaining a pore's stability over a period of time, and integrating the pore into a device that is inexpensive and easy to fabricate and use. Here, we describe our group's effort in developing a fully-integrated artificial pore on chip for molecular sensing. As we will demonstrate, our pore addresses the technological hurdles with which other nanopore strategies are confronted. Equally important, we will show that our on-chip artificial pore is a flexible platform technology that has a number of diverse applications—from label-free immunoassays to single-molecule DNA sizing.

3.2. THE BASIC DEVICE: FABRICATION AND MEASUREMENT

The on-chip artificial pore we have developed is based on the Coulter counter technique of particle sizing [7]. Coulter counters typically consist of two reservoirs of particle-laden solution separated by a membrane and connected by a single pore through that membrane.

By monitoring changes in the electrical current through the pore as individual particles pass from one reservoir to another, a Coulter counter can measure the size of particles whose dimensions are on the order of the pore dimensions. While this method has long been used to characterize cells several microns in diameter [8, 9], its relative simplicity has led to many efforts to employ it to detect nanoscale particles [10, 11].

Our on-chip artificial pore was initially fabricated on top of a quartz substrate using standard microfabrication techniques [12]. We utilized a four-terminal measurement of the current through the pore. Because we are able to control precisely the pore dimensions (which we can easily measure using optical and atomic force microscopies), we can predict quantitatively the response of our pore to various-sized particles. We have fabricated pores with lateral dimensions between 400 nm and 1 μm, and used them to detect latex colloidal particles as small as 87 nm in diameter. As we will show, our device is easily integratable with other on-chip analysis systems.

3.2.1. Fabrication of the Pore

Our device, shown in Fig. 3.1, is fabricated in multiple stages. Each stage consists of lithographic pattern generation, followed by pattern transfer onto a quartz substrate using either reactive ion etching (RIE) or metal deposition and lift-off. The first stage is the fabrication of the pore. A line is patterned on the substrate using either photolithography (PL) for line widths ≥1 μm, or electron-beam lithography (EBL) for line widths between 100 and 500 nm, and then etched into the quartz using a CHF_3 RIE. The substrate subsequently undergoes a second stage of PL and RIE to define two reservoirs that are 3.5 μm deep, separated by 10 μm, and connected to each other by the previously-defined channel. The length of the pore is defined in this second stage by the separation between the two reservoirs. The final stage consists of patterning four electrodes across the reservoirs, followed by two

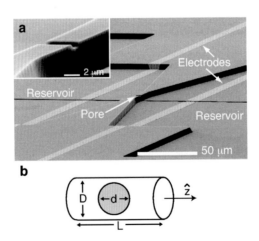

FIGURE 3.1. (a) Scanning electron micrograph of our on-chip artificial pore. The 3.5 μm deep reservoirs and the inner Ti/Pt electrodes, which control the voltage applied to the pore but pass no current, are only partially shown. The outer electrodes, which inject current into the solution, are not visible in this image. The inset shows a magnified view of this device's pore, which has dimensions $5.1 \times 1.5 \times 1.0$ μm^3. (b) A schematic diagram of a spherical particle of diameter d in a pore of diameter D and length L. [From Ref. 12.]

depositions of 50 Å/250 Å Ti/Pt in an electron-beam evaporator with the sample positioned 45 degrees from normal to the flux of metal to ensure that the electrodes are continuous down both walls of the reservoirs.

The device is sealed on top with a silicone-coated (Sylgard 184, Dow Corning Corp.) glass coverslip before each measurement. Prior to sealing, both the silicone and the substrate are oxidized in a DC plasma to insure the hydrophilicity [13] of the reservoir and pore and to strengthen the seal [14] to the quartz substrate. After each measurement, the coverslip is removed and discarded, and the substrate is cleaned by chemical and ultrasonic methods. Thus, each device can be reused many times.

3.2.2. Pore Measurement

We have measured solutions of negatively-charged (carboxyl-coated) latex colloids (Interfacial Dynamics, Inc.) whose diameters range from 87 nm to 640 nm using the device we just described. All colloids were suspended in a solution of 5x concentrated TBE buffer with a resistivity of 390 Ω-cm and pH 8.2. To reduce adhesion of the colloids to the reservoir and pore walls, we added 0.05% v/v of the surfactant Tween 20 to every solution. The colloidal suspensions were diluted significantly from stock concentrations to avoid jamming of colloids in the pore; typical final concentrations were $\sim 10^8$ particles/mL. The pore and reservoirs were filled with solution via capillary action.

The sensitivity of a Coulter counter relies upon the relative sizes of the pore and the particle to be measured. The resistance of a pore R_p increases by δR_p when a particle enters since the particle displaces conducting fluid. δR_p can be estimated [9] for a pore aligned along the z-axis (see Fig. 3.2) by

$$\delta R_p = \rho \int \frac{dz}{A(z)} - R_p \qquad (3.1)$$

where $A(z)$ represents the successive cross sections of the pore containing a particle, and ρ is the resistivity of the solution. For a spherical particle of diameter d in a pore of diameter D and length L, the relative change in resistance is

$$\frac{\delta R_p}{R_p} = \frac{D}{L} \left[\frac{\arcsin(d/D)}{(1 - (d/D^2))^{1/2}} - \frac{d}{D} \right] \qquad (3.2)$$

Eqs. (3.1) and (3.2) assume that the current density is uniform across the pore, and thus is not applicable for cases where the cross section $A(z)$ varies quickly, i.e. when $d \ll D$. For that particular case, Deblois and Bean [10] formulated an equation for δR_p based on an approximate solution to the Laplace equation:

$$\frac{\delta R_p}{R_p} = \frac{d^3}{LD^2} \left[\frac{D^2}{2L^2} + \frac{1}{1\sqrt{1 + (D/L)^2}} \right] F\left(\frac{d^3}{D^3}\right) \qquad (3.3)$$

where $F(d^3/D^3)$ is a numerical factor that accounts for the bulging of the electric field lines into the pore wall. When employing Eq. (3.3) to predict resistance changes, we find

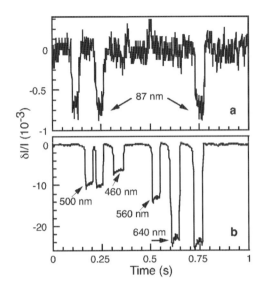

FIGURE 3.2. Relative changes in baseline current $\delta I/I$ vs time for (a) a monodisperse solution of 87 nm diameter latex colloids measured with an EBL-defined pore of length 8.3 μm and cross section 0.16 μm², and (b) a polydisperse solution of latex colloids with diameters 460, 500, 560, and 640 nm measured with a PL-defined pore of length 0.5 μm and cross section 1.2 μm². Each downward pulse represents an individual particular entering the pore. The four distinct pulse heights in (b) correspond as labeled to the four different colloid diameters. [From Ref. 12.]

an effective value for D by equating the cross sectional area of our square pore with that of a circular pore.

If R_p is the dominant resistance of the measurement circuit, then relative changes in the current I are equal in magnitude to the relative changes in the resistance, $|\delta I/I| = |\delta R_p/R_p|$, and Eqs. (3.2) and (3.3) can both be directly compared to measured current changes. This comparison is disallowed if R_p is similar in magnitude to other series resistances, such as the electrode/fluid interfacial resistance, $R_{e/f}$, or the resistance R_u of the reservoir fluid between the inner electrodes and the pore. We completely remove $R_{e/f}$ from the electrical circuit by performing a four-point measurement of the current (see Fig. 3.1a). We minimize R_u by placing the inner electrodes close to the pore (50 μm away on either side), and by designing the reservoir with a cross section much larger than that of the pore. For a pore of dimensions 10.5 μm by 1.04 μm² we measured $R_p = 36$ MΩ, in good agreement with the 39 MΩ value predicted by the pore geometry and the solution resistivity. This confirms that we have removed R_u and $R_{e/f}$ from the circuit.

Fig. 3.2 shows representative data resulting from measuring a monodisperse solution of colloids 87 nm in diameter with an EBL-defined pore (Fig. 3.1a), and from measuring a polydisperse solution containing colloids of diameters 460 nm, 500 nm, 560 nm, and 640 nm with a PL-defined pore (Fig. 3.1b). Each downward current pulse in Fig. 3.2, corresponds to a single colloid passing through the pore. For the data shown, 0.4 V was applied to the pore. In other runs, the applied voltage was varied between 0.1 and 1 V to test the electrophoretic response of the colloids. We found that the width of the downward current pulses varied

AN ON-CHIP ARTIFICIAL PORE FOR MOLECULAR SENSING

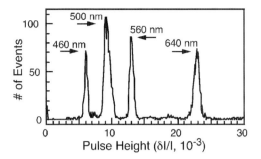

FIGURE 3.3. A histogram of pulse heights resulting from measuring the polydisperse solution shown in Figure 3.2(b). The resolution of this particular device is ±10 nm in diameter for the particles measured. [From Ref. 12.]

approximately as the inverse of the applied voltage, as is expected for simple electrophoretic motion.

Fig. 3.3 shows a histogram of ∼3000 events measured for the polydisperse solution. The histogram shows a very clear separation between the pore's response to the differently-sized colloids. The peak widths in Fig. 3.3 represent the resolution of this device, which we find to be ±10 nm in diameter for the measured colloids. This precision approaches the intrinsic variations in colloid diameter of 2–4%, as given by the manufacturer. In this run, the maximum throughput was 3 colloids/s, a rate easily achievable for all of our samples. Event rates are limited by the low concentrations needed to avoid jamming.

We used a device whose pore size was 10.5 µm by 1.04 µm^2 to measure colloids ranging from 190 nm to 640 nm in diameter. Figure 3.4 shows the comparison between the measured mean pulse heights and those predicted by Eqs. (3.2) and (3.3). As shown, there is excellent agreement between the measured and calculated values, with the measured error insignificant compared to the range of pulse heights. In addition, the measurements more

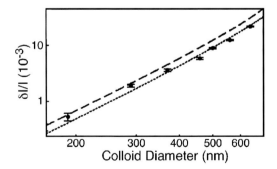

FIGURE 3.4. Comparison of measured δI/I values (circles) those predicted by Eq. (3.2) (dotted line) and Eq. (3.3) (dashed line). The measured data were taken over several runs on a single PL-defined pore of length 10.6 µm and cross section 1.04 µm^2. Error bars for the larger colloid sizes are obscured by the size of the plotted point. As the colloid diameter increases, there is a transition from agreement with Eq. (3.3) to Eq. (3.2). This reflects the fact that the derivation of Eq. (3.3) assumes the colloid diameter d is much less than the pore diameter D; conversely Eq. (3.2) relies on an assumption that holds only as d approaches D, and breaks down for smaller colloids. [From Ref. 12.]

3.2.3. PDMS-Based Pore

In the initial work we have just described, we drove the particles through the pore electrophoretically, thus requiring the particles to carry a relatively high electrostatic charge for effective electric field-driven motion. Motivated by the desire to measure particles that are not highly charged, such as viruses or protein-coated colloids, we have developed a second version of the device that utilizes hydrostatic pressure to drive the particles through the pore. Here, we describe the fabrication of a pressure-driven pore and also discuss refinements we have made in the analysis of our data. These refinements, associated with off-axis particle effects, allow us to have higher precision when we determine the colloid size.

Figure 3.5 shows a picture of our modified device: a polydimethylsiloxane (PDMS) mold sealed to a glass coverslip. The PDMS mold is cast from a master [15] and contains two reservoirs (7 μm deep, 400 μm wide) connected to an embedded pore (typically 7–9 μm long and 1 μm in diameter). The glass coverslip has platinum electrodes that extend across the width of the reservoirs and are fabricated on the glass coverslip prior to PDMS sealing. These electrodes are used to perform the four-point electronic measurement. We prepare both the PDMS slab and the coverslip using standard lithographic, micro-molding, and metal deposition techniques. Solution is added to the reservoirs via two holes cut through the PDMS slab, and capillary action is used to initially draw the solution through both reservoirs and the pore. Pressure (~1 psi) is applied to the access holes after loading the solution in order to drive the suspended latex colloids at a velocity of ~200μm/sec through the pore.

The analysis of the pulses produced during pressure-driven flow is complicated by the effects of particles that travel off the pore's central axis. Relative to particles of identical

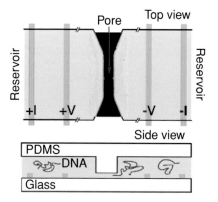

FIGURE 3.5. Schematic top and side views of our nanopore device, which consists of two 5 μm deep reservoirs connected by a lateral pore of 3 μm length and 200 nm diameter; an optical image of an actual pore sealed to a glass coverslip is incorporated into the top view. Molecules in the reservoirs are electrophoretically drawn through the pore, partially blocking the flow of ions. The current through the pore is measured using a four-terminal technique, where the voltage and current controlling platinum electrodes are as labeled. [From Ref. 37.]

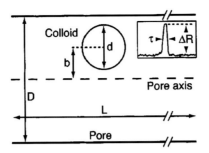

FIGURE 3.6. Schematic of the measurement geometry for a pore of diameter D and length L containing a colloid of diameter d that travels a distance b off the pore axis. Inset: Typical trace of the measured resistance vs time showing the passage of a single colloid that produces a resistance pulse of width τ and height ΔR. [From Ref. 16.]

size that travel on-axis, off-axis particles take longer to transit the pore (causing wider pulses) and produce larger electrical resistance changes. The former effect, which we refer to as the hydrodynamic off-axis effect, is simply due to the parabolic distribution of fluid velocity within the pore. The latter effect, which we refer to as the electrical off-axis effect, occurs because off-axis particles enhance the non-uniformity in the distribution of electrical current density and consequently increase further the electrical resistance. Here, we discuss two main results [16]: first, we show how off-axis particles affect data taken on populations of colloidal particles and propose a method to remove these effects. Second, we point out that a device utilizing pressure-driven flow will have an increased resolution over one using electrophoretic flow, since the algorithm we have developed to remove off-axis effects can only be performed for pressure-driven flow. As we will demonstrate, both results should increase the precision of the resistive-pulse technique in future applications.

To describe quantitatively our data, we follow the work of Berge et al. [17] who formulated phenomenological equations to describe the two aforementioned off-axis effects. For the hydrodynamic effect, they found that previous experimental data [18] on the time τ for a particle to pass through the pore are well described by

$$\tau = \frac{\tau_0}{(1 - x^2)(c_1 - c_2 x^5)} \tag{3.4}$$

where $\tau_0 = 16\eta(L/D)^2/\Delta P$ is the on-axis transit time for an infinitely small particle, η is the fluid viscosity, L is the pore length, D is the pore diameter, ΔP is the pressure drop across the pore, $x = 2b/D$ is the fractional radial position for a particle centered a distance b off of the pore axis, $c_1 = 1 - (2/3)(d/D)^2$, $c_2 = 23.36(1 - c_1)$, and d is the particle diameter (see Fig. 3.6). Berge et al. [17] then utilized Eq. 3.4 to describe empirically the variation in the change in electrical resistance ΔR with an off-axis coordinate x as

$$\Delta R = \Delta R_0 \left[1 + \alpha \left(\frac{xd}{D} \right)^3 \right] \tag{3.5}$$

where $\Delta R_0(d, D, L)$ is the change in resistance for the on-axis particle (see Ref. 18 for its functional form) and α is a constant whose value varies between 4.2 and 7.5.

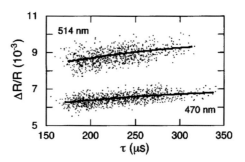

FIGURE 3.7. Comparison of measured normalized pulse heights ($\Delta R/R$) and pulse widths (τ) and the predictions of Eqs. (3.4) and (3.5). Each point represents the measured pulse height vs pulse width for the passage of a 470 nm diameter latex colloid (lower group of points) or a 514 nm diameter latex colloid (upper group of points) through a pore of length 0.4 μm and diameter 1.16 μm. For each type of colloid, the correlation between the measured heights and widths of the pulses is a result of the effect of colloids that travel off the pore axis. The measured data agree well with the predictions of Eqs. (3.4) and (3.5) for each colloid size, shown here as the solid lines. [From Ref. 16.]

In Fig. 3.7, we plot the values we measured of the normalized change in electrical resistance $\Delta R/R$ vs. τ for pulses produced by two populations of latex colloids: one population with a mean diameter of 470 nm, and one with a mean diameter of 514 nm. The data was taken using a pore that is 9.4 μm in length and 1.16 μm in diameter. For both types of colloids, there is a clear positive and nearly linear correlation between $\Delta R/R$ and τ as qualitatively expected from Eqs. 3.4 and 3.5. One interpretation of this positive correlation is that it is due to deviations in the size of individual colloids within each population, since it is clear that relatively larger colloids will both move slower and produce larger pulse amplitudes. The 470 nm diameter colloid population shown in the lower portion of the data plotted in Fig. 3.7 has a standard deviation of 12 nm as measured by the manufacturer. Eq. 3.4 predicts that the expected variation in τ of on-axis particles, due solely to differences in particle size within the population, will be ~2%. As seen in Fig. 3.7, the measured values for τ vary by much more than that (~80%). We thus conclude that the measured variations in τ can be attributed almost entirely to off-axis effects and not to differences in particle size.

Given particle and pore dimensions, we can use Eqs. 3.4 and 3.5 to find the predicted dependence of ΔR on τ due to the off-axis effects. In Fig. 3.7, we plot this result and compare it to the measured data. For both types of colloids, we find good agreement between the predicted dependence and the measurements when $\alpha = 6$ in Eq. 3.5; this value for α falls well within the range Berge et al. [17] found. The nearly linear measured correlation between ΔR and τ is then explained by the fact that variations in ΔR (caused by both electrical noise and the intrinsic size distribution of the colloid population) obscure the slight non-linearity in the predicted dependence. Based on this, we propose that off-axis effects can be effectively removed in the data analysis of a given population by first fitting a line $f(\tau)$ to the plot of ΔR vs. τ, and then calculating an adjusted value ΔR_{adj} for each event of height ΔR and width τ:

$$\Delta R_{adj} = \Delta R - [f(\tau) - f(\tau_{min})] \qquad (3.6)$$

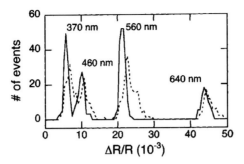

FIGURE 3.8. Histogram of the normalized pulse heights ($\Delta R/R$) measured for a solution containing four different sizes of latex colloids (of diameters 370, 460, 560, and 640 nm as labeled); each peak corresponds to the colloids of a given size. The dotted line represents the raw data while the solid line shows the same data after correcting for the effects of off-axis particles, as described by Eq. (3.6). The distribution of measured pulse heights for each type of colloid is both sharpened and more symmetric after applying the correction. For example, the application of the adjustment caused a decrease in the coefficient of variation of the pulses measured from the 560 nm colloids from 7.1% to 3.5%. [From Ref. 16.]

where $\tau_{min} = \tau_0/c_1$ is the minimum transit time measured. We thus use Eq. 3.6 as an algorithm to calculate the pulse height each colloid would have caused had it traveled on the pore's central axis.

To illustrate the increase in resolution that results from employing Eq. 3.6, we have measured a polydisperse solution containing four different sizes of latex colloids (of diameters 370 nm, 460 nm, 560 nm and 640 nm). In Fig. 3.8, we plot the distribution of measured ΔR values both before and after applying Eq. 3.6. As shown, the correction clearly sharpens the distribution for each type of colloid. For example, the coefficient of variation (standard deviation divided by mean) for pulses produced by 560 nm diameter colloids is reduced from 7.1% to 3.5%.

Previously [12], we utilized an electrophoretic driving force and found relatively little correlation between the measured pulse heights and widths. In that data, we measured linear correlation coefficients R ranging from 0.1 and 0.2 between the pulse heights and widths; this is in contrast to typical values of $R \sim 0.5$ for data obtained using pressure-driven flow. Since we expect that the electrical off-axis effect must have been present in the electrophoretically-driven data, we conclude that the electrophoretic velocity of the measured colloids does not vary significantly with the off-axis coordinate. This agrees with the fact that, in the absence of a colloid, the electric field across the pore is constant. It is possible that either inhomogenieties in the electric field due to the presence of the particle or hydrodynamic interactions between the particle and pore wall can lead to an off-axis effect on the velocity of a particle subjected to only an electrophoretic force. However, we conclude that these possibilities are insignificant when compared to the noise in our measurement.

The absence of an observable hydrodynamic off-axis effect while using electrophoretic flow means that we are unable to apply an algorithm similar to Eq. 3.6 to remove the electrical off-axis effect from the electrophoretic data. Distributions of pulse heights of a given colloid population measured using an electrophoretic driving force are therefore reduced in accuracy since they contain an intractable systematic source of error: the electrical off-axis effect. Devices using pressure driven flow, where we are able to apply the correction described in Eq. 3.6, are thus more accurate than those that use electrophoretic flow.

3.3. APPLICATIONS

3.3.1. An All-Electronic Immunoassay

Antibodies can be powerful and flexible tools because of their natural ability to bind to virtually any molecule and because of the modern ability to produce specific types in large quantities. These traits have led to the development of a number of important immunosensing techniques in which antibodies of a desired specificity are used to test for the presence of a given antigen [19–22]. For example, radioimmunoassays (RIA) have been employed in clinical settings to screen for such viruses as hepatitis [23]. An integral part of all immunosensing technologies is the ability to detect the binding of antibody to antigen. To accomplish this, most common immunoassays require the labeling of the antibody using fluorescence, radioactivity, or enzyme activity. However, the need to bind chemically a label to the antibody adds to the time and cost of developing and employing these technologies.

In this section, we show how we can use our PDMS-based pore as a new, all-electronic technique for detecting the binding of *unlabeled* antibody-antigen pairs [24]. As we discussed **Section 3.2.2**, our pore measurement is based on a particle passing through the pore and displacing conducting fluid. This, in turn, causes a transient increase, or pulse, in the pore's electrical resistance that is subsequently measured as a decrease in current. Because the magnitude of the pulse is directly related to the diameter of the particle that produced it [10, 25], we can use the pore to detect the increase in diameter of a latex colloid upon binding to an unlabeled specific antibody. We have employed this novel technique to perform two important types of immunoassays: an inhibition assay, in which we detect the presence of an antigen by its ability to disrupt the binding of antibody to the colloid; and a sandwich assay, in which we successively detect the binding of each antibody in a two-site configuration.

Previous particle-counting based immunoassays have used optical or electronic methods to detect the aggregates formed when the antibody crosslinks antigen-coated colloids [26–29]. However, relying on crosslinking as a general binding probe is limiting since it requires a free ligand with at least two binding sites. In contrast, our method is more general, since it relies only on the *added volume* of bound ligand and does not place any limitations on the ligand's functionality. While it cannot as of yet perform the kinetic analyses that surface plasmon resonance (SPR) techniques [30] are capable of, our device already represents an alternative to SPR for end-point analysis of biological reactions in that it is more rapid, inexpensive, and compact.

We perform our measurements on a chip-based microfluidic device that confers three additional advantages upon our system when compared to traditional immunoassays. First, because we have miniaturized the reservoirs leading to the pore, each measurement uses sub-microliter quantities of sample and can be performed within minutes. Second, we utilize common microfabrication and micro-molding techniques [15] to make the pore, reservoirs,

AN ON-CHIP ARTIFICIAL PORE FOR MOLECULAR SENSING

and electrodes. This allows for quick and inexpensive device construction. Third, using chip-based fabrication can extend the device's capabilities by permitting either future integration of our measurement with other microfluidic components [31, 32] such as separation units or mixers, or construction of arrays of sensors on a single chip for performing many measurements or assays in parallel.

3.3.1.1. Sample Preparation and Measurement All solutions are mixed in 0.5x PBS, pH 7.3, and contain 0.05% Pluronics F127 surfactant (a non-ionic surfactant) and 0.2 µg/mL Bovine Serum Albumen (BSA). The BSA and surfactant are added to decrease both sticking of colloids to the device walls, and non-specific adhesion of antibodies to the particles. We prepare a stock colloidal solution by mixing and twice centrifugally rinsing the colloids in the above buffer. This stock solution is then diluted by a factor of ten and mixed with the relevant antibodies and/or antigens prior to each measurement. For the sandwich assay, we attach biotinylated antibody to streptavidin colloids by incubating a high concentration of the biotinylated antibody with the stock solution, then centrifugally rinsing to remove unbound molecules. Some solutions are passed through a 0.8 µm pore size filter immediately prior to measuring so as to remove aggregates caused by the crosslinking of the colloids by the antibody.

Once each device is loaded with the solution to be analyzed, we measure the current through the pore at constant applied DC voltage (0.2–0.5 V). Figure 3.9 shows a typical measurement of the current: each downward pulse corresponds to a single colloid passing through the pore. Particle transit times are typically ~200 µs when a pressure of ~7 kPa (~1 psi) is applied. Such transit times are long enough to establish a stable square pulse shape (see inset, Fig. 3.9). We measure several hundred colloids in a given solution during a single experimental run, after which the device is either cleaned appropriately and reused or

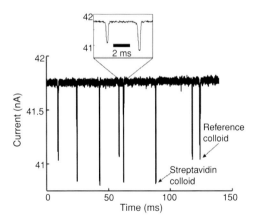

FIGURE 3.9. A typical measurement of the current across a pore as different colloids pass through it. Each downward pulse corresponds to a single colloid transiting the pore. There is a clear difference in pulse magnitude as a result of the difference in size of the streptavidin colloids as compared to the reference colloids. This difference allows us to separate the pulses for pore calibration (see text). The inset shows an expanded view of two pulses. As shown, they are well resolved in time and consequently allow an unambiguous measurement of the pulse height. The data shown was taken with an applied voltage of 0.4 V and a pressure of ~6.9 kPa. [From Ref. 24.]

discarded. Custom written software is used to extract both the height and width of each pulse in a trace. As we described in the **Section 3.2.3**, the accuracy of the measurement is increased by correcting for off-axis particles flowing through the microfluidic channel [17, 33].

3.3.1.2. Analysis Our goal is to detect an increase in the magnitude of the pulses due to the volume increase when ∼510 nm diameter streptavidin-coated latex colloids specifically bind to antibodies. As shown in Eq. 3.2, the relative height of the pulse depends on the relation of the diameter d of each colloid (∼510 nm) to the diameter D (∼900 nm) and length L of the pore. We can determine d for each streptavidin colloid measured if we know the dimensions of the pore. We directly measure L with an optical microscope. However, we cannot directly measure the pore's diameter D; instead, we perform a calibration by adding a reference colloid of known diameter (a 470 nm diameter sulfate-coated latex colloid) to each solution of streptavidin colloids. The absolute difference in diameter (470 nm to 510 nm) between the two types of colloids results in a clear difference in the pulse heights (see Fig. 3.9); consequently, we can determine easily which size colloid produced each pulse. We use the values of $\delta I/I$ arising from the reference colloids, along with the known values of L and d, to invert numerically Eq. 3.2 to thus determine the pore diameter D. Once this is accomplished, we use Eq. 3.2 once again to correlate the magnitude of each pulse to the diameter of the streptavidin colloid that produced it.

Figure 3.10a shows a histogram comparing the distribution of measured colloid diameters obtained from two different solutions: one containing only the streptavidin and the reference colloids, and one containing both types of colloids *and* 0.1 mg/mL of monoclonal mouse anti-streptavidin antibody (with an affinity for streptavidin $> 10^{10}$). As shown, there is a clear increase of 9 nm in the diameter of the streptavidin colloids in the solution containing the antibody (see also Fig. 3.10b). We attribute this increase to the volume added to the colloid upon the specific binding to the anti-streptavidin. Specificity is demonstrated by the much smaller increase in diameter (∼2.5 nm) when mixing the colloids with 0.1 mg/mL of a monoclonal isotype matched irrelevant antibody (mouse anti-rabbit; see Fig. 3.10b). This smaller increase is a result of non-specific binding of the irrelevant antibody to the colloids.

In Fig. 3.11, we show the measured change in colloid diameter as the concentration of the specific, high affinity antibody (monoclonal anti-streptavidin) is varied from 0.1 μg/mL to 100 μg/mL. As shown, the colloid diameter reaches its maximum value when the colloids are mixed with ≥ 5μg/mL of antibody. Using a Bradford protein assay [34], we determined the minimum saturating concentration of antibody for the colloid concentration in our experiment (1.2×10^9 particles/mL) to be 3.5 μg/mL, which is in good agreement with the results of our electronic pore-based immunoassay. Furthermore, the manufacturer-quoted binding capacity of the colloids indicates that each colloid has approximately 9800 streptavidin molecules on its surface. If each colloid binds to an equivalent number of antibodies, the minimum saturating concentration for a solution containing 1.2×10^9 colloids/mL will be ∼3.0 μg/mL; again, this is in good agreement with our results. As shown in Fig. 3.11, the dynamic range of our assay corresponds to antibody concentrations from 0.5 μg/mL to the saturating concentration of ∼5 μg/mL. By decreasing the colloid concentration, we can decrease the binding capacity of the solution, thus decreasing the saturating concentration of antibody. In this manner, we can expect the range of sensitivity of the device to decrease to antibody concentrations as low as 10–50 ng/mL.

AN ON-CHIP ARTIFICIAL PORE FOR MOLECULAR SENSING

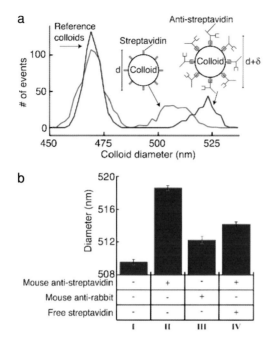

FIGURE 3.10. **A**: A histogram showing the distribution of colloid diameters measured from a solution that contains only the reference and streptavidin colloids (green line), and a solution that contains both types of colloids and 0.1 mg/mL of monoclonal anti-streptavidin antibody (red line). The specific binding of anti-streptavidin to the streptavidin colloids produces a clear increase in the diameter of the colloids. **B**: A summary of the measurements of the mean diameter of the streptavidin colloids when mixed in different solutions. A single experimental run consists of measuring several hundred colloids of each type in one solution; the plotted bars represent the mean diameter extracted from 3–5 such runs on the same solution, but using different devices. All solutions contained the streptavidin colloids and the reference colloids in a 0.5 × PBS buffer (pH 7.3). The presence of additional components in each solution is indicated by a '+' in the column beneath the plotted bar. Column **I** shows the mean diameter measured without any protein added to the solution. A 9 nm increase in colloid diameter is seen in the presence of the specific antibody to streptavidin (0.1 mg/mL mouse anti-streptavidin, column **II**); we attribute this to the volume added to the colloid due to the specific binding of the antibody. The specificity of the probe is shown by the lack of a similar diameter increase in the presence of isotype matched irrelevant antibody (0.1 mg/mL mouse anti-rabbit, column **III**); the small diameter increase in this solution can be attributed to non-specific adhesion. We also perform an inhibition assay, where the specific binding of the anti-streptavidin to the colloid is disrupted by the presence of 0.2 mg/mL free streptavidin (column **IV**)- the presence of free antigen is shown by the decrease in diameter compared with the antigen-free solution (column **II**). The error bars in this figure, and in all other figures, represent the uncertainty in determining the mean diameter based on one standard deviation of the measured distributions. The dominant source of error in our measurements is the intrinsic distribution in the streptavidin colloids' diameter, with smaller contributions from the spread in diameter of the reference colloids and the electrical noise in the current measurement. [From Ref. 24.]

We use our technique's ability to detect successfully the specific binding of unlabeled antibodies to the colloids to perform an inhibition immunoassay. We measure a 4.5 nm increase (see column IV of Fig. 3.10b) in the diameter of the streptavidin colloids when mixed with 0.1 mg/mL anti-streptavidin that had been preincubated with 0.2 mg/mL of free streptavidin. This smaller increase (relative to the solution containing only anti-streptavidin) indicates a decrease in the number of antibodies binding to each colloid. We

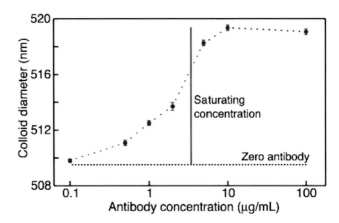

FIGURE 3.11. Measurements of the mean colloid diameter when mixed in solutions of varying monoclonal mouse anti-streptavidin concentrations. The vertical line marks the binding capacity of the colloids as determined by a Bradford protein assay. The diameter of the colloids in the absence of antibody is shown as the black dashed line. [From Ref. 24.]

primarily attribute this to the free streptavidin blocking the antibody binding sites. The measured diameter of the streptavidin-coated colloid therefore indicates the presence of free streptavidin in the solution. In general, this inhibition method can be extended to detect any antigen that can be immobilized on the colloid surface.

The 4.5 nm increase seen in column IV of Fig. 3.10b shows that some binding of antibody to the colloid does in fact occur. Based on the control measurement with an irrelevant antibody (column III of Fig. 3.10b), we attribute this increase to a combination of non-specific binding of blocked antibodies, and incomplete inhibition of the antibody by the free streptavidin. The possibility of non-specific binding does decrease the dynamic range of the measurement. However, because of the very small uncertainty in the measured mean colloid diameter, the dynamic range necessary to determine the amount of ligand bound to the colloid is still quite large.

As a second demonstration of our technique's high sensitivity to the volume added by molecules bound to a streptavidin colloid, we perform an immunoassay (summarized in Fig. 3.12) using a sandwich configuration. Here, a primary antibody that is immobilized on the colloid surface binds to a free antigen, which in turn is bound to a secondary antibody. We immobilize the primary antibody by mixing streptavidin colloids with a biotinylated antibody (Rabbit anti-Streptococcus Group A) to thus create a colloid-antibody conjugate through the streptavidin-biotin bond. As shown in Fig. 3.12, the measured conjugated colloids are 514 nm in diameter, a 5 nm increase over the 'bare' streptavidin colloids. Next, we mix the colloid-antibody conjugates with both the specific antigen to the primary antibody (extract from a culture of Streptococcus Group A), and 0.1 mg/mL of a secondary antibody (unlabeled rabbit anti-Streptococcus Group A). Measurements of this solution show the colloids further increase in diameter by 1.6 nm. This 1.6 nm increase is not seen when the colloids are mixed with the antigen alone, indicating that the binding of the secondary antibody is the principal reason for the diameter increase. The specificity of this arrangement is demonstrated by the absence of a diameter increase in the control measurements we perform

AN ON-CHIP ARTIFICIAL PORE FOR MOLECULAR SENSING 49

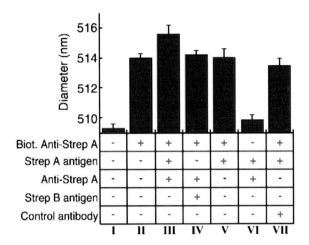

FIGURE 3.12. Summary of the mean colloid diameters measured when forming an antibody-antigen-antibody 'sandwich' on the colloid surface. All solutions contain the reference and streptavidin colloids in a 0.5 × PBS buffer (pH 7.3), along with additional components as indicated by the '+' in the column below the plotted bar. Column **I** indicates the measured diameter of the 'bare' streptavidin colloid. We measure a ∼5 nm increase (column **II**) in diameter after conjugating a biotinylated antibody (biotinylated anti-Streptococcus Group A) to the streptavidin coated colloids. A further increase of ∼1.6 nm is seen (column **III**) when adding both extract from a culture of Streptococcus Group A and a secondary antibody specific to that antigen (unlabeled anti-Streptococcus Group A); this increase indicates the formation of the sandwich on the colloid surface. The specificity of the configuration is shown by the lack of an increase in diameter when adding extract from a culture of Streptococcus Group B (which is not bound by either antibody) in place of the Group A extract (column **IV**), or an irrelevant antibody in place of the specific secondary antibody (column **VII**). When adding the specific antigen and secondary antibody to unconjugated colloids (column **VI**), we measure no significant diameter increase, indicating that non-specific adhesion of antigen-secondary antibody complexes are not the cause of the diameter increase seen in column **III**. Finally, when adding the specific antigen alone to the conjugated colloids (column **V**), we see no increase in diameter, indicating that the diameter increase in column **III** is primarily due to the binding of the secondary antibody. [From Ref. 24.]

in which either the antigen or the secondary antibody is replaced by non-specific counterparts (see Fig. 3.12).

It is intriguing that the measured 5 nm increase after attachment of the biotinylated antibody is less then the maximum 9 nm increase seen when utilizing the antibody-antigen bond (Fig. 3.10 and 12) to attach antibody to the colloid. This surprising difference is most likely due to the differing conformations of the antibody in each case; however, further work is needed to clarify this. Nonetheless, despite the smaller size increase, the ability of the device to perform the sandwich assay is still clearly demonstrated.

While we have used an antibody/antigen reaction to demonstrate the power of our technique, we emphasize that its true strength is its generality: it does *not* rely on any functional properties of the free ligand. Thus, it can be applied to any ligand/receptor pair, provided the free ligand is large enough to produce a discernible change in the size of the colloid.

Future work on the device will focus on optimizing its sensitivity in terms of both ligand size (mass) and concentration. The sensitivity is dependent on four factors: the amount of ligand bound to each colloid, the intrinsic dispersion in colloid size, the colloid geometry, and the colloid concentration. First, increasing the number of binding sites will lead to more ligands bound per colloid, and consequently a larger change in size. For the colloids used here, the parking area for each binding site is $\sim$80 nm^2; while this is close to the steric limit for antibody molecules, the use of a smaller ligand would permit more binding sites per colloid. Second, the intrinsic spread in the sizes of the streptavidin colloids is the largest source of error in our measurement. The device's sensitivity would be enhanced by using a more monodisperse population of colloids (one with a coefficient of variation in diameter of less than 2%), or even a solution of highly monodisperse nanocrystals [35]. Third, at constant binding density, the measured change in pulse height upon binding to free ligand is proportional to the surface-to-volume ratio of the colloid. Thus, we could increase the sensitivity and dynamic range of the assay by employing a smaller colloid. For example, we estimate that using a colloid 250 nm in diameter would increase the sensitivity of the assay by a factor of four in either ligand size or concentration. Thus, based on the data shown in Fig. 3.11, using a 250 nm colloid at the same particle concentration employed in this paper would make the assay sensitive to either 38 kDa ligand molecules at concentrations of 0.5 µg/mL, or antibody concentrations near 0.1 µg/mL. We mention that an even more effective strategy to increase the surface-to-volume ratio would be to use a non-spherical or porous colloid (assuming the pore size is large enough to admit the free ligand) as the substrate for the immobilized receptor. Fourth, as previously mentioned, decreasing the concentration of colloids would further increase the sensitivity since it would decrease the minimum saturating concentration of free ligand. Overall, a combination of these four strategies should result in the increased sensitivity of our assay to ligand concentrations at or below 1 ng/mL.

3.3.2. Summary

In conclusion, we have demonstrated our ability to use an electronic measurement to detect the binding of unlabeled antibodies to the surface of latex colloids. This ability is generally applicable to determining rapidly and precisely the thickness of a layer of *any* kind of biological macromolecule bound to a colloid. Here, we specifically showed that our technique can be employed to perform two widely used and important immunoassays—an inhibition assay and sandwich assay—in which either the antigen or antibody is immobilized on the colloid. In contrast to how these assays are performed today, ours requires no labeling of analytes, uses only sub-microliter volumes of sample, and can be performed rapidly and inexpensively. For instance, we have compared our technique's ability (using a sandwich configuration) to detect the presence of Streptoccocus Group A to that of a standard latex agglutination assay. We have found our method to be an order of magnitude more sensitive and over four times as fast as the agglutination assay. Overall, our device can be used to detect many different kinds of analytes, since the colloids can be easily modified to have almost any specificity (through, for example, the biotin-streptavidin interaction used here). Furthermore, our technique can be extended to multi-analyte detection not only by utilizing several microparticles with different chemical sensitivities and different mean diameter but also by employing devices consisting of arrays of pores [36]. Finally, in addition to a host of biosensing

3.3.3. Single Molecule Detection

We have pushed the length scale of our PDMS-based pore so that the pore is able to sense single molecules of unlabeled lambda-phage DNA [37]. Our success provides many opportunities for diverse single molecule detection applications.

To measure single molecules of DNA, we require pores that have dimensions 200 nm in diameter and a length of a few microns (here we used a length of 3 μm). The 200 nm diameter is achieved using electron-beam lithography. Soft lithography [15], as described in **Section 3.2.2**, is used to embed the pore and reservoirs into PDMS.

To demonstrate the sensing capabilities of our nanopore, we have measured solutions of 2.5 μg/mL lambda-phage DNA in a 0.1 M KCl, 2 mM Tris (pH 8.4) buffer. Typical traces of measured current are shown in Figure 3.13. The striking downward peaks, of height 10–30 pA and width 2–10 ms, correspond to individual molecules of DNA passing through the pore. In contrast, such peaks are absent when measuring only buffer. We further note that peaks are only present when using pores with diameters of 300 nm or less.

Previous work on colloids [10, 12] has shown that, for particles of diameter much smaller than that of the pore, the ratio of peak height to baseline current is approximately equal to the volume ratio of particle to pore: $\delta I/I \sim V_{particle}/V_{pore}$. We can estimate the volume of a single lambda DNA molecule by approximating it as a cylinder with a 2 nm radius (which includes a 1 nm ionic, or Debye, layer), and a height equal to the contour length of the molecule (~16 μm). Given the known pore volume and a total current $I = 15$ nA, we can expect a decrease in current $\delta I \sim 30$ pA when a DNA molecule fully inhabits the pore. This estimate agrees well with the upper range of measured peak heights. Further corroboration for this model comes from the fact that no peaks are observed when using larger pores (pores >300 nm in diameter). When a molecule inhabits a pore with a diameter >300 nm, the expected response in current is less than 40% of that for a 200 nm diameter pore. Therefore, at 15 nA total current, the maximum peak heights for a lambda DNA molecule will be less than 12 pA, a value not well resolvable above the noise. Our results

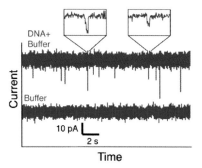

FIGURE 3.13. Typical traces of current vs. time for solutions of buffer (lower trace), and buffer with lambda phage DNA molecules (upper trace), when 0.4 V is applied across the pore. The traces are offset for clarity; the total current in each case is ~15 nA. Each downward spike in the lower trace represents a DNA molecule passing through the pore. The spikes are typically 2–10 ms in duration, and are well resolved, as shown in the insets. The variations in peak height most likely correspond to the different conformation of each molecule. [From Ref. 37.]

suggest that the measured variation in δI is most likely due to differences in molecular conformation: maximum peak heights arise when an entire molecule inhabits the pore, while smaller peak heights occur when only a portion of a molecule resides within the pore. Future experiments will focus on controlling the conformation of each molecule in order to relate the measured peak height to the length of each DNA molecule. Thus, our nanopore device may provide a simple and quick method for the coarse sizing of large DNA molecules.

The results described here represent a first step towards a host of single-molecule sensing applications. By relying on common micro-fabrication techniques, we can easily create arrays of pores for the simultaneous measurement of many different molecules [36]. Decreasing the pore size will allow us detect and size smaller molecule such as proteins or viruses. The minimum achievable pore diameter for the PDMS used here (Sylgard 184) is $\sim$150 nm, but recent work has shown that other PDMS formulations can maintain features as small as 80 nm [37]. Finally, we can add chemically specificity in two ways: first, by covalently attaching molecules of interest to the pore wall, we expect to see changes in the transit times of molecules in solution that interact with the immobilized molecules. Second, we can measure changes in the diameter of chemically-functionalized colloids upon binding of molecules in the solution, as we have already done using our electronic immunoassay described in the previous section [24]. The ease and reproducibility of micro-molding and the simplicity of our device greatly enhances the capabilities of artificial nanopores for molecular sensing.

3.4. CONCLUSIONS

We have described the fabrication and measurement of a fully-integrated on-chip artificial pore. Because we employ standard integrated circuit fabrication techniques, including photo- and electron-beam lithographies, reactive ion etching, and metal deposition, as well as employ soft-lithography, we can reproducibly fabricate a stable pore that is inexpensive and easy to use. Furthermore, we are able to scale the pore to include arrays on a chip for massively-parallel screening. We have demonstrated two applications to our pore: a label-free immunoassay and coarse-sizing of single molecules of DNA. These are only two of the many molecular-sensing applications we forsee with our artificial on-chip pore.

REFERENCES

[1] S. Howorka, S. Cheley, and H. Bayley. Sequence-specific detection of individual DNA strands using engineered nanopores. *Nature Biotech.*, 19:636–639, 2001.

[2] H. Bayley and P.S. Cremer. Stochastic sensors inspired by biology. *Nature*, 413:226–230, 2001 and references therein.

[3] J. Li, M. Gershow, D. Stein, E. Brandin, and J. Golovchenko. DNA molecules and configurations in a solid-state nanopore microscope. *Nat. Mater.*, 2:611–615, 2003.

[4] J. Li, D. Stein, C. McMullan, D. Branton, M.J. Aziz, and J.A. Golovchenko. Ion-beam sculpting at nanometer length scales. *Nature*, 412:166–169, 2001.

[5] P. Kohli, C.C. Harrell, Z. Cao, R. Gasparac, W. Tan, and C.R. Martin. DNA-functionalized nanotube membranes with single-base mismatch selectivity. *Science*, 305:984–986, 2004.

[6] R.R. Henriquez, T. Ito, L. Sun, and R.M. Crooks. The resurgence of Coulter counting for analyzing nanoscale objects. *Analyst*, 129:478–482, 2004.

[7] W.H. Coulter. U. S. Patent No. 2,656,508, issued 20 Oct. 1953.

[8] H.E. Kubitschek. Electronic counting and sizing of bacteria. *Nature*, 182:234–235, 1958.

[9] E.C. Gregg and K. David Steidley. Electrical counting and sizing of mammalian cells in suspension. *Biophys. J.*, 5:393–405, 1965.
[10] R.W. DeBlois and C.P. Bean. Counting and sizing of submicron particles by the resistive pulse technique. *Rev. Sci. Inst.*, 41:909–913, 1970.
[11] M. Koch, A.G.R. Evans, and A. Brunnschweiler. Design and fabrication of a micromachined coulter counter. *J. Micromech. Microeng.*, 9:159–161, 1999.
[12] O.A. Saleh and L.L. Sohn. Quantitative sensing of nanoscale colloids using a microchip Coulter counter. *Rev. Sci. Inst.*, 72:4449–4451, 2001.
[13] D.W. Fakes, M.C. Davies, A. Browns, and J.M. Newton. The surface analysis of a plasma modified contactlens surface by SSIMS. *Surf. Interface Anal.*, 13:233–236, 1988.
[14] M.K. Chaudhury and G.M. Whitesides. Direct measurement of interfacial interactions between semispherical lenses and flat sheets of poly(dimethylsiloxane) and their chemical derivatives. *Langmuir*, 7:1013–1025, 1991.
[15] Y.N. Xia and G.M. Whitesides. *Angewandte Chemie-International Edition*, 37:551–575, 1998.
[16] O.A. Saleh and L.L. Sohn. Correcting off-axis effects on an on-chip resistive-pulse analyzer. *Rev. Sci. Inst.*, 73:4396–4398, 2002.
[17] L.I. Berge, T. Jossang, and J. Feder. Off-axis response for particles passing through long aperatures in Coulter-type counters. *Measure. Sci. Technol.*, 1(6):471, 1990.
[18] H.L. Goldsmith and S.G. Mason. Flow of suspensions through tubes .1. Single spheres, rods, and discs. *J. Colloid Sci.*, 17:448, 1962.
[19] P.B. Luppa, L.J. Sokoll, and D.W. Chan. Immunosensors–principles and applications to clinical chemistry. *Clin. Chim. Acta*, 314:1–26, 2001.
[20] T. Vo-Dinh and B. Cullum. Biosensors and biochips: Advances in biological and medical diagnostics. *Fresen. J. Anal. Chem.*, 366:540–551, 2000.
[21] A.P. Turner. Biochemistry: Biosensors–sense and sensitivity. *Science*, 290:1315–1317, 2000.
[22] R.I. Stefan, J.F. van Staden, and H.Y. Aboul-Enein. Immunosensors in clinical analysis. *Fresen. J. Anal. Chem.*, 366:659–668, 2000.
[23] T.T. Ngo. Developments in immunoassay technology. *Methods*, 22:1–3, 2000.
[24] O.A. Saleh and L.L. Sohn. Direct detection of antibody-antigen binding using an on-chip artificial pore. *Proc. Natl. Acad. Sci.*, 100:820–824, 2003.
[25] L.Q. Gu, S. Cheley, and H. Bayley. Capture of a single molecule in a nanocavity. *Science*, 291:636–640, 2001.
[26] Y.K. Sykulev, D.A. Sherman, R.J. Cohen, and H.N. Eisen. Quantitation of reversible binding by particle counting: Hapten-antibody interaction as a model system. *Proc. Natl. Acad. Sci. U.S.A.*, 89:4703–4707, 1992.
[27] G.K. von Schulthess, G.B. Benedek, and R.W. Deblois. Experimental measurements of the temporal evolution of cluster size distributions for high-functionality antigens cross-linked by antibody. *Macromolecules*, 16:434–440, 1983.
[28] G.K. von Schulthess, G.B. Benedek, and R.W. Deblois. Measurement of the cluster size distributions for high functionality antigens cross-linked by antibody. *Macromolecules*, 13:939–945, 1980.
[29] G.K. von Schulthess, R.W. Deblois, and G.B. Benedek. Agglutination of antigen coated carrier particles by antibody. *Biophys. J.*, 21:A115–A115, 1978.
[30] W.M. Mullett, E.P. Lai, and J.M. Yeung. Surface plasmon resonance-based immunoassays. *Methods*, 22:77–91, 2000.
[31] G.M. Whitesides and A.D. Stroock. Flexible methods for microfluidics. *Phys. Today*, 54:42–48, 2001.
[32] T. Chovan and A. Guttman. Microfabricated devices in biotechnology and biochemical processing. *Trends Biotechnol*, 20:116–122, 2002.
[33] W.R. Smythe. Off-axis particles in coulter type counters. *Rev. Sci. Inst.*, 43:817, 1972.
[34] M.M. Bradford. Rapid and sensitive method for quantitation of microgram quantities of protein utilizing principle of protein-dye binding. *Anal. Biochem.*, 72:248–254, 1976.
[35] M. Bruchez, Jr., M. Moronne, P. Gin, S. Weiss, and A.P. Alivisatos. Semiconductor nanocrystals as fluorescent biological labels. *Science*, 281:2013–2016, 1998.
[36] A. Carbonaro and L.L.A Sohn. resistive-pulse sensor chip for multianalyte immunoassays. *Lab on a Chip*, 5:1155–1160, 2005.
[37] O.A. Saleh and L.L. Sohn. An artificial nanopore for molecular sensing. *NanoLetters*, 3:37–38, 2003.
[38] H. Schmid and B. Michel. Siloxane polymers for high-resolution, high-accuracy soft lithography. *Macromolecules*, 33(8):3042–3049, 2000.

4

Cell Based Sensing Technologies

Cengiz S. Ozkan, Mihri Ozkan, Mo Yang, Xuan Zhang,
Shalini Prasad, and Andre Morgan
Mechanical Engineering Department, University of California, Riverside, CA 92521, USA

Biosensor technology is the driving force in the development of biochips capable of detecting and analyzing biomolecules. A biosensor is a device that detects, records, and transmits information regarding a physiological change or the presence of various chemical or biological materials in the environment. Cell based sensing is the most promising alternative to the existing bio-sensing techniques as cells have the capability of identifying very minute concentrations of environmental agents. The use of living cells as sensing elements provides the opportunity for high sensitivity to a broad range of chemically active substances which affect the electrochemical activity of cells. This chapter provides an overview of the development of cell-based sensors for biological and chemical detection applications, along with significant advances over the last several years. Special emphasis will be given on recently developed planar microelectrode arrays for enabling extracellular recording from electrochemically active cells cultured *in vivo*. The extracellular signal spectrum can be modulated when the cells are exposed to a variety of chemical agents and this modulated signal constitutes a "signature pattern" which serves as the finger print for a specific chemical agent. Cell based sensors can change the sensing paradigm from "detect-to-treat" to "detect-to-warn".

4.1. OVERVIEW

General interest in biosensors has grown considerably since the description by Updike and Hicks [94] of the first functional enzyme electrode based on glucose oxidase deposited on an oxygen sensor. The last decade in particular has seen efforts within both academic and commercial sectors being directed towards the development of practical biosensors.

However, it is important to realize that advances in allied subject areas have been important in aiding these research activities. The field of biotechnology has contributed enormously by providing an increased understanding of immobilized bioreagents and improved techniques for immobilization, and purely technological advances in the microelectronics and communication industries have provided more refined transducer elements and devices. Today, biosensor technology is the driving force in the development of biochips for the detection of gaseous pollutants [100], biological and chemical pollutants [43], pesticides [74], and micro-organisms [85]. Biosensors combine the selectivity of biology with the processing power of modern microelectronics and optoelectronics to offer powerful new analytical tools with major applications in medicine, environmental diagnostics and the food and processing industries. A novel challenge is the development of effective and multifunctional biosensors based on fundamental research in biotechnology, genetics and information technology, such that the existing axiom of "detect -to-treat" would change to "detect -to-warn".

Conventional methods for detecting environmental threats are primarily based on enzyme [48], antibody [10, 84], or nucleic acid-based assays [20, 65, 96], which rely on chemical properties or molecular recognition to identify a particular agent [103]. The current method involved in risk assessment for humans, fail in field situations due to their inability to detect large numbers of chemical agents, characterize the functionality of agents and determine the human performance decrements [62].

4.2. CELL-BASED BIOSENSORS

Cell based sensing [81] is the most promising alternative to the existing bio-sensing techniques as cells have the capability of identifying very minute concentrations of environmental agents. In cell based sensing, mammalian cells with excitable cell membranes are used as biosensors. The membranes of mammalian are comprised of ion channels, which open or close based on the changes in the internal and external local environment of the cells. This results in the development of ionic current gradients that are responsible for the modification of the electrical conductivity. Cells express and sustain an array of potential molecular sensors. Receptors, channels and enzymes that are sensitive to an analyte are maintained in a physiologically relevant manner native to the cellular machinery. In contrast with the antibody-based approaches, cell-based sensors should optimally only respond to functional biologically active analytes. It is also important to note that there are two major difficulties involved in using cells as sensors: it requires the knowledge of microbiology or tissue culture, and the lifetime of living cells is usually more limited than enzymes [51]. Nevertheless, there are a number of compelling motivations that make it very attractive to work with living cells for sensing applications. The first and most important one is that only by using a living component it is possible to obtain functional information, i.e. information about the effect of a stimulus on a living system. This can be contrasted with analytical information, which answers the question of how much of a given substance is present. There are many circumstances in which the type of information required is really functional, and analytical tests are carried out only to estimate the functional consequences of the substances being investigated. In those cases, a measurement method using a living system is very attractive because it can yield that functional information directly, which provides real-time sensing capability.

Using a living cell as the sensing element, one can also obtain analytical information, both qualitatively and quantitatively. In its simplest form, it tells us whether a given substance is present, and in what concentration. Cells with a given type of receptors can be considered as sensors for agonists, with a sensitivity determined by the binding constant of that receptor/ligand combination [5, 78]. Another large body of work uses bacteria, often genetically engineered to respond to specific substances. Using amperometric detection, it has been possible to detect herbicides [101], benzene [87], alcohol [44], and trimethylamine gas [50]. Another detection method which has recently become popular for bacteria is bioluminescence. One of the advantages invoked for the use of cells in environmental applications is that it allows the measurement of the total bioavailability of a given pollutant rather than its free form [36, 76]. For instance, a bioluminescent bacteria detector specific for copper also detects insoluble copper sulfide [95]. This means that the analytical question really becomes a functional one, namely how bio-functional a given substance is. When used for environmental applications, a further potential advantage of biosensor devices is that they are capable of continuous monitoring, and can be made small enough to use in the field rather than in the laboratory. The main potential problem is the handling and lifetime of the living component.

Cells express and sustain an array of potential molecular sensors. Receptors, channels, and enzymes that may be sensitive to an analyte are maintained in a physiological relevant manner by a native cellular machinery. In contrast with antibody-based approaches, cell-based biosensors should optimally only respond to functional, biologically active analytes. Cell-based biosensors have been implemented using microorganisms, There are several approaches for transduction of cell signals including cell fluorescence, metabolism, impedance, intracellular potentials and extracellular potentials.

4.2.1. Cellular Microorganism Based Biosensors

Metabolism cell stress can activate microorganism pathways due to some analytes, such as pollutants [3]. The members of bacteria was sensitive to several groups of chemicals including phenols, halomethanes and several oxidants responding by increased luminescence to a different type of environmental stress. Cell biosensor specific for formaldehyde was developed using double-mutant cells of the methylotrophic yeast, where the activities of some of the enzymes in the metabolic pathway of the wild-strain cells were deliberately suppressed by introducing respective genetic blocks to optimize the selectivity and acidification rate [47]. Mutant yeast cells produced in this way were immobilized in Ca-alginate gel on the gate of a pH-sensitive field effect transistor. Another sensor approach is based on genetically engineered bacteria such as a bioluminescent catabolic reporter bacterium developed for continuous on-line monitoring of naphthalene and salicylate bioavailability and microbial catabolic activity potential in waste streams [36]. The bioluminescent reporter bacterium, Pseudomonas fluorescens HK44, carries a transcriptional nahG-luxCDABE fusion for naphthalene and salicylate catabolism. Exposure to either compound resulted in inducible bioluminescence. Engineered bacteria were used as whole cell sensor elements for detecting benzene [2], toluene [7], mercury [76] and octane [83]. The alteration of a microorganism-based biosensor response is important and genetic detection is favored by insufficient selectivity [21]. Cell-based biosensors for genetic detection derived from a biological system of interest can offer functional and physiologically relevant information.

4.2.2. Fluorescence Based Cellular Biosensors

Fluorescence based sensors are showing several signs of wide-ranging development [14, 16] for the clarification of the underlying photophysics, the discovery of several biocompatible systems, and the demonstration of their usefulness in cellular environments. Another sign is that the beneficiaries of the field are multiplying. They range from medical diagnostics through physiological imaging, biochemical investigations, environmental monitoring, and chemical analysis to aeronautical engineering.

The design of fluorescent molecular sensors for chemical species combines a receptor and a fluorophore for a "catch-and-tell" operation. The receptor module engages exclusively in transactions of chemical species. On the other hand, the fluorophore is concerned solely with photon emission and absorption. Molecular light emission is particularly appealing for sensing purposes owing to its near-ultimate detectability, "off/on" switchability, and very high spatiotemporal resolution including video imaging. The commonest approaches to combining fluorophore and receptor modules involve integrated or spaced components [4]. New fluorescence reagents based on the combination of molecular biology, fluorescent probe chemistry and protein chemistry have been developed for cell-based assays. Variants of the green fluorescent protein (GFP) with different colors would be very useful for simultaneous comparisons of multiple protein fates, developmental lineages and gene expression levels [37]. The simplest way to shift the emission color of GFP is to substitute histidine or tryptophan for the tyrosine in the chromophore, but such blue-shifted point mutants are only dimly fluorescent. The longest wavelengths previously reported for the excitation and emission peaks of GFP mutants are 488 and 511 nm, respectively.

The integrated or intrinsic sensor format relies on internal charge transfer within the excited state. The partial electronic charges so separated can interact with the target species when it is trapped by the receptor. The energy of the excited state is thereby disturbed and shows up as a blue- or red-shifted light absorption and/or emission [93]. The separation of charges within the spaced or conjugate sensor format occurs after excited state creation. This is photo-induced electron transfer (PET), which competes against fluorescence to dominate the energy dissipation of the excited state, i.e., fluorescence is switched off when the target species is absent. When it arrives, however, PET is arrested and fluorescence regains the upper hand, i.e., fluorescence is switched on. Czarnik's compound [41], de Silva's compound [15], and Calcium Green-1 from Molecular Probes [35] respond dramatically to Zn^{2+}, H^+, and Ca^{2+}, respectively.

Sensors for cell-based applications developed in this manner reveal that intracellular ionic signals are heterogeneous at the single-cell level [93]. To analyze whether this heterogeneity is preserved in downstream events, a sensitive, single-cell assay for gene expression was developed. The reporting molecule is the bacterial enzyme β-lactamase, which generates an amplified signal by changing the fluorescence of a substrate made available intracellularly.

4.2.3. Impedance Based Cellular Biosensors

The electrical properties of biological material have been studied using suitable instrumentation. Impedance techniques have been used to study organs in the body [17], explanted neural tissues [12, 39], whole blood and erythrocytes [22, 23], cultured cell suspensions [73], bacterial growth monitoring [34], and anchorage dependent cell cultures [26]. There is a great deal of relevant information regarding the characteristics of biological material to be obtained from those studies. Most significant are the frequency dependent dielectric properties of biological materials including cells which yield insight into the expected behavior within different frequency ranges.

The membranes of biological materials including cells exhibit dielectric properties. By measuring the changes in the effective electrode impedance, cultured cell adhesion, spreading and motility can be interpreted from the extracellular signal of the cells. The reliability of impedance measurements depends on the observation that intact living cells are excellent electrical insulators at low signal frequencies. When the coverage over an electrode area increases, the effective electrode impedance increases as well. Impedance measurements have been used for monitoring the behavior of an array of nonexcitable cell types including macrophages [46], endothelial cells [90] and fibroblasts [26]. Figure 4.1 shows the schematic of an impedance sensor.

It is desirable to improve the interfacial sealing at the cell-electrode junction for conducting the measurement of action potentials extracellularly [86]. Further work has been done to deduce cell-electrode interface characteristics for the development of a better understanding of extracellular action potential measurements (Lind, et al., 1991). Surface roughness effects on cell adhesion were examined by looking at smooth gold electrodes, rough platinized gold electrodes, and gold electrodes roughened by dry etching (Lind, et al., 1991). In 1995, the work was continued using *Lymnaea* neurons [6]. Impedance measurements were performed both before and during cell culture and estimates of the cell to substrate sealing impedance were made. These impedance values were then correlated with the recorded extracellular action potentials, revealing a directly proportional relationship. As the sealing impedance increased, the extracellular signal strength did as well, thereby verifying that the sealing impedance is indeed critical for improved signal to noise ratio (SNR).

4.2.4. Intracellular Potential Based Biosensors

The functional or physiological significance of the analyte to the organism can be related to the information derived from cell-based biosensors. Bioelectric signals from excitable cells have been used to relay functional information concerning cell status [28]. Membrane excitability plays a key physiological role in primary cells for the control of secretion and contraction, respectively. Thus, analytes that affect membrane excitability in excitable cells are expected to have profound effects on an organism. Furthermore, the nature of the changes in excitability can yield physiological implications for the response of the organism to analytes. Direct monitoring of cell membrane potential can be achieved through the use of glass microelectrodes. Repetitively firing neurons from the visceral ganglia of the pond snail has been used to quantitatively assess the concentration of a model analyte, serotonin [80]. Figure 4.2 portrays an example of the graded increase in firing rate seen in both the VV1 and VV2 neurons with serotonin concentration. As indicated in the figure, the traces are the different cellular response to additions of 10^{-6} M, 10^{-5} M, 10^{-4} M, and 10^{-3} M serotonin. The basic principle behind intracellular measurements is that tissue slices are prepared and are exposed to chemical analytes under test and the electrical activity from excitable cells are measured using the patch clamp technique. This

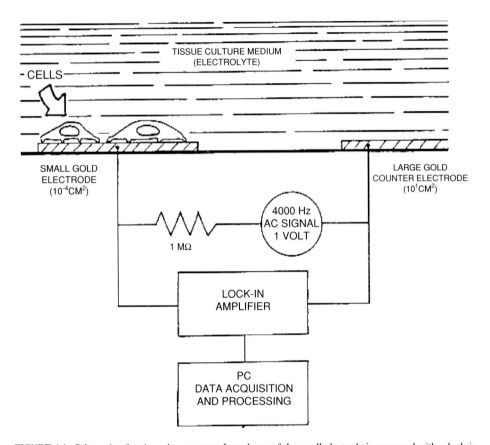

FIGURE 4.1. Schematic of an impedance sensor. Impedance of the small electrode is measured with a lock-in amplifier in series with a $1M'\Omega$ resistor to obtain an approximate constant current source. Electric cell–substrate impedance sensing (ECIS) is the technique that is used to monitor attachment and spreading of mammalian cells quantitatively and in real time. The method is based on measuring changes in AC impedance of small gold-film electrodes deposited on a culture dish and used as growth substrate. The gold electrodes are immersed in the tissue culture medium. When cells attach and spread on the electrode, the measured electrical impedance changes because the cells constrain the current flow. This changing impedance is interpreted to reveal relevant information about cell behaviors, such as spreading, locomotion and motility. They involve the coordination of many biochemical events [108].

technique illustrates the utility of excitable cells as sensors with sensitivity to chemical warfare agents; however, the invasive nature of intracellular recording significantly limits the robustness of this approach for biosensor applications. Another drawback is that excitable cells assemble into coupled networks rather than acting as isolated elements; as a result, for certain sensing applications the ability to simultaneously monitor two or more cells is essential as it permits measurements of membrane excitability and cell coupling. This is not possible using intracellular techniques. The advantage of the technique is that the physiological state of a cell can be assessed. Due to the invasiveness of the technique, it is not possible to apply it for long term measurements.

Evaluation of neuron-based sensing with serotonin

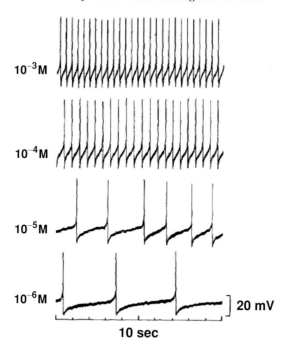

FIGURE 4.2. An example of the effects of serotonin on the spontaneous firing rate of the VV1 and VV2 neurons in a *Limnea stagnalis* snail. As indicated, the traces are the cellular response to additions of 10^{-6} M, 10^{-5} M, 10^{-4} M, and 10^{-3} M serotonin [80].

4.2.5. Extracellular Potential Based Biosensors

In recent years, the use of microfabricated extracellular electrodes to monitor the electrical activity in cells has been used more frequently. Extracellular microelectrode arrays offer a noninvasive and long-term approach to the measurement of bio-potentials [11]. Multi-electrode arrays, typically consisting of 16 to 64 recording sites, present a tremendous conduit for data acquisition from networks of electrically active cells. The invasive nature of intracellular recording, as well as voltage-sensitive dyes, limits the utility of standard electrophysiological measurements and optical approaches. As a result, planar microelectrode arrays have emerged as a powerful tool for long term recording of network dynamics. Extracellular recordings have been achieved from dissociated cells as well; that is more useful in specific chemical agent sensing applications. The current state of the art microelectrode technology comprises of 96 microelectrodes fabricated using standard lithography techniques as shown in Figure 4.3A [13]. More detailed work by Gross and his colleagues at the University of North Texas over the past 20 yrs have demonstrated the feasibility of neuronal networks for biosensor applications [28, 29]. They have utilized transparent patterns of indium–tin–oxide conductors 10 μm wide, which were photo-etched and passivated with a polysiloxane resin [30, 31]. Laser de-insulation of the resin resulted in 64 recording "craters" over an area of 1 mm^2, suitable for sampling of the neuronal ensembles achieved in culture.

Principle of extra cellular potential based biosensors

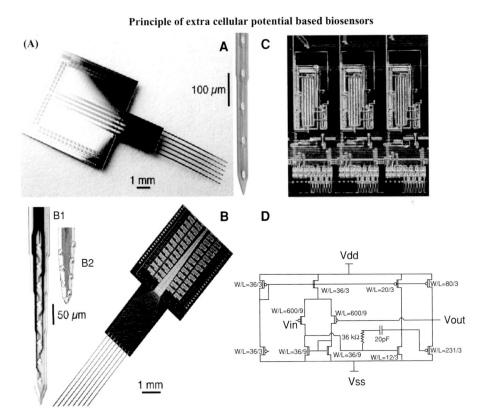

FIGURE 4.3. (**A**) Extra cellular multiple-site recording probes. *A*: 6-shank, 96-site passive probe for 2-dimensional imaging of field activity. Recording sites (16 each; 100 μm vertical spacing) are shown at higher magnification. *B*: 8-shank, 64-site active probe. Two different recording site configurations (linear, *B1* and staggered sites, *B2*) are shown as insets. *C*: close-up of on-chip buffering circuitry. Three of the 64 amplifiers and associated circuits are shown. *D*: circuit schematic of operational amplifier for buffering neural signals) [13].

Indeed, neurons cultured over microelectrode arrays have shown regular electrophysiological behavior and stable pharmacological sensitivity for over 9 months [32]. Figure 4.3B shows neuronal cultures obtained on a microelectrode array with 64 sites [109]. In fact, their precise methodological approach generates a co-culture of glial support cells and randomly seeded neurons, resulting in spontaneous bioelectrical activity ranging from stochastic neuronal spiking to organized bursting and long-term oscillatory activity [32]. Microelectrode arrays coupled with "turnkey" systems for signal processing and data acquisition are now commercially available. In spite of the obvious advantages of the microelectrode array technology for biosensing, in determining the effect of chemical analytes at the single cell level, it becomes essential to pattern the dissociated cells accurately over the microelectrodes. Single cell based sensing forms the basis for determining cellular sensitivity to a wide range of chemical analytes and determining the cellular physiological changes. Analysis of the extracellular electrical activity provides unique identification tags associated with cellular response to each specific chemical agent also known as "Signature Patterns".

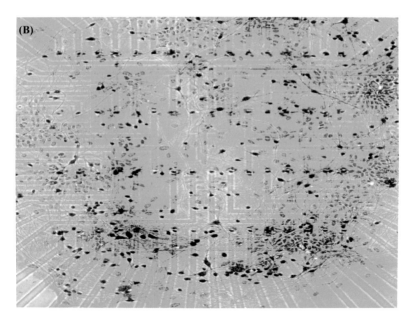

FIGURE 4.3. (*Continued*) (**B**) Neuronal cultures on a 64 microelectrode array. Laser de-insulation of the resin resulted in 64 recording "craters" over an area of 1 mm2, suitable for sampling the neuronal ensembles achieved in culture. neurons cultured over microelectrode arrays have shown regular electrophysiological behavior and stable pharmacological sensitivity for over 9 months [109].

4.3. DESIGN AND METHODS

4.3.1. Requirements for Cell Based Sensors

When developing a system for monitoring the extracellular action potential or cellular impedance of anchorage dependent cell types, it is necessary to design the sensing system with several criteria in mind: Biocompatibility, maintenance of the physiochemical environment (temperature, pH, etc.), maintenance of sterility during cell growth and sample introduction, methods of sample introduction, a transducer for monitoring the desired electrical signal, low signal path parasitics, electronics for extraction of the electrical signal, and packaging which facilitates insertion of the cell culture system in to the measurement electronics while protecting the living system from the external environment. These requirements often trade off against each other and require compromise for the best overall solution. Biocompatibility is perhaps the most important consideration when developing a cell based biosensor. If biocompatible materials are not employed through the design, the sensing element (the cells) will not survive to perform the initial signal transduction required. While biocompatibility generally means not having a toxic, harmful, or otherwise deleterious effect on biological function, there are varying degrees of it dependent on the application. Chronic studies where foreign materials are in contact with living tissue require a more diligent effort for the determination of biocompatibility than do acute studies where the tissue is in contact with the materials for a short duration. Cell culture for hybrid biosensor applications falls somewhere in between, depending on the application area. For all of the work presented herein, the requirements are for acute studies only (those where cells are cultured for less than one week). All materials that are in contact with the cellular system (comprising the cells and culture media) must be biocompatible as described above. This includes the substrate, electrodes, chamber housing, adhesives, sealants, tubing, valves, and pumps. Biocompatibility was determined for most materials by culturing the cells of interest with the cells themselves or the culture media in direct contact with the material to be tested. If the cells appeared "normal" under optical inspection and proliferated as expected, the material was deemed biocompatible for the acute studies.

4.3.2. Cell Manipulation Techniques

There are three cell patterning methods that are currently in use. The first is a topographical method, which is based on the various microfabrication schemes involved in developing microstructures that enable the isolation and long-term containment of cells over the substrates [8, 52]. Other fabrication techniques used for cell patterning and the formation of ordered networks involves the development of bio-microelectronic circuits, where the cell positioning sites function as field effect transistors (FET). This provides a non invasive interface between the cell and the microelectronic circuit [42, 59, 107]. These multi-electrode designs incorporating the topographical method have become increasingly complex, as the efficiency of cell patterning, has improved and hence fabrication has become more challenging and the devices are unsuitable for large-scale production. The other drawback is the need for an additional measurement electrode for determining the electrical activity from the electrically excitable cells. The second method is based on micro-contact printing (μCP) where simple photolithography techniques are coupled with the use of some growth permissive molecules (e.g. an aminosilane, laminin-derived synthetic peptide, Methacrylate and acrylamide polymers or poly-L-Lysine) that favor cell adhesion and growth and anti permissive molecules like fluorosilanes to form ordered cell networks [71, 72, 75, 98, 101]. The disadvantage of this technique is the presence of multiple cells on a single patterned site that results in formation of a dense network of cell processes along the patterned areas. This in turn results in difficulties in measurement as well as determination of the electrical activity associated with a specific cell. The third method is based on using biocompatible silane elastomers like polydimethylsiloxane (PDMS). Cell arrays are formed using microfluidic patterning and cell growth is achieved through confinement within the PDMS structure. This technique is hybrid in the sense that it also incorporates μCP for promoting cell adhesion [27, 89]. The drawback of this technique is its complexity. As of today no single technique has been developed that (1) efficiently isolates and patterns individual cells onto single electrodes (2) provides simultaneous electrical and optical monitoring (3) achieves reliable on-site and non-invasive recordings using the same electrode array for both positioning as well as recording.

4.3.3. Principles of Dielectrophoresis (DEP)

Dielectrophoresis, the force experienced by a polarized object in an electric field gradient, has been shown to manipulate and trap submicron particles. When particles are subjected to an electric field, a dipole moment is induced in them. In a nonuniform electric field, a polarized particle experiences a net force, which can translate the particle to high or low field regions termed positive and negative DEP, respectively. This movement depends on the polarizability of the particle relative to that of the medium. In an AC field, positive and negative DEP can be achieved by choosing the appropriate frequencies. The frequency at which there is no force acting on the particle is called the "crossover frequency".

The dieletrophoretic force acting on a spherical particle of radius r is given by

$$F_{DEP} = 2\pi r^3 \varepsilon_m \text{Re}(f_{CM}) \nabla E^2 \qquad (4.1)$$

where ε_m is the absolute permittivity of the suspending medium, E is the local (rms) electric field, ∇ is the del vector operator, and Re(f_{CM}) is the real part of the polarization factor (Clausius-Mossotti factor), defined as

$$f_{CM} = \frac{(\varepsilon_P^* - \varepsilon_m^*)}{(\varepsilon_P^* + 2\varepsilon_m^*)} \qquad (4.2)$$

In the above equation, ε_P^* and ε_m^* are the complex permittivity of the particle and the medium respectively, where $\varepsilon^* = \varepsilon - j\sigma/\omega$ and ε is the permittivity, σ is the conductivity, ω is the angular frequency of the applied field and $j = (-1)^{1/2}$.

At the crossover frequency, $f_{crossover}$, Equation (4.1) should be equal to zero. Therefore, the crossover frequency is given by

$$f_{crossover} = \frac{1}{2\pi} \sqrt{\frac{(2\sigma_m + \sigma_P)(\sigma_m - \sigma_P)}{(2\varepsilon_m + \varepsilon_P)(\varepsilon_m - \varepsilon_P)}} \qquad (4.3)$$

The principle is illustrated schematically in Figure 4.4. If a polarizable object is placed in an electric field, there will be an induced positive charge on one side of the object and an induced positive charge and an induced negative charge (of the same magnitude as the induced positive charge) on the other side of the object. The positive charge will experience a pulling force; the negative charge will experience a pushing force. In a non-uniform field, as depicted in figure 4.4B, the electric field will be stronger on one side of the object and weaker on the other side of the object. Hence, the pulling and pushing forces will not cancel, and there will be a net force on the object.

Biological cells consist of structures of materials which have different electrical properties and will be polarized in a nonuniform electrical field. The suspending medium, usually water or a dilute electrolyte is already a highly polar material. It will itself be strongly pulled toward the region of highest field intensity by the nonuniform electrical field. If the cell is to move to the region of highest field intensity, it must therefore exhibit an even higher specific polarizability. There are several ways the cellular systems can attain the higher polariazabiliy [66]. First, the cell itself is largely composed of water. Second, there are numerous polar molecules dissolved in the intracellular regions including proteins, sugars, DNA, RNA, etc., all of which can contribute to the polarization. Third, there are structured regions which can act as capacitive regions, e.g. lipid across which the electrolytes can act to produce charge distributions. Fourth, there are structured areas in the surface where ionic double layers can produce enormous polarizations. Of all the possible mechanisms, the fourth one is perhaps the most important, especially at frequencies below 10 MHz.

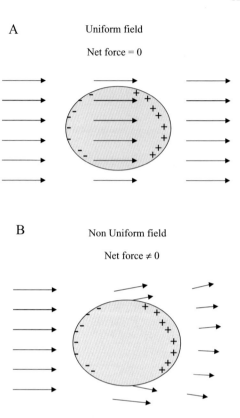

FIGURE 4.4. Schematic description of dielectrophoresis. **A** In a uniform field, the net force is zero. **B** In a nonuniform field, the net force is not zero. The direction of the arrows represent the direction of the electric field; and the length of the arrow represents the magnitude of the electric field. Poul, H.A., Dielectrophoresis: The behavior of neutral matter in nonuniform electric fields. 7, Cambridge University Press.

4.3.4. Cell Manipulation Using Dielectrophoresis (DEP)

The first application of dielectrophoresis to living cells was described by Pohl and Hawk [67]. They described what appears to have been the first purely physical means of separating live and dead cells. After that, nonuniform field effects have been shown useful in a variety of biological systems, including algae, bacteria, yeasts, mammalian blood cells, chloroplasts, mitochondria and viruses. DEP is particularly useful in the manipulation and separation of microorganisms and has been employed successfully in isolation and detection of sparse cancer cells, concentration of cells from dilute suspensions, separation of cells according to specific dielectric properties, and trapping and positioning of individual cells for characterization [97], for example, for separations of viable and nonviable yeast cells [40, 53], leukemia and breast-cancer cells from blood, and the concentration of $CD34^+$ cells from peripheral-stem-cell harvests [82], live and dead cells of the same species of small bacteria as *Listeria* [49]. Previous research in the field of DEP has already shown that small particles and living cells can be manipulated by DEP [25, 54, 57].

TABLE 4.1. Parameters for positive and negative DEP for neurons and osteoblasts [68, 104].

Cell type	Separation buffer for DEP	Conductivity of buffer solution (mS/cm)	Positive DEP frequency	Negative DEP frequency	Cross over frequency	V_{pp} (Volts)
Neurons	250 mM Sucrose/1640 RPMI	1.2	4.6 MHz	300 kHz	500 kHz	8
Osteoblasts	250 mM Sucrose/ Dubecco's modified Eagle Medium	6.07	1.2 MHz	75 kHz	120 kHz	2

4.3.5. Cell Types and Parameters for Dielectrophoretic Patterning

Mammalian cells that have electrically excitable cell membranes are suitable for cell based sensing. Prasad et al. [68] and Yang et al. [104] used rat hippocampal cells from a H19-7 cell line (ATCC, Inc.) and cells from a primary rat osteoblast culture. Their parameters for DEP isolation and positioning are summarized in Table 4.1. They have established a gradient AC field among electrodes on a microarray device and swept the applied frequency, the peak-to-peak voltage and varied the conductivity of the separation buffer to determine the optimum parameters. Cells under the absence of an electric field have a uniformly distributed negative charge along the membrane surface; on applying a gradient AC field, a dipole is induced based on the cell's dielectric properties and due to the nonuniform electric field distribution, the electrically excitable cells experience a positive dielectrophoretic force that causes their migration to the electrodes, which are the regions of high electric fields [68, 104]. This constitutes the technique for isolating and positioning the cells over the electrodes.

4.3.6. Biosensing System

The biosensing system comprises of a chip assembly and an environmental chamber to maintain a stable local environment for accurate data acquisition. The biosensing system is schematically represented in Figure 4.5.

4.3.6.1. Chip Assembly A 4 × 4 microelectrode array comprising of platinum electrodes (diameter: 80 μm, center-to-center spacing: 200 μm) spanning a surface area of 0.88 × 0.88 mm^2 on a silicon/silicon nitride substrate with electrode leads (6 μm thick) terminating at electrode pads (100 μm × 120 μm) has been fabricated using standard microlithography techniques [102]. To achieve a stable local microenvironment for sensing, the microelectrode array has been integrated to a silicone chamber (16 × 16 × 2.5 mm^3) with a microfluidic channel (50 μm, wide); to pump in the testing agent and pump out the test buffer once the sensing process has been completed. The flow rate of the buffer was 40 μL/min. The silicone chamber was provided with an opening (8 × 8 × 2.5 mm^3) and covered by a glass cover slip for in-situ monitoring. Simultaneous electrical and optical monitoring has been achieved by using a Microzoom^{TM}(Nyoptics Inc, Danville, CA)

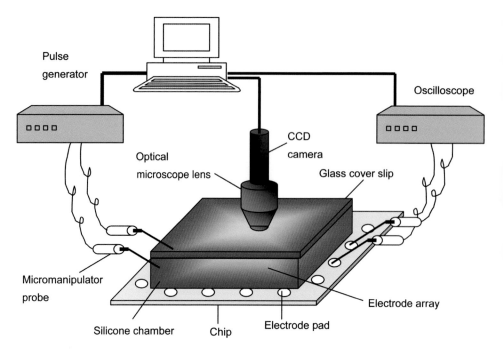

FIGURE 4.5. Schematic representation of the measurement system. It provides simultaneous electrical and optical monitoring capability [105].

optical probe station under 8 × and 25 × magnification. The electrical stimulation and measurements were achieved by using micromanipulators (Signatone, Gilroy, CA).

4.3.6.2. Environmental Chamber The optical probe station along with the chip assembly was enclosed by an acrylic chamber (S&W Plastics, Riverside, CA). The environment in the chamber is controlled so as to maintain a constant temperature of 37°C. A heat gun (McMaster, Santa Fe Springs, CA) inside the chamber heats the air in the chamber and this is linked to a temperature controller (Cole Parmer, Vernon Hills, Illinois) that stops the heat gun from functioning above the desired temperature. A 6" fan (McMaster, Santa Fe Springs, CA) inside the chamber circulates the hot air to maintain temperature uniformity throughout the chamber and is monitored by a J-type thermocouple probe attached to the temperature controller. The carbon dioxide concentration inside the chamber is maintained at 5% and is humidified to prevent excessive evaporation of the medium. This chamber with all of its components will ensure cell viability over long periods of time and stable cell physiology in the absence of the chemical agents.

4.3.7. Cell Culture

4.3.7.1. Neuron Culture The H19-7 cell line is derived from hippocampi dissected from embryonic day 17 (E17) Holtzman rat embryos and immortalized by retroviral transduction of temperature sensitive tsA58 SV40 large T antigen. H19-7 cells grow at the permissive temperature (34°C) in epidermal growth factor or serum. They differentiate to a neuronal phenotype at the non-permissive temperature (39C) when induced by basic fibroblast growth factor (bFGF) in N2 medium (DMEM-high glucose medium with supplements). H19-7/IGF-IR cells are established by infecting H19-7 cells with a retroviral vector expressing the human type I insulin-like growth factor receptor (IGF-IR). The cells are selected in medium containing puromycin.H19-7/IGF-IR cells express the IGF-IR protein. IGF-IR is known to send two seemingly contradictory signals inducing either cell proliferation or cell differentiation, depending on cell type and/or conditions. At 39°C, expression of the human IGF-IR in H19-7 cells induces an insulin-like growth factor (IGF) I dependent differentiation. The cells extend neuritis and show increased expression of NF68. This cell line does not express detectable levels of the SV40 T antigen. Following spin at $100 \times g$ for 10 minutes at room temperature; cells were re-suspended in a separation buffer (see Table 4.1). The density of the re-suspended cells (2500 cell/mL) ensured single cell positioning over individual electrodes. Separation buffer used for neurons contained 250 mM sucrose/1640 RPMI (Roswell Park Memorial Institute), with a conductivity of 1.2 mS/cm and a pH of 7.48. The separation buffer was replaced by a buffer comprising of minimum essential medium/10% Fetal Bovine Serum (FBS)/5% Phosphate buffer saline (PBS) of conductivity 2.48 mS/cm and pH of 7.4 suitable for cell viability.

4.3.7.2. Primary Osteoblast Culture Primary rat osteoblast cells were cultured to a concentration of 10,000 cells in 1 mL for sensing experiments. To achieve the patterning of a single cell over a single electrode, a 10 µL of cell culture solution was mixed with 500 µL Dulbeco modified eagle medium (DMEM; Gibco, Grand Island NY) supplemented with 10% fetal bovine serum (FBS; Gibco, Grand Island NY), 100 µg/mL penicillin, and 100 µg/mL streptomycin (P/S; Gibco, Grand Island NY). The cells were centrifuged and re-suspended in 1 mL of separation buffer consisting of 1:9 dilutions of Phosphate Buffer Saline 250 mM Sucrose (Sigma, St Louis) and de-ionized water (weight/volume). The conductivity of the separation buffer was 4.09 mS/cm and with a pH of 8.69. The separation buffer was replaced with a test buffer ((DMEM)/Fetal Bovine Serum (FBS)/Phosphate Buffer Saline (PBS)) with conductivity of 2.5 mS/cm and a pH of 8.06.

4.3.8. Experimental Measurement System

Figure 4.5 shows a schematic representation of the measurement system. It comprises of extracellular positioning, stimulating and recording units. The cells were isolated and positioned over single electrodes by setting up a gradient AC field using an extracellular positioning system comprising of a pulse generator (HP 33120A) and micromanipulators (Signatone, Gilroy, CA). The signal from the pulse generator was fed to the electrode pads of the selected electrodes using the micromanipulators. The extracellular recordings from the individual osteoblasts obtained from the electrode pads were amplified and recorded on an oscilloscope (HP 54600B, 100 MHz). The supply and measurement systems were integrated using a general purpose interface bus (GPIB).

4.4. MEASUREMENTS

4.4.1. Long Term Signal Recording in vivo

In essence, signals can be obtained from the microelectrodes which are related to the action potentials (Figure 4.6). The extracellular electrodes record a current which provides

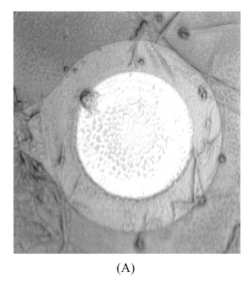

(A)

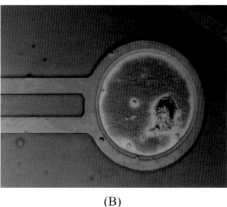

(B)

FIGURE 4.6. Single Neuron positioned on the surface of microelectrode. **A.** single neuron adhered to the edge of microelectrode due to dielectric potential trap (DEP); **B**. single neuron well spread over the surface of a microelectrode [106].

a voltage in the external load impedance. The magnitude and temporal characteristics of an action potential so recorded depends on local conditions, e.g. when axons cross the bare surface of an electrode, the resulting signal-to-noise ratios (SNR) are high. The electrodes produce signals that resemble action potentials in shape when the electrode sealing to the cell is good, i.e. impedance to earth is very high. The relative impedances of the electrode paths determined both the magnitude and the form of the signal.

The form of the extracellular signal changed with the condition of the sealing of cells over the electrodes. Poor sealing between cells and electrodes resulted in signals with low S/N ratios and the quality of the signal recorded from a neuron with a good sealing over the electrode improved due to an increase in the sealing (interfacial) impedance. Figure 4.7

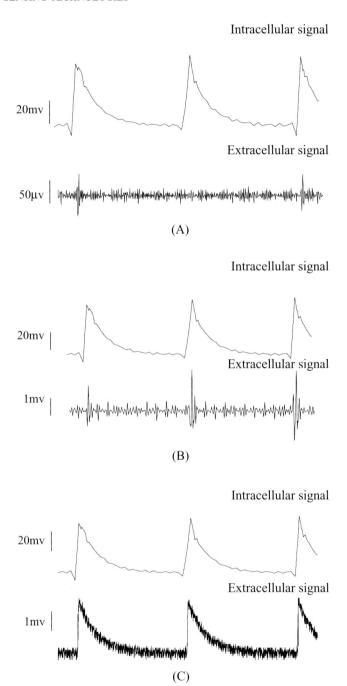

FIGURE 4.7. Effect of cell-electrode interfacial sealing conditions on the quality of signals recorded via the microfabricated electrodes. **A**. Poor sealing conditions. **B**. Better sealing conditions. **C**. Excellent sealing conditions. Cells were well spread over the surface of the microelectrodes, and the shape of the extracellular signal spectrum is similar to that of the intracellular signal [106].

illustrates three such examples in which the top traces are intracellular signals and the lower traces are the corresponding extracellular recordings. In the first example, the sealing condition was not good and the amplitude of the recorded extracellular signal was very small (50–80 μV). In the second and third examples, the cell had spread well over the electrode and a larger amplitude extracellular signal was recorded (2–5 mV). A clear relationship must exist between the amplitude of the signal and the degree of sealing over the electrode. The other feature to note is that as the sealing conditions become better, the signal shape becomes similar to that of the intracellularly recorded changes and not the differential signal (Figure 4.7 (c)).

The capped microelectrode array is a highly stable recording environment primarily because the electrodes do not invade the cell membrane and do not vibrate or slip relative to the neural components. However, the number of active electrodes with sufficiently high SNRs can vary from culture to culture and are influenced by the neuronal cell density, glial density and the size of the adhesion island over the recording matrix. They don't seem to be greatly affected by the age of the culture patterned on the electrodes. A statistical interpretation of the active electrodes with mean and maximum SNRs as a function of culture age are provided in Figure 4.8. The value of maximum and mean SNRs were around

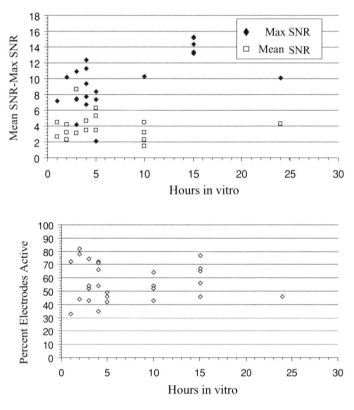

FIGURE 4.8. Long term *in vivo* studies of signal-to-noise ratio (SNR) [106].

CELL BASED SENSING TECHNOLOGIES

10 and 4 respectively. It was also observed that there was no obvious trend of a decrease in the value of the SNRs even after many hours of *in vivo* sensing.

4.4.2. Interpretation of Bioelectric Noise

The amplitude distribution of a biological noise signal usually yields little information about the membrane events that give rise to the observed noise. This difficulty arises for two reasons. First, the calculated shape of an amplitude distribution may be characteristic of noise generated by more than one mechanism. Second, the shape of the distribution may alter the frequency of the underlying membrane event itself. This means a single noise generating process can give rise to signals with widely differing amplitude distributions. In order to determine the probable membrane mechanisms underlying a biological noise signal, it is necessary to analyze the signal with respect to its frequency composition. This analysis is carried out using the methods of fluctuation statistics and fast Fourier transformation (FFT).

In the low frequency range (f = 1 $\sim$ 10 Hz), the presence of Johnson noise in cell membranes is usually obscured by other forms of biological noise which display greater intensity (Figure 4.9). The shape of 1/f noise can be seen in this frequency band (1 $\sim$ 10 Hz) which contributes more than the other noise sources. After the 100 Hz boundary, a steady "platform" is observed in the frequency spectra where Johnson noise is the dominating content of the biological noise which is independent of frequency and the amplitude of the 1/f noise becomes negligible with increasing the frequency. Actually, in the high frequency range (f > 100 Hz), Johnson noise is always shadowed by the appearance of a capacitative current noise signal which arises from voltage fluctuations in the recording apparatus.

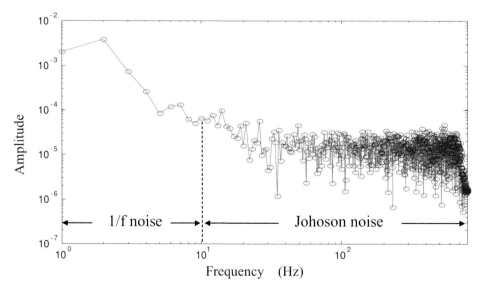

FIGURE 4.9. Frequency domain of noise signal from a single neuron coupled to a microelectrode [106].

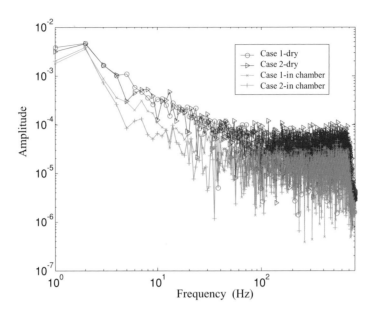

FIGURE 4.10. Effect of environmental parameter to noise. The diameter of microelectrode is 50 μm [106].

4.4.3. Influence of Geometry and Environmental Factors on the Noise Spectrum

The effect of microelectrode dimensions and the environmental conditions of the microchamber on the noise spectrum is shown in Figures 4.10 and 4.11. Noise measurements were conducted for two types of microelectrode arrays with diameters of 50 μm and 80 μm respectively. Figure 4.10 shows that immersing the microelectrode into the media solution and obtaining a good sealing of the microchamber reduced the noise level approximately by a factor of 1.5 compared to the noise level for dry and open condition of the microchamber. When the diameter of microelectrode is increased from 50 μm to 80 μm, the noise level was reduced by an additional factor of 2 (Figure 4.11). The Johnson noise content was filtered beyond the 100 Hz regime and the drop of the 1/f noise is seen clearly in the overall frequency domain. This is because in the low frequency domain, the main content of the noise signal is composed of the 1/f noise which is inversely proportional to the diameter of the microelectrode. Hence, geometry factors are more dominant in the low frequency range.

Figure 4.12 shows the frequency spectra finger print for single neuron to ethanol sensing (9 ppm). SNR is improved from 8 to around 17 after the denoising process. Statistical analysis of interspike interval histogram (ISIH) finger print of neuron spike to ethanol after denoising is shown in Figure 4.12. The interspike interval τ is around $0.015 \sim 0.017$ seconds and the relative firing frequency is in the $58.8 \sim 66.67$ Hz range which corresponds to the frequency spectral finger print.

Four groups of SNR measurements for different microelectrode dimensions (50μm or 80μm) and environmental conditions (dry or in-chamber) are presented in Figure 4.13. It can be seen that, the case of 80μm diameter and in-chamber condition indicated the best SNR after denoising. Hence, the SNR of a signal spectrum for cell based chemical sensing can be improved by choosing the optimum geometrical factors and environmental conditions.

4.4.4. Signal Processing

Changes in the extracellular potential shape have been used to monitor the cellular response to the action of environmental agents and toxins. The extracellular electrical activities of a single osteoblast cell are recorded both in the presence and absence of chemical agents and the modulation in the electrical activity is determined. However, the complexity of this signal makes interpretation of the cellular response to a specific chemical agent rather difficult. It is essential to characterize the signal both in time domain and frequency domain for extracting the relevant functional information. The use of power spectral density analysis as a tool for classifying the action of a chemically active agent was investigated and

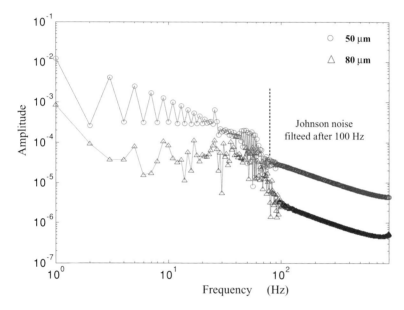

FIGURE 4.11. Noise amplitude as a function of frequency and elecrode dimensions [106].

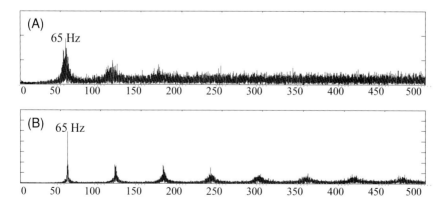

FIGURE 4.12. Frequency spectra finger print for single neuron sensing of ethanol. **A**. before denoising **B** after denoising [106].

found to offer a more suitable technique for data analysis. The power spectrum of the extracellular potential is a better indicator of the cell response than the monitored peak-to-peak amplitude.

Additionally, by examining the Root Mean Square (RMS) power in different frequency bands, it is possible to approximate the power spectral density analysis performed numerically herein. Using FFT analysis, the shifts in the signal's power spectrum were analyzed. FFT analysis extracts the modulation in the frequency of the extracellular potential burst rate and hence is termed as "frequency modulation" and generates the SPV (Signature Pattern Vector). The "Eigen Vectors" corresponding to the modulated firing rate of the osteoblast cell are determined from the SPV. However the FFT process is a transformation based on the whole scale, i.e. either absolutely in time domain, or absolutely in frequency domain. Therefore, it is impossible to extract the local information in time domain. Thus, WT (Wavelet Transformation) analysis was performed to extract the information from the local time domain. WT is a time-scale (time-frequency) analysis method whereby multi-resolution analysis of the parameters is achieved. This can express the local characterization of signals both in time and frequency domains, hence, extracting functional information from the extracellular potential, such as the response time and the limits of detection. As this analysis relies on the determination of the modulation of the amplitude of the signal due to the effect of the chemical agents it is termed as "Amplitude Modulation".

4.4.5. Selection of Chemical Agents

It is essential to obtain the effect of a broad spectrum of chemical agents ranging from highly toxic and physiologically damaging to relatively less toxic to determine and evaluate the time window of response of a particular cell type for a specific known agent based on varying concentrations and finally determine the limit of detection for a specific chemical

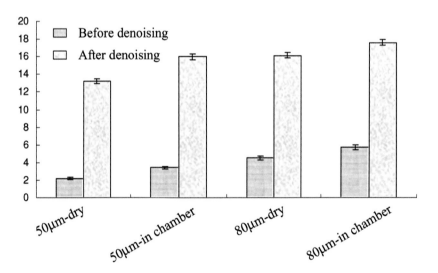

FIGURE 4.13. SNR comparison for different dimension and environmental conditions before and after denoising [106].

agent. All the experiments were conducted based on the hypothesis that a unique SPV would be generated for each cell type for a specific chemical. This was hypothesized as it has been scientifically proven that different chemicals bind to different ion channel receptors thus, modifying the electrical response of the cell in a unique manner [7, 76]. Here, the responses of single osteoblast cells were presented to the effect of the following chemical agents: Ethanol, Hydrogen peroxide, Ethylene diamene tetra acetic acid (EDTA), and Pyrethroids, for n = 15.

4.4.5.1. Ethanol Ethanol produces anesthetic effects but in a milder form as compared to pentobarbitone and ketamine, though the mechanism of action is essentially assumed to be the same [79]. We hypothesized that determination of single cell ethanol sensitivity would help us identify the lowest threshold concentration, for the family of chemicals whose physiological response mechanism would mimic that of ethanol. The concentration ranges tested for ethanol were from 5000 ppm to 15 ppm. The detection limit for ethanol using this technique was determined to be 19 ppm.

4.4.5.2. Hydrogen Peroxide It is one of the major metabolically active oxidants present in the body and leads to apoptosis. Hydrogen peroxide also leads to the degradation of cells. As the behavior of hydrogen peroxide *in-vivo* is similar to the behavioral responses obtained from exposure to carcinogenic chemicals such as rotenone, it was estimated that hydrogen peroxide would make an ideal candidate for sensing studies [33]. The range of concentration of hydrogen peroxide varied from 5000 ppm to 20 ppm. The sensitivity limit for a single osteoblast due to the action of hydrogen peroxide was determined to be 25 ppm.

4.4.5.3. Pyrethroids They are active ingredients in most of the commercially used pesticides. Pyrethroids share similar modes of action, resembling that of DDT. Pyrethroids are expected to produce a "knock down" effect *in-vivo*; the exact *in-vitro* response at a cellular level has not yet been understood. Hence, they are ideal candidates for the analysis of this genre of chemicals. The concentration range of pyrethroids varied from 5000 ppm to 850 ppb. The detection limit for pyrethroids was determined to be 890 ppb.

4.4.5.4. Ethylene Diamene Tetra Acetic Acid (EDTA) EDTA belongs to a class of synthetic, phosphate-alternative compounds that are not readily biodegradable and once introduced into the general environment can re-dissolve toxic heavy metals. Target specificity of EDTA in a single osteoblast cell has not been electrically analyzed to date. The concentration ranges for EDTA varied from 500 ppm to 175 ppm. The limit of detection in this case for a single osteoblast was determined to be 180 ppm.

In all of the experimental cycles, a unique response was obtained from the osteoblast cells to specific chemical agents. The detection limits for a single osteoblast cell for various chemical agents were also determined. The firing rate of a single osteoblast cell in the absence of a chemical agent was determined to be 668 Hz after the FFT analysis of the recorded extracellular electrical activity. On performing FFT analyses on the modified extracellular electrical activity in the presence of specific chemical agents at varying concentrations, specific burst frequencies were obtained that can be used as identification tags for recognizing the chemical agents (Eigen Vectors).

78   CENGIZ S. OZKAN ET AL.

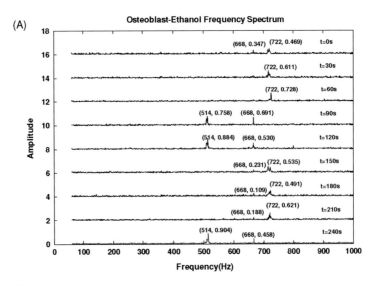

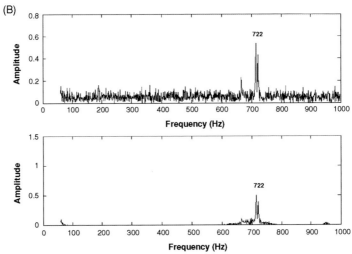

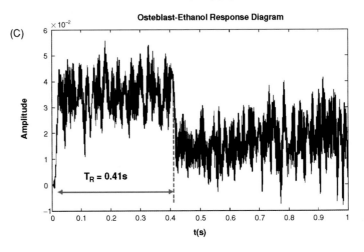

CELL BASED SENSING TECHNOLOGIES

To simulate real time field sensing conditions, "cascaded sensing" was performed using a single osteoblast cell to detect the response to two chemical agents introduced in a cascade form. The results obtained for cascaded ethanol-hydrogen peroxide sensing are presented later in this chapter.

4.4.6. Control Experiment

In order to determine the SPV corresponding to a specific chemical, the initial activity pattern vector for each cell type was determined. Using the process of dielectrophoresis, a single cell was positioned over a single electrode and its initial electrical activity was recorded.

4.4.7. Chemical Agent Sensing

4.4.7.1. Ethanol Sensing Using Single Osteoblast Single osteoblast cells were positioned over individual electrodes. The sensing agent was then introduced onto the microelectrode array using the microfluidic inlet channel. The initial concentration of ethanol used was 5000 ppm and the modified electrical activity was recorded. The concentration of ethanol was decremented in a stepwise manner and in each case the modified electrical activity was recorded. The lowest concentration of ethanol sensed by a single osteoblast was 19 ppm. The lowest detectable concentration of ethanol (19 ppm) by this technique is far more sensitive than the detectable limits as obtained from the optical waveguide technique [77] (35 ppm: 0.4×10^{-6} M) that is considered to be one of the most sensitive detection techniques to date [26].

The analysis was performed on the acquired data pertaining to the modified extracellular potential to yield the SPV. The instant at which the chemical is added to the chip system is denoted by t = 0 sec. Figure 4.14(A) represents the SPV for a single osteoblast due to the action of ethanol at 19 ppm. Osteoblasts have an unmodulated firing rate of 668 Hz. This corresponds to the frequency of firing of osteoblasts in the absence of a chemical agent. There are two Eigen vectors (514 Hz and 722 Hz) in the SPV corresponding to the modulated firing rate of the osteoblast. During the first phase of the sensing cycle (t = (0, 60) sec) the modulated firing rate is focused at 722 Hz. During the second phase of the sensing cycle (t = (60, 120) sec) the modulated firing rate shifts towards the lower frequency value (514 Hz). During the third phase of the sensing cycle (t = (120, 180) sec), the modulated frequency shifts back to the original higher frequency bursting (668 Hz and 722 Hz) as observed in the first phase. As the concentration of ethanol is very low; the cell quickly recovers and on re-introducing the chemical at t = 180 sec; the SPV starts to repeat itself (t = (180, 240 sec).

The WT analysis is performed on the acquired data to yield the local time domain characteristics in order to extract the first modulated maxima corresponding to the first

FIGURE 4.14. **A.** Signature pattern vector of single osteoblast due to the action of ethanol at 19 ppm. **B.** Wavelet transformation analysis to determine the first Eigen vector of a single osteoblast due to the action of ethanol at 19 ppm. (c) Response time of a single osteoblast due to the action of ethanol [105].

Eigen vector of the response. Figure 4.14(B) indicates the extraction of the first Eigen vector using WT at an ethanol concentration of 19 ppm. The response time for an osteoblast is also determined using WT analysis. The response time is defined as the time taken for the functional sensing element -osteoblast, to respond to the specific input- chemical agent, and reach its first extreme value. Figure 4.14(C) indicates the response time for the osteoblast at an ethanol concentration of 19 ppm. We determined that the response time of the osteoblast to a specific chemical until the detection limit remained constant irrespective of the concentrations of the chemical agent. The response time of a single osteoblast to ethanol was determined to be 0.41 sec and is denoted by T_R.

4.4.7.2. Hydrogen Peroxide Sensing Using Single Osteoblast Single osteoblast cells were isolated and positioned over individual electrodes in a manner previously described. The initial concentration of hydrogen peroxide used was 5000 ppm and the modified electrical activity was recorded. The concentration of hydrogen peroxide was decremented in a stepwise manner and in each case; the modified electrical activity was recorded. The lowest concentration of hydrogen peroxide sensed by a single osteoblast was 25 ppm. FFT analysis was performed on the acquired data pertaining to the modified extracellular potential to yield the SPV. The instant at which the chemical is added to the chip system is denoted by t = 0 sec. Figure 4.15(A) represents the SVP.

There are three Eigen vectors (257 Hz, 565 Hz, and 873 Hz) in the SPV corresponding to the modulated firing rate of the osteoblast. The frequency of 668 Hz corresponds to the osteoblast firing rate in the absence of a chemical agent. During the sensing cycle, low-frequency subsidiary peaks (129 Hz, 334 Hz, 257 Hz and 437 Hz) are expressed. We hypothesize that these occur due to probable nonspecific interactions between the chemical agent and the sensing system. The hypothesis is based on the fact that the control burst frequency for a single osteoblast is 668 Hz and the Eigen vector range for the Eigen vectors due to the interaction of the chemical agents have been observed to vary within ± 30% of the control value.

The WT analysis is performed on the acquired data to yield the local time domain characteristics in order to extract the first modulated maxima corresponding to the first Eigen vector of the response (spectrum not shown). The response time for an osteoblast is determined using WT analysis by evaluating the time required to achieve the first maximum after the application of hydrogen peroxide. Figure 4.15(B) indicates the response time for the osteoblast at a hydrogen peroxide concentration of 25 ppm. This technique produces a sensitivity of $2.94 \times 10-8$ M (25 ppm) in comparison to the sensitivity of $1.2 \times 10-6$ M: 42 ppm produced by the optical waveguide technique [4]. We determined that the response time remained constant for varying concentrations of hydrogen peroxide. The response time of a single osteoblast due to hydrogen peroxide was determined to be 0.71 sec for its varying concentrations.

4.4.7.3. Pyrethroid Sensing Using Single Osteoblast The initial concentration of pyrethroid used was 5000 ppm and the modified electrical activity was recorded. The concentration of pyrethroid was decremented in a stepwise manner and in each case the modified electrical activity was recorded. The lowest concentration of pyrethroid sensed by a single osteoblast was 890 ppm. The sensitivity limit obtained via this technique is far more sensitive than that obtained through the waveguide detector technique (≈950 ppb:

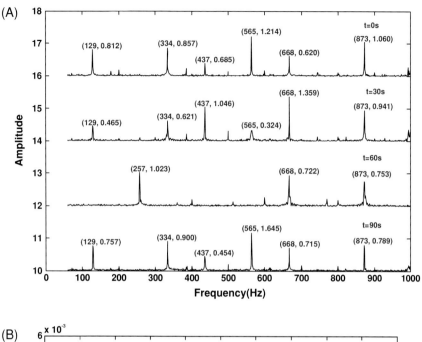

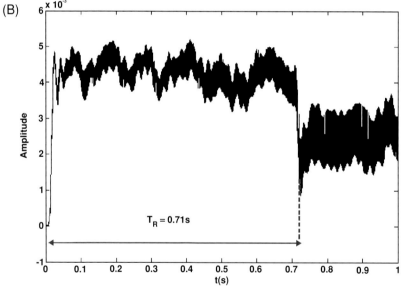

FIGURE 4.15. **A**. Signature pattern vector of single osteoblast due to the action of hydrogen peroxide at 25 ppm. **B**. Response time of a single osteoblast due to the action of hydrogen peroxide [105].

0.1×10^{-6} M) [88]. FFT analysis was performed on the acquired data pertaining to the modified extracellular potential to yield the SPV. The instant at which the chemical is added to the chip system is denoted by t = 0 sec. Figure 4.16(A) represents the SPV.

There are two Eigen vectors (257 Hz, and 873 Hz) in the SPV corresponding to the modulated firing rate of the osteoblast. The frequency of 668 Hz corresponds to the osteoblast firing rate in the absence of a chemical agent. During the first half of the cycle, there are subsidiary peaks corresponding to 129 Hz and 565 Hz corresponding to the non-specific interactions of the chemical agent within the sensing system.

The local time domain characteristics are obtained by performing WT analysis on the acquired data. The first modulated maximum is extracted and this corresponds to the first Eigen vector of the response (spectrum not shown). The response time for an osteoblast is determined using WT analysis by evaluating the time required to achieve the first maximum after the application of pyrethroid. Figure 4.16(B) indicates the response time for the osteoblast at a pyrethroid concentration of 890 ppm. We determined that the response time remained constant for varying concentrations of pyrethroid. The response time of a single osteoblast to pyrethroid was determined to be 0.23 sec and this value remained constant irrespective of the concentrations of pyrethroid.

4.4.7.4. EDTA Sensing Using Single Osteoblast The initial concentration of EDTA used was 5000 ppm and the modified electrical activity was recorded. The concentration of EDTA was decremented in a stepwise manner and in each case. The modified electrical activity was recorded. The lowest concentration of EDTA sensed by a single osteoblast was 180 ppm. The sensitivity of this technique is far superior to that obtained from previous studies which resulted in a detection limit of $4.6 \times 10-6$ M: 210 ppm [92]. FFT analysis was performed on the acquired data pertaining to the modified extracellular potential to yield the SPV. The instant at which EDTA is added to the chip system is denoted by t = 0 sec. Figure 4.17(A) represents the SPV.

The initial peak in the frequency spectrum is observed at 514 Hz corresponding to the first Eigen vector. This is obtained at t = 0 sec, after the immediate application of EDTA. Osteoblast cells then regain their control of the firing rate, corresponding to 667 Hz. The next two Eigen vectors of 258 Hz and 872 Hz are obtained in the time interval (t = (60, 90) sec). Subsidiary low frequency peaks are observed at 129 Hz, 334 Hz, and 437 Hz and high-frequency peaks are observed at 514 Hz and 565 Hz due to the non specific interactions.

The local time domain characteristics and functional information are obtained by performing WT analysis on the acquired data. The first modulated maximum is extracted and this corresponds to the first Eigen vector of the response (spectrum not shown). The response time for an osteoblast is determined using WT analysis by evaluating the time required to achieve the first maximum after the application of EDTA. Figure 4.17(b) indicates the response time for the osteoblast at a pyrethroid concentration of 180 ppm. We determined that the response time remained constant for various concentrations of EDTA. The response time of a single osteoblast to EDTA was determined to be 0.14 sec and this value remained constant irrespective of the concentrations of EDTA.

4.4.7.5. Comparison of Detection Limits and Response Times It was found that a single osteoblast cell was the most sensitive to ethanol (19 ppm) whereas it was the least

CELL BASED SENSING TECHNOLOGIES

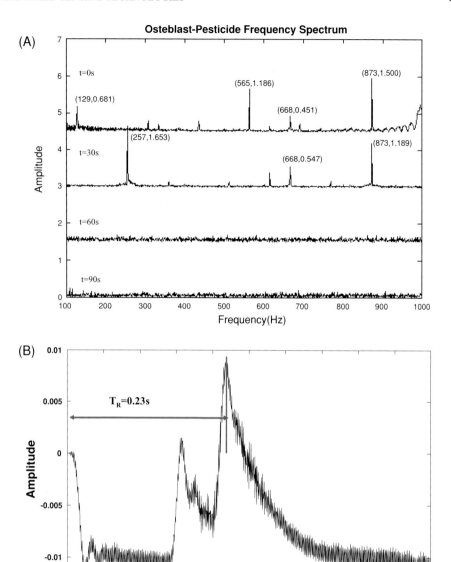

FIGURE 4.16. **A**. Signature pattern vector of single osteoblast due to the action of pyrethroids at 890 ppm. **B**. Response time of a single osteoblast due to the action of pyrethroids [105].

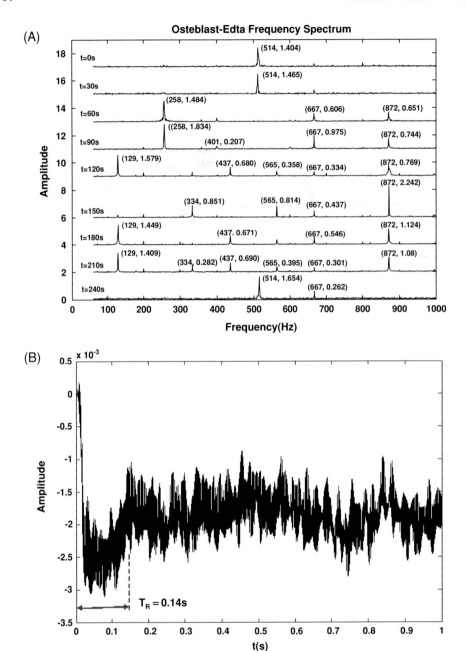

FIGURE 4.17. **A**. Signature Pattern Vector of single osteoblast due to the action of EDTA at 180 ppm. **B**. Response time of a single osteoblast due to the action of EDTA [105].

TABLE 4.2. Comparison of chemical concentrations and response times.

	Ethanol	Peroxide	EDTA	Pyrethroids
Response time(s)	0.41	0.71	0.14	0.23
Concentration	19 ppm	25 ppm	180 ppm	890 ppm

sensitive to pyrethroid (890 ppm). Also, single osteoblast cells respond the fastest to EDTA (0.14 sec) whereas they take the maximum time to respond to hydrogen peroxide (0.71 sec). This data is summarized in Table 4.2. Figure 4.18 is a graphical representation of the response times obtained for each specific chemical agent. The graph shows the repeatability of the response. The response time for each chemical was determined by testing a specific agent in three cycles and each cycle comprised of three runs.

4.4.7.6. Effect of Varying Concentration of Chemical Agents It was observed for all the chemical agents that the amplitude of the response decreased as the concentration of the chemical agent in the local microenvironment increased. WT analysis was performed where local time domain characterization of the amplitude was performed as a function of concentration. This analysis identified the amplitude shifts corresponding to the varying concentration. WT analysis indicated that at a higher concentration (1000 ppm), there was a large decrement in the amplitude of the time domain signal of the extracellular potential. For low levels of concentration near the detection limit, the decrement of the amplitude was much smaller, by a factor of about 80%. Figures 4.19(A) and (B) represent the variation in amplitude due to low (180 ppm) and high (1000 ppm) concentrations of EDTA. It was also observed that there is no noticeable difference in the response times due to varying concentrations for a specific chemical agent.

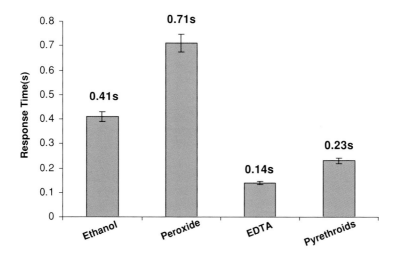

FIGURE 4.18. Representation of response times for specific chemical agents [105].

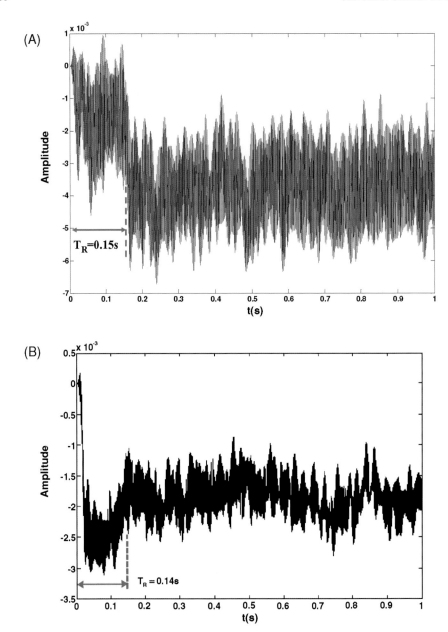

FIGURE 4.19. **A**. Variation in amplitude of single osteoblast due to the action of EDTA at 180 ppm. **B**. Variation in amplitude of single osteoblast due to the action of EDTA at 1000 ppm [105].

4.4.7.7. Cascaded Sensing of Chemical Agents Using Single Osteoblast To simulate real time field conditions, the selectivity of single osteoblast sensors was tested. The ability of the sensor was examined to identify specific chemical agents when introduced in cascade by exhibiting the SPV corresponding to each chemical agent. Here, cascaded sensing of ethanol and hydrogen peroxide was described by single osteoblast cells. After determining the detection limits for both of the chemical agents, first, ethanol at 19 ppm was introduced into the chip sensor and the modified extracellular potential was recorded. As observed previously the osteoblast cell then regains its initial spectrum after undergoing modulation. Hydrogen peroxide at 25 ppm was then introduced into the chip sensor and the modulated response was recorded. FFT analysis of the acquired data indicates that the SPV obtained in the cascaded sensing exactly correlated with the SPVs obtained from individual sensing of ethanol and hydrogen peroxide. The Eigen vectors corresponding to ethanol (514 Hz and 722 Hz) and those corresponding to hydrogen peroxide (257 Hz, 576 Hz and 852 Hz) can be correlated to those obtained during individual chemical sensing. There is a slight shift in two of the Eigen vectors of hydrogen peroxide from 565 Hz to 576 Hz and from 873 Hz to 852 Hz which can be accounted for by the interaction between ethanol and hydrogen peroxide. Figures 4.20(A) and (B) represent the SPVs of a single osteoblast cell due to the cascaded action of ethanol and hydrogen peroxide respectively (for $n = 15$).

4.5. DISCUSSION AND CONCLUSION

The surge of interest in bioanalysis over the last decade has resulted in an ingenious of new proposals and a series of solutions to earlier problems. We have seen the reduction to commercial reality of some of the more speculative proposals from earlier years. Among the most rapidly advancing of these fronts is the area of biosensing, whether it is single analyte detection methods or multiarray-based biochip technology. The 1990s have seen the development of biosensors for many different analyses, and even seen them begin to advance to clinical and in some cases commercially available technologies [55, 60].

Cell-based biosensors constitute a promising field that has numerous applications ranging from pharmaceutical screening to environmental monitoring. Cells provide an array of naturally evolved receptors and pathways that can respond to an analyte in a physiologically relevant manner. Enzymes, receptors, channels, and other signaling proteins that may be targets of an analyte are maintained and, as necessary, regenerated by the molecular machinery present in cells. The array of signaling systems characteristic of cell-based sensors yields generic sensitivity that is a distinguishing feature in comparison to other molecular biosensor approaches. In addition, cell-based sensors offer an advantage of constituting a function-based assay that can yield insight into the physiologic action of an analyte of interest. Three important issues that constitute barriers for the use of cell-based sensors have been presented and discussed. There are certainly other areas that will require attention, as cell-based sensors move from the laboratory environment; namely, cell delivery and/or preservation technologies. As further progress is made to address fundamental challenges, cell-based biosensors and related cellular function based assays will undoubtedly become increasingly important and useful. It is of popular belief that such function-based assays will become an indispensable tool for monitoring in environmental, medical, and defense applications. Strategies relying on a single population of excitable cells appear most well

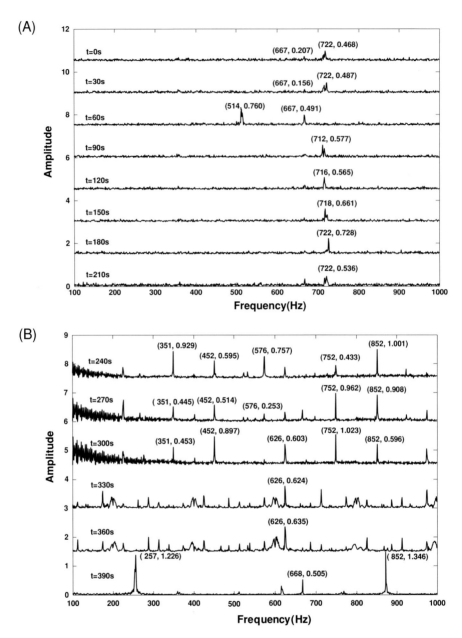

FIGURE 4.20. **A**. Signature pattern vector of single osteoblast due to the cascaded action of ethanol-hydrogen peroxide at 19 ppm and 25 ppm respectively. **B**. Signature Pattern Vector of single osteoblast due to the cascaded action of ethanol-hydrogen peroxide at 19 ppm and 25 ppm, respectively [105].

suited for measurements of acute and direct effects of receptor agonist/antagonists. Compounds that fall within this category include ion channel modulators, metals, ligand–receptor blockers, and neurotransmitters. In fact, the detection of acute and direct effects of compounds may be sufficient and relevant for certain operational situations, such as a battlefield environment or the floor of an assembly plant, where cognitive function is absolutely critical. The prospect of detecting all physiologically active analytes using a single cell or tissue type is improbable. It is possible that particular analytes may undergo biotransformation, resulting in a secondary or tertiary compound of substantial physiologic effect. In spite of that drawback single cell based sensors are highly reliable. This has been shown by the newly developed single cell based sensing technique. This technique functions on the principle of integrating a fundamental biological tool like dielectrophoresis to biochip technology. Single cell arrays of the same biological state and differentiation can be developed using this method. Simultaneous sensing can be achieved, which reduces false alarms. Unique identification tags have been generated for identifying specific chemical analytes using this technique. These are known as Signature Patterns. The greatest advantage of this technique is its high sensitivity and speed of response. Chemical analytes of concentrations in the order of parts per billion have been detected. To determine the veracity and reliability of the sensor simultaneous fluorescence detection techniques have also been implemented at the detection limit obtained from the single cell based sensor. The physiological behavior corroborates the sensing. This establishes the viability of this technique for potential commercial implementation.

Finally, the threat of biological weapons has become a major concern to both the civilian and military populations. All of the present biological warfare and environmental agent rapid detection systems, in field use or under prototype development, rely on structural recognition approaches to identify anticipated agents. Cell based sensor technology utilizing biochip capability can be thought to be one potentially reliable solution. It is now only a matter of time before this technology will impinge on a wide range of commercial situations.

REFERENCES

[1] T. Akin, K. Najafi, R.H. Smoke, and R.M. Bradley. *IEEE Trans. Biomed. Eng.*, 41:305, 1994.
[2] B.M. Applegate, S.R. Kermeyer, and G.S. Sayler. *Appl. Environ. Microbiol.*, 64:2730, 1998.
[3] S. Belkin, D.R. Smulski, S. Dadon, A.C. Vollmer, T.K. Van Dyk, and R.A. Larossa. *Wat. Res.*, 31:3009, 1997.
[4] R.A. Bissell, A.P. de Silva, H.Q.N. Gunaratne, P.L.M. Lynch, G.E.M. Maguire, and K.R.A.S. Sandanayake. *Chem. Soc. Rev.*, 21:187–195, 1992.
[5] L. Bousse, R.J. McReynolds, G. Kirk, T. Dawes, P. Lam, W.R. Bemiss, and J.W. Parce. *Sens. Actu. B*, 20:145, 1994.
[6] L.J. Breckenridge, R.J.A.Wilson, P. Connolly, A.S.G. Curtis, J.A.T. Dow, S.E. Blackshaw, and C.D.W. Wilkinson. *J. Neurosci. Res.*, 42:266, 1995.
[7] R.S. Burlage, A.V. Palumbo, A. Heitzer, and G. Sayler. *Appl. Microbiol. Biotechnol.*, 45:731, 1994.
[8] J.C. Chang, G.J. Brewer, and B.C. Wheeler. *J. Biomed. Microdev.*, 2(4):245, 2000.
[9] P. Clark, P. Connolly, A.S.G. Curtis, J.A.T. Dow, and C.D.W. Wilkinson. *J. Cell Sci.*, 99:73, 1991.
[10] B.A. Cornell, V.L. Braach-Maksvytis, L.G. King, P.D. Osman, B. Raguse, L. Wieczorek, and R.J. Pace, *Nature*, 387:580, 1997.
[11] P. Connolly, G.R. Moores, W. Monaghan, J. Shen, S. Britland, and P. Clark. *Sens. Actu.*, B6:113, 1992.

[12] K.S. Cole and H.J. Curtis. *J. Gen. Physiol.*, 22:649, 1939.
[13] J. Csicsvari, D.A. Henze, B. Jamieson, K.D. Harris, A. Sirota, P. Bartho, K.D. Wise, and G. Buzsaki. *J. Neurophysi.*, 90:1314, 2003.
[14] A.W. Czarnik and J.P. Desvergne. *Chemosensors for Ion and Molecule Recognition.* Kluwer, Dordrecht, The Netherlands, 1997.
[15] A.P. De Silva and R.A.D.D. Rupasinghe. *J. Chem. Soc. Chem. Commun.*, 166:14, 1985.
[16] A.P. De Silva, H.Q.N. Gunaratne, T. Gunnlaugsson, A.J.M. Huxley, C.P. McCoy, J.T. Rademacher, and T.E. Rice. *Chem. Rev.*, 97:1515, 1997.
[17] A.M. Dijkstra, B.H. Brown, A.D. Leathard, N.D. Harris, D.C. Barber, and L. Edbrooke. *J. Med. Eng. Technol.*,. 17:89, 1993.
[18] C.S. Dulcey, J.H. Georger, A. Krauthamer, Jr., D.A. Stenger, T.L. Fare, and J.M. Calvert. *Science*, 252:551, 1991.
[19] D.J. Edell, V.V. Toi, V.M. McNeil, and L.D. Clark. *IEEE Trans. Biomed. Eng.*, 39:635, 1992.
[20] C.F. Edman, D.E. Raymond, D.J., Wu, E.G. Tu, R.G. Sosnowski, W.F. Butler, M. Nerenberg, and M.J. Heller. *Nucleic Acids Res.*, 25:4907, 1997.
[21] G.A. Evtugyn, E.P. Rizaeva, E.E. Stoikova, V.Z. Latipova, and H.C. Budnikov. *Electroanalysis*, 9:1–5, 1997.
[22] H. Fricke and S. Morse. *J. Gen. Physiol.*, 9:153, 1926.
[23] H. Fricke and H.J. Curtis. *J. Gen. Physiol.*, 18:821, 1935.
[24] P. Fromherz, A. Offenhausser, T. Vetter, and J. Weis. *Science*, 252:290, 1991.
[25] G. Fuhr, H. Glasser, T. Muller, and T. Schnelle. *Biochim. Biophys. Acta.*, 1201:353, 1994.
[26] L. Giaever and C.R. Keese. *IEEE Trans. Biomed. Eng.*, 33:242, 1986.
[27] L. Griscom, P. Degenaar, B. LePioufle, E. Tamiya, and H. Fujita. *Sens. Actu. B*, 83(1–3):15, 2002.
[28] G.W. Gross, B.K. Rhoades, and R. Jordan. *Sen. Actu. B*, 6:1, 1992.
[29] G.W. Gross, B.K. Rhoades, H.M.E. Azzazy, and M.C. Wu. *Biosens. Bioelectron.*, 10:553, 1995.
[30] G.W. Gross, W. Wen, and J. Lin. *J. Neurosci. Meth.*, 15:243, 1985.
[31] G.W. Gross, B.K. Rhoades, D.L. Reust, and F.U. Schwalm. *J. Neurosci. Meth.*, 50:131, 1993.
[32] G.W. Gross. Internal dynamics of randomized mammalian neuronal networks in culture. In *Enabling Tedmologiesfor Cultured Neural Networks.* Academic Press, Vol. 277, 1994.
[33] W. Gopel, J.L. Hesse, and J.N. Zemel. *A Comprehensive Survey of Sensors, Trends in Sensor Technology/ Sensor Markets.* (ed.), Elseiver, Netherlands, 1995.
[34] L.L. Hause, R.A. Komorowski, and F. Gayon. *IEEE Trans. Biomed. Eng.*, BME-28:403, 1981.
[35] R.P. Haugland. *Handbook of Fluorescent Probes and Research Chemicals*, 6th Ed. Molecular Probes, Eugene, OR, 1996.
[36] A. Heitzer, K. Malachowsky, J.E. Thonnard, P.R. Bicnkowski, D.C. White, and G.S. Sayler. *Appl. Environ. Microbiol.*, 60:1487, 1994.
[37] R. Heim and R.Y. Tsien. *Curr Biol.*, 1:178–182, 1996.
[38] A.W. Hendricson, M.P. Thomas, M.J. Lippmann, and R.A. Morrisett. *J. Pharmacol. Exp. Ther.*, 307(2):550, 2003.
[39] A.L. Hodgkin and A.F. Huxley. *J. Physiol.*, 117:500, 1952.
[40] Y. Huang, R. Holzel, R. Pethig, and X.B. Wang. *Phys. Med. Biol.*, 37(7):1499–1517, 1992.
[41] M.E. Huston, K.W. Haider, and A.W. Czarnik. *J. Am. Chem. Soc.*, 110:4460, 1988.
[42] M. Jenker, B. Muller, and B. Fromherz. *Biol. Cybernetics.*, 84:239, 2001.
[43] I.S. Kampa and P. Keffer. *Clin. Chem.*, 44:884, 1998.
[44] Y. Kitagawa, M. Ameyama, K. Nakashima, E. Tamiya, and I. Karube. *Analyst*, 112:1747, 1987.
[45] G.T.A. Kovacs. *IEEE Trans. Biomed. Eng.*, 1992.
[46] M. Kowolenko, C.R. Keese, D.A. Lawrence, and I. Giaever. *J. Immunol. Methods*, 127:71, 1990.
[47] Y.I. Korpan, M.V. Gonchar, N.F. Starodub, A.A. Shul'ga, A.A. Sibirny, and A.V. El'skaya. *Anal. Biochem.*, 215:216, 1993.
[48] S. Lacorte, N. Ehresmann, and D. Barcelo. *Environ. Sci. Technol.*, 30:917, 1996.
[49] H.B. Li and R. Bashir. *Sens. Actu. B*, 86:215, 2002.
[50] N. Li, H. Endo, T. Hayashi, T. Fujii, R. Takal, and E. Watanahe. *Biosens. Bioelectron.*, 9:593, 1994.
[51] B. Luc. *Sens. Actu. B*, 34:270, 1996.
[52] M.P. Maher, J. Pine, J. Wright, and Y.C. Tai. *J. Neurosci. Meth.*, 87:45, 1999.
[53] G.H. Markx, M.S. Talary, and R. Pethig. *J. Biotechnol.*, 32:29, 1994.

[54] G.H. Markx and R. Pethig. *Biotech. Bioeng.*, 45:337, 1995.
[55] M. Malmquist. *M Biochem. Soc. T.*, 27:335, 1999.
[56] H.M. McConnell, J.C. Owicki, J.W. Parce, D.L. Miller, G.T. Baxter, H.G. Wada, and S. Pitchford. *Science*, 257:1906, 1992.
[57] T. Müller, A. Gerardino, T. Schnelle, S.G. Shirley, Bordoni, Gasperis, G. De, R. Leoni, and G.J. Fuhr. *Phys. D: Appl. Phys.*, 29:340, 1996.
[58] K. Najafi. *Sens. Actu. B*, 1:453, 1990.
[59] A. Offenhausser, C. Sprossler, M. Matsuzawa, and W. Knoll. *Biosens. Bioelectron.*, 12(8):819, 1997.
[60] A. Ota and S. Hybridoma. *Ueda*, 17:471, 1998.
[61] J.C. Owicki and J.W. Parce. *Biosen. Bioelectron.*, 7:255, 1992.
[62] J.C. Owicki, J.W. Parce, K.M. Kercso, G.B. Sisal, V.C. Muir, J.C. Venter, C.M. Fraser, and H.M. McConnell. *Proc. Natl. Acad. Sci.*, 87:4007, 1990.
[63] B.M. Paddle. *Biosens. Bioelectron.*, 11:1079, 1996.
[64] J.W. Parce, J.C. Owicki, K.M. Kercso, G.B. Sigal, H.G. Wada, V.C. Muir, L.J. Bousse, K.L. Ross, B.I. Sikic, and H.M. McConnell. *Science*, 246:243, 1989.
[65] D. Pollard-Knight, E. Hawkins, D. Yeung, D.P. Pashby, M. Simpson, A. McDougall, P. Buckle, and S.A. Charles. *Ann. Biol. Clin.*, 48:642, 1990.
[66] H.A. Pohl. *Dielectrophoresis, the Behavior of Neutral Matter in Nonuniform Electric Fields*. Cambridge University Press, 1978.
[67] H.A. Pohl and I. Hawk. *Science*, 152:647, 1966.
[68] S. Prasad, M. Yang, X. Zhang, C.S. Ozkan, and M. Ozkan. *Biomed. Microdev.*, 5(2):125, 2003.
[69] W.G. Regehr, J. Pine, and D.B. Rutledge. *IEEE Trans. Biomed. Eng.*, 35:1023, 1998.
[70] B.K. Rhoades and G.W. Gross. *Brain Res.*, 643:310, 1994.
[71] J. Ruhe, R. Yano, J.S. Lee, P. Koberle, W. Knoll, and A. Offenhausser. *J. Biomater. Sci. Polym. Ed.*, 10(8):859, 1999.
[72] T.G. Ruardij, M.H. Goedbloed, and W.L.C. Rutten. *Med. Biol. Eng. Comput.*, 41(2):227, 2003.
[73] H.P. Schwan. C.A. Academic Press, New York, Vol. 5, pp. 147, 1957.
[74] G. Schmuck, A. Freyberger, H.J. Ahr, B. Stahl, and M. Kayser. *Neurotoxicology*, 24:55, 2003.
[75] M. Scholl, C. Sprossler, M. Denyer, M. Krause, K. Nakajima, A. Maelicke, W. Knoll, and A. Offenhausser. *J. Neurosci. Meth.*, 104:65, 2000.
[76] O. Selifonova, R.S. Burlage, and T. Barkay. *Appl. Environ. Micobiol.*, 59:3083, 1993.
[77] D.E. Semler, E.H. Ohlstein, P. Nambi, C. Slater, and P.H. Stern. *J. Pharmacol. Exp. Ther.*, 272:3:1052, 1995.
[78] J.B. Shear, H.A. Fishman, N.L. AIIbritton, D. Garigan, R.N. Zare, and R.H. Scheller. *Science*, 267:74, 1995.
[79] J. Singh, P. Khosala, and R.K. Srivastava. *Ind. J. Pharmacol.*, 32:206, 2000.
[80] R.S. Skeen, W.S. Kisaalita, and B.J. Van Wie *Biosens. Bioelectronic.*, 5:491, 1990.
[81] D.A. Stenger, G.W. Goss, E.W. Keefer, K.M. Shaffer, J.D. Andreadis, W. Ma, and J.J. Pancrazino. *Trends in Biotechnol.*, 19:304, 2001.
[82] M. Stephens, M.S. Talary, R. Pethig, A.K. Burnett, and K.I. Mills. *Bone Narrow Transpl.*, 18:777, 1996.
[83] P. Sticher, M.C.M. Jaspers, K. Stemmler, H. Harms, A.J.B. Zehnder, and vand der Meer. *J.R. Appl. Environ. Microbiol.*, 63:4053–4060, 1997.
[84] A.A. Suleiman and G.G. Guilbault. *Analyst.*, 119:2279, 1994.
[85] F.J. Swenson. *Sens. Actu. B.*, 11:315, 1992.
[86] D.W. Tank, C.S. Cohan, and S.B. Kater. *IEEE Conf. on Synthetic Microstructures*, Airlie House, Arlington, Virginia, IEEE, New York, 1986.
[87] H.M. Tan, S.P. Chcong, and T.C. Tan. *Biosens. Bioelectron.*, 9:1, 1994.
[88] T. Takayasu, T. Ohshima, and T. Kondo. *Leg. Med.*, (Tokyo). 3:(3):157, 2001.
[89] P. Thiebaud, L. Lauer, W. Knoll, and A. Offenhausser. *Biosens. Bioelectron.*, 17(1–2):87, 2002.
[90] C. Tiruppathi, A.B. Malik, P.J. Del Vecchio, C.R. Keese, and I. Giaever. *Proc. Natl. Acad. Sci.*, 89:7919, 1992.
[91] R.Y. Tsien. *Annu. Rev. Neurosci.*, 12:227–253, 1989.
[92] D.W. Tank, C.S. Cohan, and S.B. Kater. *IEEE Conf. on Synthetic Microstructures*. Airlie House, Arlington, Virginia, IEEE, New York, 1986.
[93] R.Y. Tsien. *Annu. Rev. Neurosci.*, 12:227–253, 1989.
[94] S.J. Updike and G.P. Hicks. *Nature*, 214:986, 1967.

[95] D. Van der Lelie, P. Corbisier, W. Baeyens, S. Wnertz, L. Diels, and M. Mergeay. *Res. Microbiol.*, 145:67, 1994.
[96] H.J. Watts, D. Yeung, and H. Parkes*Anal. Chem.*, 67:4283, 1995.
[97] X. Wang, Y. Huang, P.R.C. Gascoyne, and F.F. Becker. *IEEE Tran. Ind. Appl.*, 33:660, 1997.
[98] B.C. Wheeler, J.M. Corey, G.J. Brewer, and D.W. Branch. *J. Biomech. Eng.*, 121:73, 1999.
[99] J.P. Whelan, L.W. Kusterbeck, G.A., Wemhoff, R. Bredehorst, and F.S. Ligler. *Anal. Chem.*, 65:3561, 1993.
[100] P. Wilding, J. Pfahler, H.H. Ban, J.N. Zemcl, and L.J. Kricka. *Clin. Chem.*, 40:43, 1994.
[101] C. Wyart, C. Ybert, L. Bourdieu, C. Herr, C. Prinz, and D.J. Chattenay. *J. Neurosci. Meth.*, 117(2):23, 2002.
[102] M. Yang, X. Zhang, K. Vafai, and C.S. Ozkan. *J. Micromech. Microeng.*, 13:864, 2003.
[103] M. Yang, X. Zhang, and C.S. Ozkan. *Biomed. Microdev.*, 5:323, 2003.
[104] M. Yang, S. Prasad, X. Zhang, M. Ozkan, and C.S. Ozkan. *Sensor Letters,* 2:1, 2004.
[105] M. Yang, S. Prasad, X. Zhang, A. Morgan, M. Ozkan, and C.S. Ozkan. *Sens. Mater.*, 15(6):313, 2003.
[106] M. Yang, X. Zhang, Y. Zhang, and C.S. Ozkan. *Sens. Actu. B*, (in print), 2004.
[107] G. Zeck and P. Fromherz.. *Proc. Natl. Acad. Sci. U.S.A.*, 98(18):10457, 2001.
[108] I. Giaever and C.R. Keese. *Proc. Natl. Acad. Sci. U.S.A.*, 88:7896, 1991.
[109] J. Pancrazio, S.A. Gray, Y.S. Shubin, N. Kulagina, D.S. Cuttino, K.M. Shaffer, K. Kisemann, A. Curran, B. Zim, G.W. Gross, and T.J. O'Shaughnessy. *Biosens. Bioelectron.*, 18(11):1339, 2003.

5

Fabrication Issues of Biomedical Micro Devices

Nam-Trung Nguyen

School of Mechanical and Production Engineering, Nanyang Technological University, 50. Nanyang Avenue, Singapore 639798

5.1. INTRODUCTION

Biomedical micro devices (BMMD) are microsystems, which can be used in surgery, biomedical diagnostics, and therapeutic management [28]. These devices allow precise surgical procedures with spatial control in the micrometer range. The minimal invasive approach enables faster recovery for patients through shorter access pathways and reduced operation trauma. BMMDs for biomedical diagnostics and therapeutic management utilizes microfluidic technology that allows faster screening of common diseases as well as painless and effective drug delivery. The successful development and introduction of these technologies in health care will have a great impact on the living quality of patients, and significantly lower the total cost of medical treatment.

Conventional fabrication method evolving from microelectronics were used for fabricating micro electromechanical systems (MEMS). MEMS-technology was successfully commercialized in products such as micro sensors and micro actuators. However, the extension of silicon-based devices to biomedical applications have certain limitations. The majority of developed devices have been realized on silicon and glass, because the fabrication technologies for them are matured and widely available [2, 28]. Almost all conventional micromachining techniques such as wet etching, dry etching, deep reactive ion etching, sputter, anodic bonding, and fusion bonding were used for fabricating BMMDs. Key components of a BMMD such as flow channels, flow sensors, chemical detectors, separation capillaries, mixers, filters, pumps, and valves have been developed based on silicon technology [26]. These devices have many advantages over their macro counterparts. They significantly improve the efficiency with smaller samples, fast response time, and higher analytical performance. However, the major drawback of silicon-based BMMDs is their cost. BMMDs for diagnostics and drug delivery are often used as disposable tools in medical treatments due to contamination hazards. Furthermore, compared to their microelectronic and micro electromechanical counterparts, BMMDs such as microfluidic devices have a relatively large size. The single use and the high material cost as well as the high processing cost in clean rooms make silicon-based BMMDs less attractive for the mass market.

Furthermore, silicon as a substrate material is not very compatible to biomedical applications. Many disposable biomedical gadgets were made of polymers. Thus, fully polymeric BMMDs promise to solve the problems of biocompatibility, and in addition offer a much lower total cost compared to their silicon counterparts. In the past few years, a number of devices were fabricated successfully using different polymers such as polydimethlsioxane (PDMS), polystyrene (PS), polyethylene (PT), polyethyleneterephthalate (PET), polycarbonate (PC), SU-8 and polymethylmethacrylate (PMMA). This chapter focuses on polymeric micro technologies, which promise the mass fabrication of BMMDs at a reasonable cost.

In contrast to conventional MEMSs, BMMDs have direct interactions with a biological environment. These interactions lead to implication for the proper operation of the devices. For instance, proteins and micro organisms tend to adsorb to synthetic surfaces [11]. The adsorbed layer creates malfunction in sensing applications and reduces the overall life span of the device. The use of BMMDs in biological environment not only affects the device itself, but also triggers a cellular response in the host [3]. Thus, the use of materials and the fabrication technology for BMMDs should consider the biocompatibility tailored to a application. Responsible agencies such as the Food and Drug Administration (FDA) only approve BMMDs for specific purposes and not the devices themselves in isolation. This chapter should read the different aspects of biocompatibility in material choice and fabrication technologies.

While numerous research works focused on the fabrication and the functionality of BMMDs, little was done on packaging and interconnection problems of these devices. Critical issues are biocompatibility of the package and microfluidic interconnects. The biological environment leads to special requirements in interfacing BMMDs with the surrounding wold. These requirements should be considered carefully in the design and the fabrication of BMMDs.

5.2. MATERIALS FOR BIOMEDICAL MICRO DEVICES

5.2.1. Silicon and Glass

Silicon is undoubtedly the most popular material in MEMS applications. Silicon micromachining technologies are established and well known [23, 26]. Single crystalline silicon wafers with high purity and different orientations are commercially available at a reasonable cost. The two basic micromachining techniques are bulk micromachining and surface micromachining. The deposition of a number of functional layers such as polysilicon, silicon dioxide, silicon nitride, metals and several organic layers is also well established. A variety of microdevices can be designed and fabricated by combining these techniques. In fact, all conventional MEMSs are silicon-based.

FABRICATION ISSUES OF BIOMEDICAL MICRO DEVICES   95

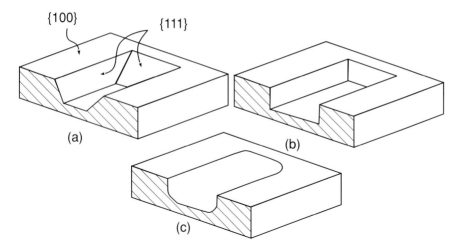

FIGURE 5.1. Etch profiles with different bulk micromaching techniques: (a) Anisotropic wet etching of silicon, (b) Deep reactive ion etching (DRIE) of silicon or wet etching of photo sensitive glasses, (c) Isotropic etching in silicon or glass.

Glass consists of silicon oxide and oxides of metals. The amount of silicon oxide varies and depends on the glass type (68% in soda-lime glass, 81% in boronsilicate glass and almost 100% in fused silica) [4]. Analytical instruments in the past are widely based on glass wares. The reasons for the use of glass include high mechanical strength, high chemical resistance, high electrical insulation and a wide optical transmission range. Thus, glass was the first choice for fabricating BMMDs in diagnosis applications such as electrophoresis [6, 9, 30, Jacobson et al., 1994] electrochromatography [Jacobson et al., 1994] and DNA separation [37]. Most of the glasses can be etched in a buffered hydrofluoric acid (HF) solution [32] using a photo resist etch mask such as AZ 4620 [19]. The typical result of isotropic etching in glass is shown in Fig. 5.1c.

Glass can also be structured with photo lithography. Commercially available glasses such as Foturan [13] and FS21 [29] belong to this glass type. These glasses consist of silicon oxide, aluminum oxide and lithium oxide. The photo sensitivity is activated by doping the glass with oxides of elements such as Ag, Ce, Sn, Sb [29] For machining, the glass is exposed to ultraviolet radiation. The amorphous glass crystallizes under UV-exposure. The crystallized area can be removed selectively by a subsequent etch process in hydrofluoric acid. Wet etching of photo sensitive glass can result in straight walls similar to deep reactive etching (DRIE) of silicon, Fig. 5.1b.

5.2.2. Polymers

As already mentioned in the introduction, silicon-based and glass-based BMMDs have the drawbacks of higher cost and biocompatibility problems. Regarding a cheaper mass production of BMMDs, polymers offer a real alternative to silicon-based and glass-based substrates. Polymers are macromolecular materials, which are formed through polymerization reactions. In this reaction, the monomer units connect each other either in linear chains or in three-dimensional network chains and form a macromolecule. Based on their properties synthetic polymers can be categorized as thermoplastics, elastomers, and thermosets [5].

Polymers as functional materials fulfill a number of requirements of BMMDs [31]:

- Polymers are suitable for micro machining, the different micro fabrication techniques are discussed in section 5.3.
- Many polymers are optically transparent. Due to the requirement of optical detection methods such as fluorescence, UV/Vis absorbance, or Raman method, the device material should allow a wide optical transmission range and have a minimum autofluorescence.
- Many polymers are chemically and biologically compatible. The materials used in BMMDs should be inert to a wide range of solvents.
- Most polymers are good electrical insulators. An electrically insulating substrate is required in applications with a strong electrical field such as electrophoretic separation. In addition, good thermal properties are also desired due to thermal load resulting from Joule heating.
- The surface chemistry of polymers can be modified for a certain application.

The following sections discuss the properties of some typical polymers, which are frequently used in recently published works on BMMDs.

5.2.2.1. PMMA PMMA is known under trade names such as Acrylic, Oroglass, Perspex, Plexiglas, and Lucite. PMMA is available commercially in form of extrusion sheets. For micromachining purposes such as X-ray exposure, PMMA can be applied on a handling substrate by different ways: multilayer spin coating, bonding of a prefabricated sheet, casting, and plasma polymerization.

PMMA is one of the thermoplastic polymers. Thermoplastic polymers are usually linear-linked and will soften when heated above glass transition temperature. It can be reheated and reshaped before hardening in its form for many times. Thermoplastic polymers can be crystalline or amorphous. In general, transparent polymers are non-crystalline and translucent polymers are crystalline. PMMA has a non-crystalline structure. Thus, it has optical properties with a 92% light transmittance. PMMA also offers other excellent properties such as low frictional coefficient, high chemical resistance, and good electrical insulation. Thus PMMA is a good substrate for microfludic devices especially in biomedical applications.

The surface properties of PMMA can be modified chemically to suit its application [10] PMMA can be machined in many ways: X-ray exposure and subsequent developing, hot embossing, and laser machining (see section 5.3).

5.2.2.2. PDMS PDMS is a polymer that has an inorganic siloxane backbone with methyl groups attached to silicon. The prepolymers and curing agents of PDMS are both commercially available. PDMS is suitable for BMMDs with microchannels for biological samples in aqueous solutions. PDMS presents the following excellent properties [24]:

- PDMS can be micro machined using replica molding. The elastomeric characteristic allows PDMS to conform to nonplanar surfaces. PDMS can be released from a mold with delicate micro structures without damaging them and itself.

- PDMS is optically transparent down to a wavelength of 280 nm, thus this material is suitable for devices utilizing UV detection schemes.
- PDMS is biocompatible, mammalian cells can be cultured directly on this material and PDMS devices can be implanted in a biological environment.
- PDMS can bond itself to a number of materials reversibly. Irreversible seal can be achieved by covalent bonds, if the contact surface is treated with oxygen plasma.
- PDMS surface is hydrophobic. An oxygen plasma treatment makes the surface hydrophilic and negatively charged, thus suitable for electrokinetic applications. Oxidized PDMS can also absorb other polymers, which modify the surface properties.

5.2.2.3. SU-8 Microstructures can be transferred to most polymers by hot embossing or replica molding. Direct photo lithography is possible with PMMA using X-ray. However, the X-Ray source and the corresponding mask are expensive and not suitable for mass production. SU-8 is a thick film resist, which allows photo lithography with high aspect ratios using conventional exposure equipments with near-UV wavelengths from 365 nm to 436 nm. Film thickness up to 2 mm and aspect ratios better than 20 can be achieved with SU-8.

SU-8 photoresist consists of three basic components:

- An epoxy resin, which has one or more epoxy groups. An epoxy group is referred to as the oxygen bridge between two atoms. During the polymerization process, epoxy resins are converted to a thermoset form of a three-dimensional network.
- A solvent such as gamma-butyrolacetone (GBL). The SU-8 2000 family (MicroChem Corp., USA) uses cyclopentanone (CP) as the solvent.
- A photoinitiator such as triarylium-sulfonium salt.

The unexposed SU-8 can be dissolved with solvent-based developers. The commercially available developer for SU-8 is propylene-glycol-methyl-ether-acetate (PGMA) (MicroChem Corp., USA).

SU-8 is chemically stable and is resistant to most acids and other solvents. The good mechanical properties allow the use of SU-8 directly as moveable components, section 5.3.1.1

5.3. POLYMERIC MICROMACHINING TECHNOLOGIES

5.3.1. Lithography

Lithography is the most important techniques for fabricating micro structures. Several lithography techniques were established during the development of microelectronics. Based on the type of the energy beam, lithography can be categorized as photolithography, electron lithography, X-ray lithography, and ion lithography [34]. The lithography process only allows transferring two-dimensional lateral structures. The desired pattern is transferred from a mask to the resist. In microelectronics and silicon-based MEMS-technology, the patterned resist is in turn the mask for transferring the pattern further into a functional layer. In polymeric micromachining, the structured resist can be used directly as the functional material or as a mold for replica molding of other polymers.

A conventional lithography process consists of three basic steps:

- Positioning: lateral positioning and gap adjusting between the mask and the substrate,
- Exposure: exposure to the energy beam, transferring pattern by changed properties of the exposed area,
- Development: selective dissolution or etching of the resist pattern.

5.3.1.1. Lithography of Thick Resists Lithography of PMMA requires collimated X-ray with wavelength ranging from 0.2 nm to 2 nm. These high-energy beams are only available in synchrotron facilities. X-ray lithography also requires special mask substrates such as beryllium and titanium. The absorber material of a X-ray mask are heavy metals such as gold, tungsten , or tantalum. For a higher aspect ratios of the PMMA structures, high X-ray energy and consequently a thicker absorber layer are required. During the exposure to X-ray, the polymer chains in the exposed area are broken. Thus the exposed area can subsequently etched away by a developer. A typical developer consists of a mixture of 20 vol% tetrahydro-1, 4-oxazine, 5 vol% 2-aminoethanol-1, 60 vol% 2-(2-butoxy-ethoxy) ethanol, and 15 vol% water [7].

In contrast to the expensive X-ray lithography of PMMA, SU-8 only requires conventional UV-exposure equipment. A standard SU-8 process consists of several steps:

- Spin coating: SU-8 is commercially available with a variety of viscosities. At the same spin speed, a higher viscosity will result in a thicker film. The film thickness can also be adjusted by the spin speed.
- Soft bake: Before exposure the solvent is evaporated in the soft bake process. This process can be carried out in a convection oven or on a hot plate. Ramping from 65 °C to 95 °C is recommended for this step.
- Exposure: Since the optical absorption of SU-8 increases sharply bellow the wavelength of 350 nm, the exposure should be carried out at wavelengths higher than this value. I-line equipment with mercury lamp is suitable for this step. The thicker the SU-8 layer, the higher is the required exposure dosage.
- Post exposure bake: After the exposure step, the SU-8 layer is selectively cross-linked by a thermal process. A two-step ramp between 65 °C to 95 °C is recommended to minimize the film stress and possible cracks. Rapid cooling after the thermal treatment should be avoided.
- Developing: The unexposed areas of the SU-8 film can be dissolved by immersion in a solvent-based developer.
- Hard bake: Another baking process after developing allows the remaining SU-8 to further cross-link and harden.
- Remove: Since polymerized SU-8 is resistant to most acids and other solvents, it's very difficult to remove a cross-linked film after exposure. Measures such as etching in a strong acid solution, reactive ion etching in oxygen plasma or laser ablation can be used to remove polymerized SU-8.

5.3.1.2. SU-8 on PMMA Technique SU-8 can be used for forming microchannels on silicon and glass. A number of silicon-based and glass-based techniques were summarized

FABRICATION ISSUES OF BIOMEDICAL MICRO DEVICES   99

in [26]. In this section, the fabrication of SU-8 on a polymer substrate is presented. The technique shows the possibility of fabrication of fully polymeric BMMDs.

In preparation for the process, a 3-mm-thick PMMA sheet was cut into 100-mm-diameter circular wafer using CO_2-laser machining (section 5.3.4). The two films protecting the PMMA were kept intact to prevent dust and oil contamination during the cutting process. These protecting films were peeled off just before the SU-8 process. The wafer was cleaned with isopropyl alcohol (IPA) and deionized (DI) water. Strong solvents such as acetone should not be used for the cleaning process due to possible damage of the PMMA surface. Next, the wafer was dried in a convection oven at 90 °C for 30 minutes. PMMA has a low glass transition temperature at around 106 °C, so all baking temperatures should be kept under 90 °C.

Next, four milliliters of SU-8 2050 (Microchem Corp, USA) was dispensed onto the PMMA wafer. Spin-coating the resist at 500 rpm for 15 s, followed by 3000 rpm for 15 s produced a 50 mm thick film. This film acted as the base for the next SU-8 structural layer. The resist was soft-baked in the convection oven at 65 °C in 2 minutes, then at 90 °C in 15 minutes, and allowed to cool down to the room temperature of 24 °C, Fig. 5.2a. Subsequently, the resist was blanket-exposed to UV light (EV620 Mask aligner, EV Group) with an energy density of 525 mJ/cm^2. The resist underwent hard baking at 65 °C in 2 minutes and at 90 °C in 5 minutes. Subsequently, a relaxation step at 65 °C, in 2 minutes was performed to release thermal stress in the SU-8 film, Fig. 5.2b.

In the next step, four milliliters of SU-8 2100 was dispensed onto the first SU-8 layer. The spinning speed was ramped up in 5 seconds to 500 rpm, held for 5 seconds, ramped up in 10 seconds to 2100 rpm, held for 22 seconds, and ramped down to full stop in 20 seconds.

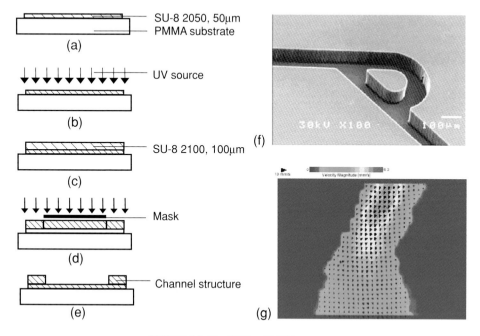

FIGURE 5.2. The SU-8 on PMMA process.

This recipe gave a 100-mm-thick film. The resist was soft baked for 10 minutes at 65 °C and for 40 min at 90 °C, Fig. 5.2c. After cooling down to room temperature, the second SU-8 layer was exposed with an energy density of 525 mJ/cm^2 through a photo mask defining the desired structures, Fig. 5.2d. Next, a two-step hard bake was performed at 65 °C for 5 minutes and at 90 °C for 20 minutes. An intermediate step at 65 °C for 2 minutes was introduced to release the thermal stress in the SU-8 film. The SU-8 was developed in the PGMEA developer for 10 minutes. The wafer was then blown dry with nitrogen, Fig. 5.2e. Figure 5.2f shows a Tesla-valve fabricated with the above technology. After sealing with a second PMMA wafer, covered with 5-μm-thick SU-8, the channel structure was tested using micro particle image velocimetry (micro-PIV). The results show and excellent seal with high-quality microchannels, Fig. 5.2g. Circular wafer is preferred for the SU-8-on-PMMA process because it reduces the excessive edge bead associated with a rectangular wafer during the coating process. In our first experiments, rectangular wafers were used. The rectangular shape resulted in very thick beads at the four edges. The edge bead was even thicker at the four corners. This excessive edge bead prevented the photo mask to properly contact with the majority of the resist surface during exposure, causing poor lithography resolution.

In some of the first experiments, the structural SU-8 was directly coated on to the PMMA wafer. After developing, a layer of undeveloped SU-8 still remained on the wafer. The residual SU-8 could not be cleaned off by IPA, DI water, or even acetone. An ultrasonic bath treatment did remove some of the residual, but part of the SU-8 structure was also damaged. It can be concluded that the residual could not be removed by normal procedures. The solution was to coat a base SU-8 layer as described in the previous section.

The thermal expansion coefficients (TEC) of PMMA and SU-8 are 60 ppm/°C and 52 ppm/°C, respectively. These matching properties result in less crack in the SU-8 film, and a better film adhesion on wafer. Thus, it is possible to use PMMA as the handling substrate for the polymeric surface micromachining techniques presented in section 5.3.2.

5.3.2. Polymeric Surface Micromachining

Polymeric surface micromachining technique is similar to its silicon-based counterpart. A functional layer is structured on top of a sacrificial layer. Removing the sacrificial layer results in a freely moveable structure. Polymers can work as both sacrificial layer and functional layer. With SU-8 as the functional layer, polymers such as polystyrene [22] or metals such as chromium [27] were used as sacrificial layers. In the following section a process with silicon directly as sacrificial material as well as the handling substrate is demonstrated by the fabrication of a polymeric micropump [35] and a polymeric microgripper.

5.3.2.1. Polymeric Micropump The micropump consists of different layers made of PMMA and SU-8. Double-sided adhesive tapes act as bonding layers. All layers have the form of a disc with 1-cm diameter. Easy assembly can be achieved using alignment holes machined in all layers and alignment pins. The PMMA-parts and the adhesive tapes are cut and drilled by the CO_2-laser (section 5.3.4).

Because of the required high accuracy, the micro checkvalves were fabricated in a 100-μm-thick SU-8 layer (SU8 2100, Microchem Corp.) by photo lithography. The process is similar to that described in section 5.3.1.2. Here, a silicon wafer was used as both the handler wafer and the sacrificial material.

FABRICATION ISSUES OF BIOMEDICAL MICRO DEVICES 101

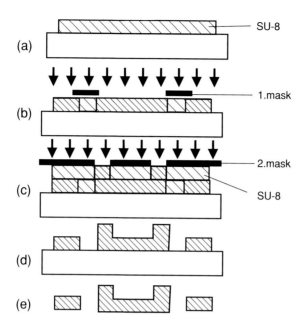

FIGURE 5.3. Steps of a two-layer polymeric surface micromachining process for making a microvalve.

The SU-8 process started with coating of the first SU-8 layer, Fig. 5.3a. The first lithography mask contains the valve disc and the valve springs, Fig. 5.3b. A second SU-8 layer was coated and structured using another mask to form the sealing ring on the valve disc, Fig. 5.3c. Both SU-8 layers were developed and hard-baked in the same process, Fig. 5.3d. Etching the silicon substrate in a KOH solution at room temperature releases the SU-8 valve discs, Fig. 5.3e. Tiny circular holes incorporated on the SU-8 disc avoid micro cracks and work as etch access for faster release. Figure 5.4a shows the fabricated SU-8 micro check valve.

The pump actuator is a commercially available piezoelectric bimorph disc that is bonded on the assembled pump stack by the same adhesive tape. The assembly process was carried out at a room temperature of 25 °C. The inlet and outlet of the micropump are stainless steel needles of 600-μm outer diameter. The needles are glued on the inlet/outlet holes of the pump body. The assembled pump is depicted in Fig. 5.4b. The micro checkvalves and the micropump were successfully tested with water. Figure 5.5a shows the characteristics of microvalves with different spring lengths. A valve with longer spring arms will be softer, thus allowing a higher flow rate at the same pressure. The behavior of the microvalves also affect the characteristics of the micropumps as shown in Fig. 5.5b. Micropumps with softer valves can deliver a higher flow rate at the same actuating frequency and voltage.

5.3.2.2. Polymeric Microgripper Microgrippers have been one of the typical applications of MEMS-technology. Microgrippers were developed for systems, which can handle microparts or manipulate cells. For the latter application, bio-compatibility and gentle handling are often required. The microgripper reported here was fabricated using polymeric surface micromachining with SU-8 as the structural material and silicon as the sacrificial

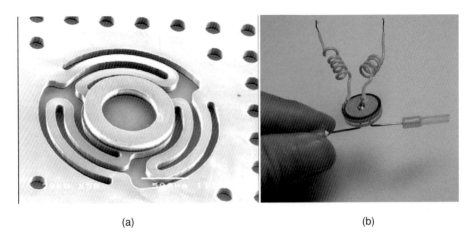

FIGURE 5.4. Fabrications results: (a) the microvalve (b) the assembled micropump.

material. A titanium/platinum layer work as the heater for the gripper. The relatively low operating temperature of less then 100 °C and the gentle gripping force make the gripper suitable for applications with living cells and bacteria.

The thick-film resist SU-8 has an unique property, that it does not soften at elevated temperatures. Higher temperatures cause better cross-links and make SU-8 even harder. Thus, SU-8 is suitable for the use with thermal actuators. Since SU-8 is not conductive, a thin titanium/platinium on top of the SU-8 structure was used as the heater. Due to the large ratio between the thickness of the metal layer (hundred nanometers) and the SU-8 part (one hundred microns), vertical bending due to thermomechanical mismatch is negligible compared to lateral bending. With an Young's modulus of 4.02 GPa [21] SU-8 is almost

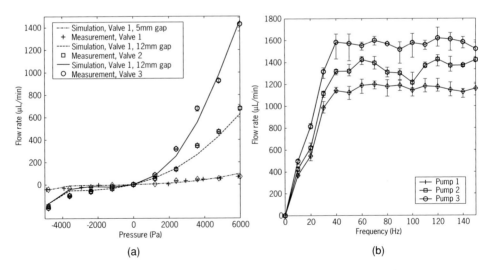

FIGURE 5.5. Characteristics of the microvalves (a) and the micropumps (b).

FABRICATION ISSUES OF BIOMEDICAL MICRO DEVICES 103

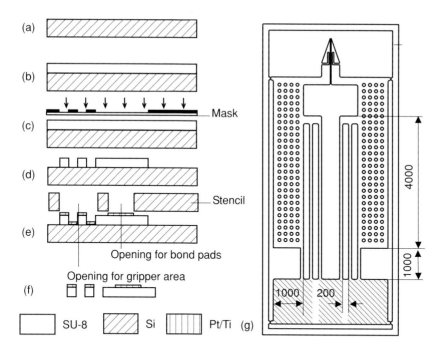

FIGURE 5.6. The polymeric microgripper: (a) the fabrication steps (b) the design.

40 times softer than silicon, while its thermal expansion coefficient of $52 \pm 5 \times 10^{-6}/K$ [21] is superior to that of silicon ($2.4 \times 10^{-6}/K$ [12]). Thus, SU-8 microgripper with thermal actuator offers a much lower operating temperature, lower power and more gentle handling force than its silicon counterparts.

The fabrication process consists of three basic steps: fabrication of the SU-8 gripper, deposition of the titanium/platinum layer, and release. The SU-8 body was fabricated with the polymeric surface micromaching techniques described in section 5.3.2.1. Silicon was used directly as sacrificial material, Fig. 5.6a. The process started with spin-coating SU-8 2100 photoresist (Microchem Corp., USA) on silicon, Fig. 5.6b. This SU-8 layer was then soft baked and exposed to UV light using a mask defining the SU-8 part, Fig. 5.6c. The intended thickness of this SU-8 film was 100 μm. After hard baking, the SU-8 layer was allowed to cool down to room temperature. They were then developed with propylene glycol methyl ether (PGMEA), Fig. 5.6d.

In preparation for the second step, a stencil was dry-etched though a silicon wafer using DRIE. The stencil only defines the bonding pads for the heaters. The heater structures themselves are defined masklessly by the SU-8 structure. Thus the entire gripper body was exposed to the subsequent evaporation processes. Next, the stencil wafer was positioned to the handler wafer containing the developed SU-8 parts. The two wafers are fixed using adhesive tapes. Subsequently, a 50-nm thick titanium layer was evaporated through the stencil. Titanium works as the adhesion layer between SU-8 and the subsequent platinum layer. A 70-nm thick platinum layer was evaporated on top of the titanium layer, Fig. 5.6e.

In the final step, the SU-8 microgrippers covered by the metal double layer were released in 30% KOH solution. Etch access created by many circular holes on the discs allows fast under etching. After releasing the grippers were rinsed in DI (deionized water) water, Fig. 5.6f.

The gripper was designed for the normally closed operation mode. That means the gripper is not actuated while holding an object. Actuation is only needed during the gripping and release actions. This design minimizes the thermal load on the object. Figure 5.6g shows the design and the corresponding geometry parameters of the polymeric microgripper. The gripper is suspended on a frame that supports the fragile structures during the release and assembly process and is removed before use. The gripper consists of two symmetrical arms. The tip has a gap of 30 μm. The L-shaped slit on the tip limits further the gripping force allowing gentle handling of the object. Each gripper arm consists of three flexures. Two small flexures with a width of 100 μm act as the "hot" arms of the thermal actuator. The large flexure works as the "cold" arm of the actuator. Holes with an 100-μm diameter are incorporated in the large flexure to allow easy etch access for the later release. The circular holes and the rounded corner arrest the possible stress in SU-8 during the fabrication and avoid cracks in the gripper. With this design the gripper is attached to the base with 6 flexures, which warrant mechanical stability for the relatively long gripper arms. Figure 5.7a shows the fabricated gripper.

Figure 5.7b depicts the measured displacement/voltage characteristics of the gripper. The circles are the measured data. The solid line is the second order polynomial fitting function. The typical quadratic behavior can be observed.

5.3.3. Replication Technologies

Replication technologies allow the mass fabrication of BMMDs at a low cost. Since most of the BMMDs have a relatively large size compared to conventional MEMS-devices

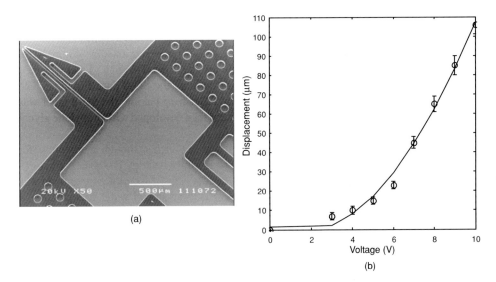

FIGURE 5.7. The fabricated microgripper (a) and its characteristics (b).

FABRICATION ISSUES OF BIOMEDICAL MICRO DEVICES 105

and consequently the small number of devices on a wafer. Thus, silicon technologies could be very expensive for BMMDs.

The basic idea behind replication technologies is the fabrication of a master mold with the "expensive" technology and the low-cost replication in polymers. However, the major drawbacks of replication technologies are [1]:

- Since the master is to be removed from the molded structures, free standing structures with undercuts can not be fabricated. A combination with polymeric surface micromachining (section 5.3.2) could be a solution for this problem.
- Only few micromachining technologies can meet the required smoothness of the master mold.
- Due to contamination and fast diffusion in micro scale, release agents used in macro scale can not be used for the release process in microscale.

The master mold can be fabricated with a number of techniques. Conventional machining techniques such as drilling, cutting, milling, and turning can be used for this purpose for structures down to several tens microns. Bulk silicon micromachining can be used for structures with high aspect-ratios. Metal mold can be electro platted with the help of a structured thick resists such as SU-8 and PMMA. For instance, the fabrication of nickel mold from structured PMMA was established and called as LIGA (Lithographie, Galvanoformung, Abformung) (German acronym for lithography, electroplatting, and molding). Following, three replication techniques are discussed in details: soft lithography, hot embossing, and injection molding.

5.3.3.1. Soft Lithography Soft lithography is a direct pattern transfer technique. The technique is based on an elastomeric stamp with patterned relief structures on its surface. There are two basic techniques for transferring the micro patterns: micro contact printing and replica molding [38]. In many BMMDs, the elastomeric part can be used directly as the functional material. Following the fabrication of microchannels with PDMS are described.

To start with, PDMS is mixed from prepolymers. The weight ratio of the base and the curing agent could be 10:1 or 5:1. The solid master is fabricated from SU-8, Fig. 5.8a. Glass posts can be placed on the SU-8 master to define the inlets and reservoirs.

Next, the PDMS mixture is poured into the master and stands for a few minutes in order to self-level, Fig. 5.8b. The whole set is then cured at relatively low temperature from 60 °C to 80 °C for several hours. After peeling off and having surface treatment with

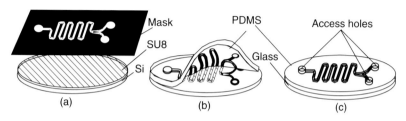

FIGURE 5.8. Fabrication of micro channels with soft lithography: (a) fabrication of a SU-8 master, (b) making a PDMS replica, (c) surface treatment in oxygen plasma and bonding to glass.

low-temperature oxygen plasma. The structured PDMS membrane is brought into contact with clean glass, silica, or another piece of surface-activated PDMS, Fig. 5.8c.

The sealed channel can withstand pressures up to five bars. Without surface treatment, PDMS also forms a watertight seal when pressed against itself, glass, or most other smooth surfaces. These reversible seals are useful for detachable fluidic devices, which are often required in research and prototyping. Inlet tubes and outlet tubes can be embedded in the PDMS device [16].

Three-dimensional structures can be formed by lamination of many PDMS sheets. In this case. methanol is used as surfactant for both bonding and self-alignment. The surface tension at superimposed holes in the PDMS sheets self-aligns them. Methanol prevents instant bonding between two PDMS sheets after plasma treatment. After evaporating methanol on a hot plate, the laminated stack is bonded.

5.3.3.2. Hot Embossing Hot embossing was widely used for the fabrication of simple microchannels. The technique uses a master mold and a flat polymer substrate. The polymer substrate is heated above the glass transition temperatures, which are typically on the order of 50 °C to 150 °C. Embossing force (0.5 to 2 kN/cm^2) is then applied on the substrate under vacuum conditions [1]. Before release, the master and the substrate are cooled under the applied embossing force. The entire hot-embossing process takes about few minutes.

5.3.3.3. Injection Molding Injection molding is a standard process for fabricating polymer parts. Using a micromachined mold insert, this technique can be extended to the fabrication of BMMDs. Injection molding uses polymer in granular form. The polymer is first transported into a cylinder with a heated screw, where the granules are melted. The melt is forced into the mold insert with a high pressure (600 to 1000 bars). The molding temperature depends on the type of the polymer (about 200 °C for PMMA and PS, about 280 °C for PC) [1]. For micro devices, the mold isert needs to be heated close to the glass transition temperature of the polymer. The entire injection molding process takes about 1 to 3 minutes.

5.3.4. Laser Machining

Laser machining is a localized, non-contact machining technique. Machining applications of laser include drilling, cutting, engraving, marking and texturing. Almost all types of material such as metals, ceramics, plastics, and wood can be used with laser machining. Most significantly, laser machining can remove materials in small amount with a small heat-affected zone. Micromachining with controlled accuracy can be achieved. A further attractive advantage of laser machining compared to other micromachining techniques is the possibility of low-cost rapid prototyping.

UV-lasers were used to realize microstructures in polymers. For more detailed discussion on the mechanism on UV-laser ablation on PMMA the reader can refer to [33]. Although UV-laser is a good choice for laser ablation, its cost is much higher than that of CO_2-laser. CO_2-laser has a relatively long characteristic wavelength of 10.6 μm. Thus, the ablation process depends more on thermal energy. The microchannels shown in Fig. 5.9 were fabricated by the Universal M-300 Laser Platform of Universal Laser Systems Inc. (http://www.ulsinc.com). The system uses a 25-Watt CO_2-laser, the maximum beam moving

FABRICATION ISSUES OF BIOMEDICAL MICRO DEVICES 107

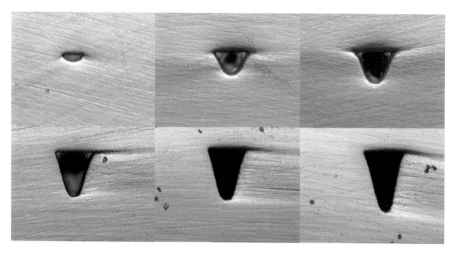

FIGURE 5.9. Typical shapes of microchannels fabricated with CO_2-laser in PMMA.

speed is about 640 mm/s. When the laser beam driven by stepper motors moves across the substrate surface, it engraves a microchannel into the substrate. As mentioned above, the ablation process of CO_2-laser is determined by thermal energy. Therefore the cross section of the microchannel depends on the energy distribution of the laser beam, its moving speed, the laser power, and the thermal diffusivity of substrate material. The energy of the laser beam has a Gaussian distribution, thus the cross section of the channel also has a Gaussian shape. Typical cross sections of Gaussian-shaped microchannels can be seen in Fig. 5.9. Klank et al. [17] found a linear relation between the channel depth and the laser power as well as the number of scanning passes. Beside these two parameters, the influence of the beam speed on the cross-section geometry is shown in Fig. 5.10b.

For the results in Fig. 5.10a the beam speed is fixed at 4%, while the laser power is varied from 3% to 10.5%. The channel depths increase linearly with the laser power as observed by [17]. However the relation between the channel width and the laser power is not linear. For the results in Fig. 5.10b the laser power was kept at a constant value of 7%, while the beam speed is varied from 1.1% to 8%. We can observe that the channel widths and channel depths are inversely proportional to the beam speed.

Figure 5.11 shows the typical velocity distribution in a Gaussian-shaped microchannel. The Poiseuille number Po represents the fluidic resistance of a microchannel and is defined as:

$$\text{Po} = -\frac{1}{\mu}\frac{dp}{dx}\frac{D_h^2}{2u}, \qquad (5.1)$$

where μ is the viscosity of a fluid, dp/dx is the pressure gradient along the channel, D_h is the hydraulic diameter, and u is the average velocity. The Poiseuille number only depends on the channel shape. Gaussian-shaped channels have a Poiseuille number ranging from 9 to 15 depending on the aspect ratio $\alpha = H/W$ between the depth H and the width W,

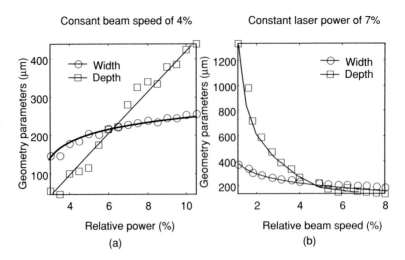

FIGURE 5.10. Geometry parameters of the microchannels as functions of laser power and laser speed. The relative values are based on a maximum laser power of 25 W and a maximum beam speed of 640 mm/s. Circles and squares are measurement results, lines are fitting curves.

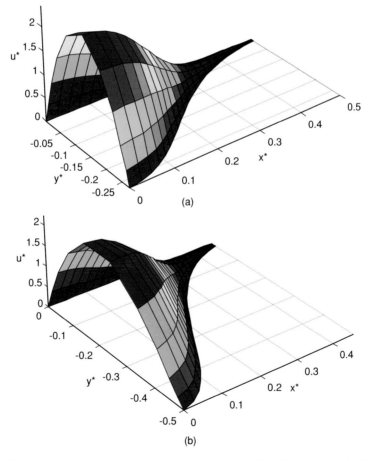

FIGURE 5.11. Simulated velocity distribution of a pressure driven flow Gaussian-shaped microchannels: (a) Aspect ratio of $\alpha = 0.25$, (b) Aspect ratio of $\alpha = 0.5$.

while circular channels and slit channels have higher Poiseuille numbers Po = 16 and Po = 24, respectively.

5.4. PACKAGING OF BIOMEDICAL MICRO DEVICES

The challenges in packaging of BMMDs are the many types of interconnects on the same micro device. In addition to electric interconnects, fluidic interconnects are also needed. The packaging technology should assure the interfaces between the electric domain, fluidic domain and the external environment. Three basic packaging concepts for BMMDs are the multi-chip module, the monolithic horizontal integration and the stacked modular system [18].

- Multi-chip module (MCM) concept is similar to the concept of a printed circuit board (PCB). The MCM concept bonds unpackaged chips to a carrier substrate, which has both electrical wires and fluidic channels. The assembly technique for this concept can be adapted from the surface-mounted technology of traditional electronics.
- Stacked modular system is based on the modular concept of MCM. The components can be stacked as modules to form a complex system.
- Monolithic horizontal integration realizes all components in the same substrate in the same fabrication process. This concept is similar to the monolithic integration of microelectronic components. The advantage of this abroach are the small total size, minimum dead volume and leak-free fluidic interconnects. The drawback is the complex process involved and the many masks needed. Most of the BMMDs have device-specific fabrication processes, which may not be compatible to the rest of the fabrication.

Following, some typical packaging techniques related to polymeric BMMDs are discussed in details.

5.4.1. Thermal Direct Bonding

Similar to diffusion bonding, the direct bonding of the polymeric parts is based on thermal reaction. The strength of most thermoplastics will change with temperature. With the temperature increase the molecules have higher kinetic energy, which breaks the bondage between the monomers. The damage of the molecular chains in a polymer depends on the extent of the absorbed energy.

There are two significant temperature points in the thermal behavior of most thermoplastics: the glass transition temperature and the start of the random chain scission phase. At the glass transition point, the plastics will lost the strength at the normal temperature, but still can keep it solid shape. In the random chain scission phase the bondage between the monomers will be damaged rapidly, and the plastics will lost its solid shape. In a thermal bonding process, this phase will damage some old bondage and form new bondage between the polymer substrates.

The bonding temperature can be selected just above glass transition temperature (106 °C for PMMA), so the substrates and the structures on the surface can keep their original shapes. However, bonding at glass transition temperature has the problem of low bonding energy, thus the bondage between surfaces is not very strong. Bonding at a higher temperature will cause the polymer structures to lost its original shapes. Fortunately, the thermal degradation of a polymer starts much later than the glass transition temperature (150 °C for PMMA), and the weak head group of monomers will be damaged at a higher temperature (near 180 °C for PMMA). From this temperature on the speed of the degradation of the polymer is very fast and it will lose its original shapes. A temperature slightly above the thermal degradation point could be chosen for thermal bonding of polymers. For instance, a temperature of 165 °C is a good bonding temperature for PMMA. At this temperature a good balance between keeping shape and high bond strength can be achieved. Following the bonding process of two PMMA-wafers is described.

To start with, the PMMA wafers are cleaned with IPA and rinsed in DI-water in an ultrasonic bath. The actual bonding process can be carried out in a commercial wafer bonder for silicon wafers. The process is similar to fusion bonding between silicon wafers. First, the wafers were heated up to 165 °C. This temperature remains for the next 30 minutes. Fast cooling may cause residual stress in the wafer stack. Thus an anneal process should be carried out after the high temperature process. The anneal temperature for PMMA is about 80 °C.

To avoid bubbles at the bond interface, a pressure of about 1 bar is to be applied on the wafer stack. Another measure can improve the bonding quality is to design channels on the chips to trap and vent out the gas.

5.4.2. Adhesive Bonding

Adhesive bonding uses an intermediate layer to bond 2 wafers. The advantages of adhesive bonding are the low process temperature and the ability to bond different polymeric materials. The intermediate layer can be an adhesive, which can be cured with temperature or UV-exposure. The example in section 5.3.1.2 used a thin SU-8 layer as the adhesive.

Liquid adhesive may block microchannels and other micro components. An alternative to coating an adhesive layer is the use of a pre-fabricated double-sided adhesive. In the example of a stand-alone micropump in section 5.3.2.1, a 50-μm-thick adhesive tape (Arclad 8102, Adhesive Research Inc., Clen Rock, PA) were used as the intermediate bonding layer. The lamination of the adhesive layers allows bonding at room temperature, which is of advantage for fully polymeric devices with different material layers.

5.4.3. Interconnects

One of the biggest challenge in the fabrication of BMMDs is the development of low-cost fluidic interconnects to the macroscopic world. A BMMD in general needs interconnects for power supply, information signals, and material flow to communicate with its environment. The material flow is the fluid flow, which is processed in a BMMD. While traditional microelectronics already offers a number of solutions for the first two types of interconnects, fluidic interconnects still pose a big challenge to the design and fabrication of BMMDs. Fluidic interconnects for BMMDs should meet some general requirements:

- Easy to handle,
- Low dead volume,
- High-pressure resistance,
- Small sizes,
- Chemical and biological compatibility.

In the following sections, fluidic interconnects are categorized by the coupling nature as press-fit interconnects and glued interconnects.

5.4.3.1. Press-Fit Interconnects Press-fit interconnects utilize elastic forces of coupling parts to seal the fluidic access. Due to the relatively small sealing forces, this type of interconnect is only suitable for low-pressure applications. For fluidic coupling out of a BMMD, tubing interconnects are needed. Figure 5.12a shows a horizontal tubing interconnect fabricated by wet etching of silicon. The external polymeric tubing is press-fitted to the silicon tubing [8]. Vertical tubing interconnects can be etched in silicon using DRIE, Fig. 5.12b. These vertical tubes can hold fused silica capillaries, which are perpendicular to the device surface. If external capillaries are to be inserted directly into etched openings, plastic couplers can be used to keep them, Fig. 5.12c [8].

The molded coupler shown in Figure 5.12d is fabricated from two bonded silicon wafers. The fused silica capillary is embedded in the plastic coupler. The capillary with the coupler is then inserted into the fluidic opening. Annealing the device at elevated temperatures allows the plastic to melt. After cooling down to room temperature, the plastic coupler seals the capillary and the opening hermetically [25]. Plastic couplers can also be compression-molded. The thermoplastic tube is inserted into the opening. Under pressure and temperatures above the glass transition temperature, the plastic melts and fills the gaps between the coupler and silicon, Fig. 5.12e.

In many cases, the elastic force of the couplers can not withstand high pressures. One solution for high pressure interconnects is the use of a mesoscale casing, which has conventional O-rings for sealing fluidic interconnects. If a microfluidic system has multiple fluidic ports, many small O-rings are required. In this case, it is more convenient to have integrated O-rings in the device [39]. First, the cavities for the O-rings are etched in silicon with DRIE. After depositing an oxide/nitride layer, silicone rubber is squeezed into the cavities . The fluidic access is opened from the backside by DRIE. Subsequently, the oxide/nitride layer is etched in buffered hydrofluoric acid and SF_6 plasma. The silicone rubber O-ring remains on top of the opening. If a capillary is inserted into the opening, the rubber O-ring seals it tightly, Fig. 5.12f.

Figure 5.12g shows a concept with standard polyetheretherketone (PEEK) tubing for high performance liquid chromatography (HPLC) [18]. The PEEK tubing is machined to fit the ferrule and the O-ring as shown in Fig. 5.12g. The tube is inserted through the hollow screw. Tightening the screw presses the O-ring against the fluidic port of the BMMD. The O-ring can be replaced by a custom molded ring as shown in Fig. 5.12g [18].

5.4.3.2. Glued Interconnects In many cases, press-fit interconnects are fixed with adhesives. Besides holding function, adhesives offer good sealing by filling the gap between the external tubes and the device opening. Figure 5.13a shows a typical glued fluidic interconnects. The glued surface can be roughened to improve adhesion. A combination of surface roughening, compression molding, and adhesive bonding is used to make tight fluidic interconnects, Fig. 5.13b [36].

If epoxy is to be avoided, metal alloy such as Kovar can be used as the sealing material. Kovar is an alloy consisting of 29% nickel, 17% coper, and the balance of iron. Because of its relatively low thermal expansion coefficient, Kovar is often used for glass to metal

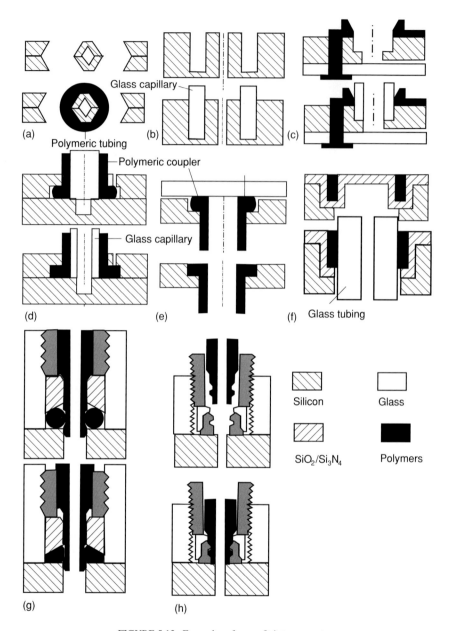

FIGURE 5.12. Examples of press-fit interconnects.

seals in electronics packaging. Figure 5.13d shows a solution with glass seal and Kovar tubes [20]. The Kovar tubes are fitted into the fluidic access. Glass beads are placed around them. A carbon fixture is used as the mold for the glass melt. Glass sealing is accomplished after annealing the assembly at 1,020 °C [20].

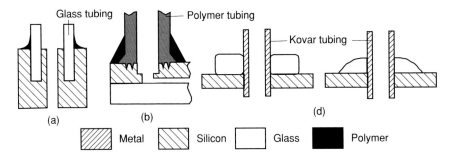

FIGURE 5.13. Examples of press-fit interconnects.

5.5. BIOCOMPATIBILITY OF MATERIALS AND PROCESSES

5.5.1. Material Response

The most common material responses to the biological environment are swelling and leaching. The simplest material response is a mass transfer across the tissue/material interface. Fluid diffuses from the host tissue into the device material, causing it to swell. The changes in dimension may cause microcracks on the material's surface, which in turn alter the mechanical properties of the device. Leaching is another reaction caused by fluid transfer. The fluid that had previously diffused into the device material can move back into the biological environment, and carries material particulates suspended within. Removed particulates damage both the device and surrounding tissues.

5.5.2. Tissue and Cellular Response

Reactionary tissue response begins with inflammation at the device/tissue interface. The symptoms of inflammation are classically reddening, swelling, heating, and pain. Chemical signals released by the damaged tissue trigger the inflammation and attract white blood cells as body response. The device is covered with macrophages and foreign body giant cells leading to a fibrous encapsulation. The encapsulation can affect the functionality of the device [40].

If the healing process occurs as described above, the device can be called biocompatible. Alternatively, a long-lasting inflammation can be caused by chemical or physical properties of the device material or by motion of the device itself. Constant local cell damages make an inflammation reaction continue to be released. If the device material causes cells to die, it is called cytotoxic. If the device material is inert, its particulates cannot be digested by macrophages. The material response to the biological environment can cause chemical changes in the device material, which can be cytotoxic and damage cells.

5.5.3. Biocompatibility Tests

There are two general test methods for biocompatibility:

- In vitro (in laboratory glassware);
- In vivo (in a live animal or human).

The usual procedure starts with in vitro tests, which can weed out the clearly dangerous, but may not expose all problems that may occur in the complex living system. In vitro tests should also be carried out to evaluate the adhesion behavior of cells, DNA, and other polymers. Furthermore, cytotoxicity of the device material should be tested for applications such as tissue engineering. Some animal tests may be required before the device moves to a human test stage. There are standardized tests to follow for the in vitro assessment.

In [40], typical materials of BMMDs such as metallic gold, silicon nitride, silicon dioxide, silicon, and SU-8 were evaluated using the cage implant system in mice. The materials were placed into stainless-steel cages and were sampled at 4, 7, 14, and 21 days, representative of the stages of the inflammatory response. The adherent cellular density of gold, silicon nitride, silicon dioxide, and SU-8TM were comparable and statistically less than silicon. These analyses identified the gold, silicon nitride, silicon dioxide, SU-8 as biocompatible with reduced biofouling.

5.6. CONCLUSIONS

Several fabrication issues were discussed in this chapter. BMMDs can be fabricated with polymeric micromachining technologies at a lower cost. Device examples in this article demonstrated, that BMMDs can be fully made of polymers. Polymers can work as substrate materials (section 5.3.1.2), sacrificial materials, and functional materials (section 5.3.2). Metals were successfully deposited on polymers using micromachined stencils (section 5.3.2.2). Polymeric BMMDs promise a simple design of interconnects (section 5.4.3) and a better degree of biocompatibility (section 5.5).

REFERENCES

[1] H. Becker and Gärtner. *Electrophoresis*, 21:12, 2000.
[2] D.J. Beebe, G.A. Mensing, and G.M. Walker. *Annu. Rev. Biomed. Eng.*, 4:261, 2002.
[3] J. Black. *Biological Performance of Materials*. Marcel Dekker, New York, 1992.
[4] A. Braithwaite and F.J. Smith. *Chromatographic Methods*. 5. ed., Blackie Academic & Professional, London, 1996.
[5] M.G. Cowie. *Polymers: Chemistry and Physics of Modern Materials*. Intertext, Aylesbury, 1973.
[6] C.S. Effenhauser, A. Manz, and H.M. Widmer. *Anal. Chem.*, 65:2637, 1993.
[7] V. Ghica and W. Glashauser (*Verfahren für die spannungsfreie Entwicklung von bestrahlten Polymethylmethacrylate-Schichten*. German Patent, #3039110, 1982).
[8] C. Gonzalez, S.D. Collins, R.L. Smith. *Sens. Actu. B*, 49:40, 1998.
[9] D.J. Harrison, A. Manz, Z.H. Fan, H. Lüdi, and H.M. Widmer. *Anal. Chem.*, 64:1926, 1992.
[10] A.C. Henry, T.J. Tutt, M. Galloway, Y.Y. Davidson, C.S. McWhorter, S.A. Soper, and R.L. McCarley. *Anal. Chem.*, 72:5331, 2000.
[11] T.A. Horbett. *Protein Adsorption on Biomaterials: Iterfacial Phenomena and Applications*. American Chemical Society, Washington D.C., 2002.
[12] Q.A. Huang and N.K.S. Lee. *J. Micromech. Microeng.*, 9:64, 1999.
[13] D.Hülsenberg. *Microelectron. J.*, 28:419, 1997.
[14] S.C. Jacobson, L.B. Kountny, R. Hergenröder, A.W. Moore, and J.M. Ramsey. *Anal. Chem.*, 66:3472, 1994.
[15] S.C. Jacobson, R. Hergenröder, L.B. Coutny, and J.M. Ramsey. *Anal. Chem.*, 66:2369, 1994.
[16] B.H. Jo et al. *J. Micromech. Microeng.*, 9:76, 2000.
[17] H. Klank, J.P. Kutter, and O. Geschke. *Lab on a Chip*, 2:242, 2002.
[18] P. Krulevitch, W. Benett, J. Hamilton, M. Maghribi, and K. Rose. *Biomed. Microdev.*, 4:301, 2002.
[19] C.H. Lin, G.B. Lee, Y.H. Lin, and G.L. Chang. *J. Micromech. Microeng.*, 11:726, 2001.

[20] A.P. London. Development and Test of a Microfabricated Bipropellant Rocket Engine. Ph.D. Thesis, Massachusetts Institute of Technology, 2000.
[21] H. Lorenz et al. *J. Micromech. Microeng.*, 7:121, 1997.
[22] C. Luo, J. Garra, T. Schneider, R. White, J. Currie, and M. Paranjape. A New Method to Release SU-8 Structures Using Polystyrene for MEMS Applications. In *Proceedings of the 16th European Conference on Solid-StateTranducers*, 2002.
[23] M.J. Madou. *Fundamentals of Microfabrication: The Science of Miniaturization*. (2. ed., CRC Press, Boca Raton, 2002.
[24] J.C. McDonald, D.C. Duffy, J.R. Anderson, D.T. Chiu, H. Wu, O.J.A. Schueller, and G.M. Whitesides. *Electrophoresis*, 21:27, 2000.
[25] E. Meng, S.Wu, and Y.-C. Tai. *Micro Total Analysis System 2000*. Kluwer Academic Publishers, Netherlands, 2000.
[26] N.T. Nguyen and S.T. Wereley. *Fundametals and Applications of Microfluidics*. Artech House, Boston, 2002.
[27] N.T. Nguyen, T.Q. Truong, K.K. Wong, S.S. Ho, L.N. Low. *J. Micromech. Microeng.*, 14:69, 2003.
[28] D.L. Polla, A.G. Erdman,W.P. Robbins, D.T. Markus, J. Diaz-Diaz., R. Rizq, Y. Nam, H.T. Brickner, A. Wang, and P. Krulevitch. *Annu. Rev. Biomed. Eng.*, 2:551, 2000.
[29] R. Salim, H. Wurmus, A. Harnisch, and D. Hülsenberg. *Microsys. Technol.*, 4:32, 1997.
[30] K. Seiler, D.J. Harrison, and A. Manz. *Anal. Chem.*, 65:1481, 1993.
[31] S.A. Soper, S.M. Ford, S. Qi, R.L. McCarley, K. Kelley, and M.C. Murphy. *Anal. Chem.*, 72:643A, 2000.
[32] G. Spierings. *J. Mater. Sci.*, 28:6261, 1993.
[33] R. Srinivasan. *Appl. Phys.*, 73:6, 1993.
[34] L.F. Thompson, C.G. Willson, and M.J. Bowden. *Introduction to Microlithography*. Americal Chemical Society, Washington D.C., 1994.
[35] T.Q. Truong and N.T. Nguyen. *J. Micromech. Microeng.*, 14:632, 2004.
[36] W. Wijngarrt et al. The First Self-Priming and Bi-Directional Valve-Less Diffuser Micropump for Both Liquid and Gas. In *Proceedings of MEMS'00 the 13th IEEE International Workshop on MEMS*, 2000.
[37] A.T. Wolly and R.A. Mathies. *Anal. Chem.*, 67:3676, 1995.
[38] Y. Xia and G.M. Whitesides. *Annu. Rev. Mater. Sci.*, 28:153, 1998.
[39] T.J. Yao et al. Micromachined Rubber O-Ring Micro-Fluidic Couplers. In *Proceedings of MEMS'00 the 13th IEEE International Workshop on MEMS*, 2000.
[40] G. Voskericiana, M.S. Shivea, R.S. Shawgoc, H. von Recumd, J.M. Andersona, M.J. Cimac, and R. Langere. *Biomaterials*, 24:1959, 2003.

6
Intelligent Polymeric Networks in Biomolecular Sensing

Nicholas A. Peppas[1,2,3] and J. Zachary Hilt[4]

[1] *Center of Biomaterials, Drug Delivery, and Bionanotechnology Molecular Recognition, Department of Chemical Engineering*
[2] *Department of Biomedical Engineering*
[3] *Department of Pharmaceutics, The University of Texas, Austin, TX 78712-0231 U.S.A.*
[4] *Department of Chemical and Materials Engineering, University of Kentucky, Lexington, KY 40506-0046 U.S.A.*

Since the development of the first biological sensor over 40 years ago [1], the biosensor field has continuously evolved. Today, biosensors are applied in a wide range of uses, including environmental analysis, medical diagnostics, bioprocess monitoring, and biowarfare agent detection. The success of the biosensor is dependent on the ability to rapidly, sensitively, and selectively recognize various biomolecules, with relative importance dependent on the application.

The tailored recognition of the desired biomolecule by the sensing element is the first step in the biosensing process, and the second step is the translation of the interaction into a measurable effect via the transduction element (Figure 6.1). In sensor platforms, a wide variety of transduction methods have been employed, such as gravimetric [2–5], optical [6, 7], and electrochemical [8] transducing elements.

For biosensor applications, the sensing element is typically natural bioreceptors, such as antibody/antigen, enzymes, nucleic acids/DNA, cellular structures/cells, due to their evolved high affinity and specificity [9, 10]. Biomimetic sensing elements, such as those based on polymer networks, can be advantageous over their biological counterparts because they can be designed to mimic biological recognition pathways and at the same time exhibit other abiotic properties that are more favorable, such as greater stability in harsh environments [11, 12].

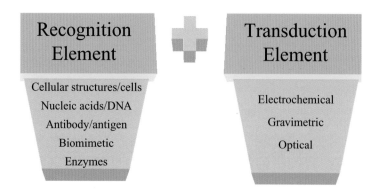

FIGURE 6.1. Illustration of the critical components of a biosensor device and some examples of each.

In particular, the development of micro- and nanoscale sensor platforms has greatly enhanced the applicability of the resultant devices. For instance, the field of clinical diagnostics presents numerous opportunities where micro- and nanoscale biosensor technology can be exploited [13, 14]. For successful patient treatment, medical diagnostics depends on the rapid and precise detection of signature biomolecules for a condition. With the development of lab-on-a-chip and other miniature point-of-care, the speed and precision with which health care is administered has been radically enhanced. These micro or miniaturized total analysis systems (μ-TAS) integrate microvalves, micropumps, microseparations, microsensors, and other components to create miniature systems capable of analysis that typically requires an entire laboratory of instruments. Since introduced as a novel concept for chemical sensing devices [15], reviews have been published that have focused on the application of μ-TAS as innovative point-of-care diagnostic [16, 17].

By developing micro- and nanoscale sensor elements and the corresponding sensor platforms, point-of-care diagnostic devices can be fabricated that not only have a significant impact in ex-vivo sensing applications, but can also be applied to in-vivo and in-vitro applications, where micro- or nanoscale dimensionality is imperative. These miniaturized sensors require small sample and/or reagent volumes, tend to be less invasive, and can be faster and more sensitive relative to macroscale technologies.

In this chapter, we point out how intelligent polymer networks can be used as fundamental sensing elements in biosensor devices focusing on the advancements towards micro- and nanoscale sensor platforms that will lead to improved and novel analysis and impact a wide variety of fields, including environmental analysis and medical diagnostics.

6.1. INTELLIGENT POLYMER NETWORKS

A polymer network is a three dimensional structure formed via physical or chemical crosslinking of polymer chains, creating one giant macromolecule where all monomer units are connected to each other within the polymer phase. The major classes of macromolecular structures are illustrated in Figure 6.2. Polymer networks are prepared by reacting functional monomers and crosslinkers beyond the critical extent of reaction, referred to as the gelation point, where the transition between free polymer chains and an insoluble polymer network

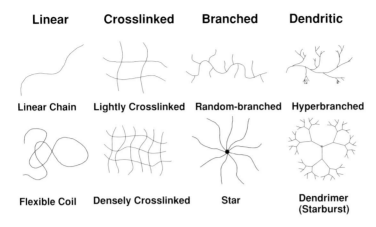

FIGURE 6.2. Illustration of the major classes of macromolecular architecture.

occurs. Various initiation mechanisms, using photochemical, thermal, and redox initiation, are commonly utilized to synthesize polymer networks.

By tailoring their molecular structure, polymer networks can be created that interact with their environment in a pre-programmed, intelligent manner. In biosensor applications, these networks can be advantageous relative to there biological counterparts, since they can be designed to recognize and respond to a biological entity or environmental condition, while exhibiting other properties more favorable for sensor applications, such enhanced stability. Intelligent polymer networks have been created based on a variety of polymer systems, including environmentally responsive hydrogels, biohybrid hydrogel networks, and biomolecularly imprinted polymers.

6.1.1. Hydrogels

Hydrogels are hydrophilic polymeric networks that swell in water or biological fluids without dissolving. The swelling characteristics are the result of crosslinks (tie-points or junctions), permanent entanglements, ionic interactions, or microcrystalline regions incorporating various chains [18–21]. They have been used in various biomedical applications including linings of artificial organs, contact lenses, and biosensors. In the last twenty years, hydrogels have been researched as prime materials for pharmaceutical applications, predominantly as carriers for delivery of drugs, peptides or proteins. They have been used to regulate drug release in reservoir-type, controlled release systems, or as carriers in swellable matrix systems [22].

Hydrogels can be classified as neutral, anionic or cationic networks. In Table 6.1, some representative functional monomers and their relevant properties are listed for reference. The network swelling behavior is governed by a delicate balance between the thermodynamic polymer-water Gibbs free energy of mixing and the Gibbs free energy associated with the elastic nature of the polymer network. In ionic hydrogels, the swelling is governed by the thermodynamic mixing, elastic-retractive forces, and also by the ionic interactions between charged polymer chains and free ions. The overall swelling behavior and the associated response or recognition are affected by the osmotic force that develops as the

TABLE 6.1. Common functional monomers used in hydrogel synthesis.

Functional Monomer	Abbrev.	Structure	Ionic Character
Acrylamide	AAm		Neutral
2-Hydroxyethyl methacrylate	2-HEMA		Neutral
Poly(ethylene glycol)$_n$ dimethacrylate	PEGnDMA		Neutral
Poly(ethylene glycol)$_n$ methacrylate	PEGnMA		Neutral
Acrylic acid	AA		Anionic
Methacrylic acid	MAA		Anionic
2-(Diethylamino) ethyl methacrylate	DEAEMA		Cationic
2-(Dimethylamino) ethyl methacrylate	DMAEMA		Cationic

charged groups on the polymer chains are neutralized by mobile counterions [23]. Electrostatic repulsion is also produced between fixed charges and mobile ions inside the gel, affecting the over swelling of the ionic gel. The equilibrium swelling ratios of ionic hydrogels are often an order of magnitude higher than those of neutral gels because of the presence of intermolecular interactions including coulombic, hydrogen-bonding, and polar forces [24].

6.1.2. Environmentally Responsive Hydrogels

There has recently been increased research in the preparation and characterization of materials that can intelligently respond to changing environmental conditions. By tailoring the functional groups along the polymer backbone, hydrogels can be made sensitive to the conditions of the surrounding environment, such as temperature, pH, electric field, or ionic strength. Because the actuation process is governed by water uptake, these hydrogel systems are attractive for any aqueous applications, and this has led to extensive research being focused on developing biosensor devices based on these hydrogels. Recent reviews have highlighted the extensive research focused on developing new and applying current environmentally sensitive hydrogels, specifically those sensitive to temperature, pH, and specific analytes [11, 25–27].

Certain hydrogels may exhibit environmental sensitivity due to the formation of polymer complexes. Polymer complexes are insoluble, macromolecular structures formed by the non-covalent association of polymers with the affinity for one another. The complexes form due to association of repeating units on different chains (interpolymer complexes) or on separate regions of the same chain (intrapolymer complexes).

Polymer complexes can be stereocomplexes, polyelectrolyte complexes, and hydrogen bonded complexes. The stability of the associations is dependent on such factors as the nature of the swelling agent, temperature, type of dissolution medium, pH and ionic strength, network composition and structure, and length of the interacting polymer chains. In this type of gel, complex formation results in the formation of physical crosslinks in the gel [28]. As the degree of effective crosslinking is increased, the network mesh size and degree of swelling is significantly reduced.

6.1.3. Temperature-Sensitive Hydrogels

Certain hydrogels undergo volume-phase transition with a change in the temperature of the environmental conditions. The reversible volume change at the transition depends on the degree of ionization and the components of the polymer chains. There is usually a negligible or small positive enthalpy of mixing which opposes the process. However, there is also a large gain in the entropy which drives the process. This type of behavior is related to polymer phase separation as the temperature is raised to a critical value known as the lower critical miscibility or solution temperature (LCST). Networks showing lower critical miscibility temperature tend to shrink or collapse as the temperature is increased above the LCST, and these gels to swell upon lowering the temperature below the LCST.

Temperature sensitive hydrogels are classified as either positive or negative temperature-sensitive systems, depending on whether they are contracted below or above a critical temperature, respectively. The majority of the research on thermosensitive hydrogels has focused on poly(N-isopropyl acrylamide) (PNIPAAm), which is a negative temperature-sensitive hydrogel exhibiting a phase transition around 33°C. PNIPAAm and other thermosensitive hydrogels have been studied for variety of applications, including in drug delivery and tissue engineering [26, 29].

6.1.4. pH-Responsive Hydrogels

In networks containing weakly acidic or basic pendent groups, water sorption can result in ionization of these pendent groups depending on the solution pH and ionic composition. The gels then act as semi-permeable membranes to the counterions influencing the osmotic balance between the hydrogel and the external solution through ion exchange, depending on ion-ion interactions. For ionic gels containing weakly acidic pendent groups, the equilibrium degree of swelling increases as the pH of the external solution increases, while the degree of swelling increases as the pH decreases for gels containing weakly basic pendent groups.

Numerous properties contribute to the swelling of ionic hydrogels. Peppas and Khare [23] discussed the effect of these properties including the ionic content, ionization equilibrium considerations, nature of counterions, and nature of the polymer. An increase in the ionic content of the gel increases the hydrophilicity leading to faster swelling and a higher equilibrium degree of swelling. Anionic networks contain acidic pendant groups, such as carboxylic acid, with a characteristic pK_a, while cationic networks contain basic pendant groups, such as amine groups, with a characteristic pK_b. In the case of anionic networks, ionization of these acid groups will occur once the pH of the environment is above the characteristic pK_a of the acid group, leading to the absorption of water into the polymer to a greater degree causing swelling. This actuation process is shown in Figure 6.3. In an ampholyte, which contains both acidic and basic groups, the isoelectric pH determines the transitional pH of swelling of the gel.

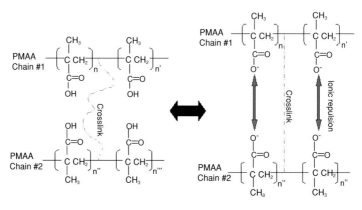

FIGURE 6.3. Schematic of the pH dependent swelling process of an anionic hydrogel: specifically, a crosslinked poly(methacrylic acid) (PMAA) is illustrated.

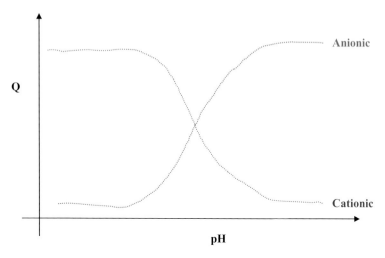

FIGURE 6.4. Equilibrium volume swelling, Q, of ionic hydrogels as a function of pH.

Ionization equilibrium considerations also affect the swelling behavior of ionic hydrogels. Fixed charges in the network lead to the formation of an electric double layer of fixed charges and counterions in the gel. Due to Donnan equilibrium, the chemical potential of the ions inside the gel is equal to that of the ions outside the gel in the swelling medium. Donnan exclusion prevents the sorption of co-ions because of electroneutrality resulting in a higher concentration of counterions in the gel phase than in the external swelling agent. The efficiency of co-ion exclusion, or an increase in the Donnan potential, increases with decreasing solution concentration. Increasing ionic content of the gel also increases the efficiency of co-ion exclusion [30]. A schematic of the characteristic pH response curves for ionic hydrogels is included as Figure 6.4. Examples of some commonly studied ionic polymers include poly(acrylic acid), poly(methacrylic acid), polyacrylamide, poly(diethylaminoethyl methacrylate), and poly(dimethylaminoethyl methacrylate).

6.1.5. Biohybrid Hydrogels

By incorporating biological elements into hydrogel systems, researchers have created hydrogels that are sensitive to specific analytes. For instance, research groups have immobilized enzymes within the network structure of the hydrogel. In the presence of specific chemicals, the enzyme triggers a reaction which changes the microenvironment of the hydrogel. Changes in the local microenvironment (such as pH or temperature) lead to gel swelling or collapse. These systems are completely reversible in nature.

One method is to immobilize into pH-sensitive hydrogel networks enzymes that act on a specific analyte leading to by-products that affect the environmental pH. An example from our studies [31] was the inclusion of activated glucose oxidase into pH-sensitive cationic hydrogels (see Figure 6.5). The glucose oxidase converts glucose into gluconic acid lowering the pH of the local environment, which then causes the hydrogel network to swell in the case of a cationic gel. This system was proposed for a responsive drug delivery system that

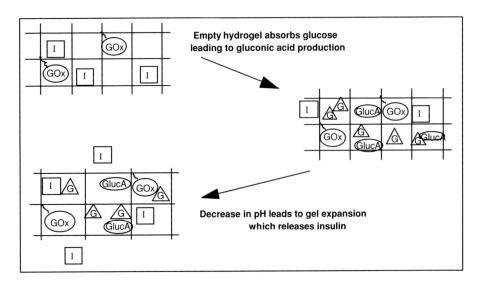

FIGURE 6.5. Actuation mechanism of biohybrid cationic hydrogels with illustration as delivery device. (A) the gel is a crosslinked network with glucose oxidase immobilized in it and insulin is physically entrapped into the system, (B) in the presence of glucose gluconic acid is produced, (C) increase in mesh size results in release of insulin.

would swell and release insulin in response to an increase in glucose concentration, but this biohybrid system has equal merit as a sensing element.

Another approach to impart analyte specificity, based on competitive binding, is loading a hydrogel, having the desired analyte as pendant groups on its chain, with a corresponding entity that selectively binds the analyte. The entity will bind the pendent analytes and form reversible crosslinks in the network, which will then be broken with the competitive binding that takes place in the presence of the analyte. Park et al. [32] immobilized concanavalin A (Con A), a lectin (carbohydrate-binding proteins), in a hydrogel that contained pendant glucose in its network. The Con A non-covalently bound to the glucose pendent groups to form crosslinks, and with increasing concentration of glucose in the environment, the crosslinks were reversibly broken allowing the network to swell.

6.1.6. Biomolecular Imprinted Polymers

Although the above mentioned techniques have been proven effective in producing analyte sensitive hydrogels, they rely on proteins, which inherently lead to limitations in the system. The major limitations of natural receptors, such as proteins, are their high cost, possible antigenicity, and low stability. An alternative to these techniques is to use synthetic biomimetic methods to create hydrogels that will bind and respond to specific analytes. These biomimetic polymer networks are advantageous because they can be tailored to bind any molecule with controlled selectivity and affinity, provided that certain interactions exist. In a recent review, several methods utilizing molecular imprinting processes, a template mediated polymerization, were proposed to design analyte responsive hydrogels that can

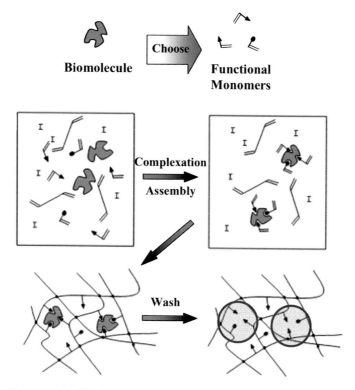

FIGURE 6.6. Schematic of the biomimetic approach to molecular imprinting. Recognition sites are highlighted with shaded circles.

respond, in theory, to any desired analyte [11]. A schematic of the biomimetic molecular imprinting methodology is included as Figure 6.6.

There are some significant characteristics to consider in the design of biomimetic polymer networks via a molecular imprinting technique. To achieve a relatively easy on/off binding event, a non-covalent recognition process is favored. Therefore, supramolecular interactions, such as hydrogen bonding, electrostatic interactions, hydrophobic interactions, and van der Waals forces, are employed to achieve recognition. For the formation of the network, it is imperative that the functional monomers, crosslinker, and template are mutually soluble. In addition, the solvent must be chosen wisely, so that it does not interact and destabilize the self-assembled functional monomer and template.

6.1.7. Star Polymer Hydrogels

Star polymers are structurally intriguing materials that have been synthesized and characterized in the past fifteen years. Star polymers are hyperbranched polymers with large number of arms emanating from a central core, as illustrated in Figure 6.7. Star-shaped polymers are usually prepared by anionic or cationic polymerizations, which provide a relatively compact structure in the form of star-shaped polymers of nanometers size. In recent years, researchers have applied star polymers and other dendritic structures in a variety

FIGURE 6.7. Structure of a Star Polymer Gel.

of biomedical and pharmaceutical applications, including tissue engineering, gene delivery, and artificial recognition elements [33–37]. The advantage of star polymers in creating responsive and recognitive networks is the presence of a large number of functional groups in a small volume.

In our lab, we have investigated the use of star polymers as advanced materials for pH-sensitive and recognitive polymer networks [37]. By imprinting star polymer building blocks with a D-glucose template, we demonstrated the ability to synthesize novel polymeric networks that were able to distinguish between the template and a similar sugar, D-fructose. In addition, star polymers were copolymerized with ionizable methacrylic acid to synthesize polymer networks that exhibited sensitivity to changes in the pH. These networks, which are hydrophilic leading to applicability in aqueous environments, are promising as sensing elements for use in biological environments.

6.2. APPLICATIONS OF INTELLIGENT POLYMER NETWORKS AS RECOGNITION ELEMENTS

A number of research groups have utilized hydrogels as functional components in biomedical applications, such as in biomaterials and biosensors. Ito and collaborators [38–40] have patterned pH-sensitive and thermo-sensitive hydrogels and proposed these microstructures for use in various microdevices. In numerous studies done by Matsuda and coworkers [41–45], hydrogels have been micropatterned using a surface polymerization technique induced by an iniferter (acts as an initiator, a transfer agent, and a terminator) to create surface regions with different physicochemical properties for direction of cell adhesion and behavior. In our own laboratory, we have patterned poly(ethylene glycol)-containing hydrogels onto polymer substrates that had incorporated iniferters in their networks to provide bonding to create novel surfaces for possible application in biosensors and in biomaterials for selective adhesion of cells and proteins [46].

Several other groups have focused on developing microdevices utilizing the mechanical response of environmentally sensitive hydrogels for microactuation. Beebe et al. [47, 48] and Zhao et al. [49] have micropatterned pH-sensitive hydrogels inside microfluidic channels to create flow controls that sense the environmental conditions and then actuate in response. In similar work, pH-sensitive hydrogels were patterned to form a biomimetic valve capable of directional flow control [50]. Similarly, Madou and coworkers [51] have utilized a blend of redox polymer and hydrogel to create an "artificial muscle" that can act as an electro-actuated microvalve for possible application in controlled drug delivery.

Of particular interest, environmentally sensitive hydrogels have been applied as sensing elements for development of novel sensor platforms, utilizing various transducing elements. Specifically, hydrogels have been applied as recognition matrices where the hydrogel provided stability for an entrapped biological component and/or its properties were monitored with changing environmental conditions. In addition, hydrogels were utilized as actuation elements where the mechanical response of the network was applied to actuate various transduction elements.

6.2.1. Sensor Applications: Intelligent Polymer Networks as Recognition Matrices

Several research groups have patterned hydrogels containing immobilized oxidoreductase enzymes, such as glucose oxidase, lactate oxidase, and alcohol oxidase, onto electrodes using photolithography to create biosensors for monitoring various analyte levels [52–55]. For example, Jimenez et al. [55] fabricated enzymatic microsensors that employed polyacrylamide as a entrapment matrix for immobilization of enzyme recognition elements, including glucose oxidase and urease. To create biosensors selective for glucose or urea, the polymer matrix containing the desired enzyme was patterned onto a pH-sensitive ion selective field-effect transistor (FET). In similar work, Sevilla et al. [56] fabricated a SO_2 sensor by coating a pH-sensitive FET with a polyurethane-based hydrogel containing a hydrogen sulfite electrolyte, not an enzyme, entrapped within its network structure. A pH change as a result of the interaction of the sulfur dioxide with the hydrogel matrix was successfully measured.

In other work, Sheppard Jr. et al. [57–59] developed miniature conductimetric pH sensors based on the measurement of the conductivity of pH-sensitive hydrogels that were photo-lithographically patterned onto planar interdigitated electrode arrays [57–59]. The sensor detection was based on the measurement of changes in the electrical conductivity of the hydrogel membrane that resulted with its swelling/collapsing. In related work, Guiseppi-Elie et al. [60] demonstrated chemical and biological sensors that applied conducting electroactive hydrogel composites as recognition elements and utilized electrochemical detection.

In addition, various optical based detection schemes have utilized to hydrogel systems. Leblanc et al. [61] have synthesized hydrogel membranes with a short peptide sequence immobilized within the network as a membrane fluorescent sensor. In this work, the peptide sequence was chosen due to its high binding affinity for Cu^{2+} and contained a dansyl fluorophore for signal transduction. It was demonstrated that the gel film exhibited a fluorescent emission that was selectively and reversibly quenched by copper ions in its aqueous environment. In work by another researcher, pH-sensitive hydrogels were photolithographically patterned for possible application in a fluorescence sensor array [62]. Arregui et al. [63] integrated neutral hydrogels with optical fibers to demonstrate an optical based humidity sensor. Pishko et al. [64] have encapsulated living cells within hydrogel microstructures and demonstrated finite viability of the cells. These systems have possible application as optical biosensor arrays of individually addressable cell-containing hydrogels for drug screening or pathogen detection.

6.2.2. Sensor Applications: Intelligent Polymer Networks as Actuation Elements

Utilizing the actuation response of hydrogels, Grimes et al. [65,66] demonstrated wireless pH sensors based on integrating pH-responsive hydrogels with magnetoelastic thick films. The sensor device functioned by monitoring change in resonance frequency due to applied mass load of the magnetoelastic sensor device by remote query. Recently, Han et al. [67] demonstrated a constant-volume hydrogel osmometer as a novel sensor platform. The concept was illustrated with a device where a pH-responsive hydrogel was confined between a rigid semipermeable membrane and the diaphragm of a miniature pressure sensor. Changes in the osmotic swelling pressure of the hydrogel resulting from changes pH were accurately measured via the pressure sensor. Although the device was of macroscale in dimensions, the design can be easily miniaturized for microscale sensor development. Other groups have demonstrated sensor platforms at the macroscale for pH [68] and CO_2 [69] using pressure sensors to transduce the swelling response of hydrogel systems. These systems also have the ability to be miniaturized, which would greatly enhance their applicability.

Recently, microelectromechanical systems (MEMS) sensor platforms, specifically those based on microcantilevers, have been applied in a wide variety of applications due their miniature size and ultrahigh sensitivity. In our work, environmentally responsive hydrogels have been integrated with silicon microcantilevers to develop an ultrasensitive bioMEMS sensor platform (see Figure 6.8). Specifically, a pH microsensor was demonstrated based on

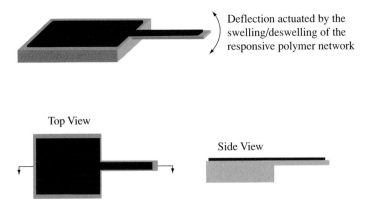

FIGURE 6.8. Schematic of MEMS sensor platform based on microcantilever patterned with an intelligent polymer network.

a methacrylic acid based pH-responsive hydrogel [70, 71]. This was the first demonstration of a microscale MEMS sensor device where actuation is controlled by an intelligent polymer network. In similar work, Thundat et al. [72] have recently demonstrated a variation on our novel sensor platform by integrating hydrogels responsive to CrO_4^{2-} with commercial silicon microcantilevers to create CrO_4^{2-} sensors. More recently, another variation has been demonstrated where hydrogels containing benzo-18-crown-6 coated on microcantilevers to create Pb^{2+} sensors [73].

6.3. CONCLUSIONS

Intelligent polymer networks are exceptional materials for application as novel sensing elements, since the response of the macromolecular network can be precisely tailored through the molecular design of the polymer network functionality and three dimensional structure. These intelligent networks have numerous advantages over natural bioreceptors, such as antibodies and enzymes, due to their low price and robustness. In particular, biomimetic networks can be advantageous over their biological counterparts because they can be designed to mimic biological recognition pathways.

Previously, sensor platforms have been demonstrated using intelligent polymer networks as sensing elements and a variety of transducing elements, such as gravimetric, optical, and electrochemical, but these have mostly been at the macroscale. The further development of micro- and nanoscale sensors that utilize intelligent polymer networks as functional components, such as those based on microcantilevers, will drastically enhance the diagnostic capabilities in a variety of fields. For instance, the field of clinical diagnostics presents numerous opportunities where micro- and nanoscale biosensor technology can be exploited by developing lab-on-a-chip and other miniature point-of-care devices that enhance the speed and precision with which health care is administered. The future of biomolecular sensing will be profoundly impacted by the increased integration of intelligent polymer materials as recognition elements, particularly in micro- and nanoscale sensors.

REFERENCES

[1] L. Clark, Jr., and C. Lions. Electrode systems for continuous monitoring in cardiovascular surgery. *Ann. N.Y. Acad. Sci.*, 102:29, 1962.
[2] C.K. O'Sullivan and C.G. Guilbault. Commercial quartz crystal microbalances–theory and applications. *Biosens. Bioelectron.*, 14:663, 1999.
[3] B.A. Cavic, G.L. Hayward, and M. Thompson. Acoustic waves and the study of biochemical macromolecules and cells at the sensor-liquid interface. *Analyst*, 124:1405, 1999.
[4] E. Benes, M. Groschl, W. Burger, and M. Schmid. Sensors based on piezoelectric resonators. *Sensor. Actuat. A—Phys.*, 48:1, 1995.
[5] M. Sepaniak, P. Datskos, N. Lavrik, and C. Tipple. Microcantilever transducers: A new approach to sensor technology. *Anal. Chem.*, 74:568A, 2002.
[6] O. Wolfbeis. Fiber-optic chemical sensors and biosensors. *Anal. Chem.*, 74:2663, 2002.
[7] J. Homola, S.S. Yee, and G. Gauglitz. Surface plasmon resonance sensors: Review. *Sensor. Actuat. B—Chem.*, 54:3, 1999.
[8] E. Bakker and M. Telting-Diaz. Electrochemical sensors. *Anal. Chem.*, 74:2781, 2002.

[9] S. Subrahmanyam, S. Piletsky, and A. Turner. Application of natural receptors in sensors and assays. *Anal. Chem.*, 74:3942, 2002.
[10] M. Byfield and R. Abuknesha. Biochemical aspects of biosensors. *Biosens. Bioelectron.*, 9:373, 1994.
[11] M.E. Byrne, K. Park, and N.A. Peppas. Molecular imprinting within hydrogels. *Adv. Drug Deliver. Rev.*, 54:149, 2002.
[12] J.Z. Hilt, M.E. Byrne, and N.A. Peppas. Configurational biomimesis in drug delivery: Molecular imprinting of biologically significant molecules. *Adv. Drug Deliver. Rev.*, 11:1599, 2004.
[13] R. McGlennen. Miniaturization technologies for molecular diagnostics. *Clin. Chem.*, 47:393, 2001.
[14] T. Vo-Dinh and B. Cullum. Biosensors and biochips: Advances in biological and medical diagnostics. *Fresen. J. Anal. Chem.*, 366:540, 2000.
[15] A. Manz, N. Graber, and H. Widmer. Miniaturized total chemical-analysis systems—a novel concept for chemical sensing. *Sensor. Actuat. B—Chem.*, 1:244, 1990.
[16] A. Tudos, G. Besselink, and R. Schasfoort. Trends in miniaturized total analysis systems for point-of-care testing in clinical chemistry. *Lab on a Chip*, 1:83, 2001.
[17] Y. Liu, C. Garcia, and C. Henry. Recent progress in the development of mu TAS for clinical analysis. *Analyst*, 128:1002, 2003.
[18] N.A. Peppas. *Hydrogels in Medicine and Pharmacy*. CRC Press, Boca Raton, FL, 1986.
[19] R. Langer and N.A. Peppas. Advances in biomaterials, drug delivery, and bionanotechnology. *AIChE J.*, 49:2990, 2003.
[20] N.A. Peppas, Y. Huang, M. Torres-Lugo, J.H. Ward, and J. Zhang. Physicochemical, foundations and structural design of hydrogels in medicine and biology. *Ann. Revs. Biomed. Eng.*, 2:9, 2000.
[21] A.M. Lowman and N.A. Peppas. Pulsatile drug delivery based on a complexation/decomplexation mechanism. In S.M. Dinh, J.D. DeNuzzio, and A.R. Comfort, (eds.), *Intelligent Materials for Controlled Release, ACS Symposium Series*, ACS, Washington, DC, Vol. 728, pp. 30–42, 1999.
[22] N.A. Peppas and P. Colombo. Analysis of drug release behavior from swellable polymer carriers using the dimensionality index. *J. Control. Rel.*, 45:35, 1997.
[23] A.R. Khare and N.A. Peppas. Release behavior of bioactive agents from pH-sensitive hydrogels. *J. Biomat. Sci., Polym. Ed.*, 4:275, 1993.
[24] L. Brannon-Peppas and N.A. Peppas. Equilibrium swelling behavior of pH-sensitive hydrogels. *Chem. Eng. Sci.*, 46:715, 1991.
[25] Y. Qiu and K. Park. Environment-sensitive hydrogels for drug delivery. *Adv. Drug Deliv. Revs.*, 53:321, 2001.
[26] B. Jeong, S.W. Kim, and Y.H. Bae. Thermosensitive sol-gel reversible hydrogels. *Adv. Drug Deliv. Revs.*, 54:37, 2002.
[27] T. Miyata, T. Uragami, and K. Nakamae. Biomolecule-sensitive hydrogels. *Adv. Drug Deliv. Revs.*, 54:79, 2002.
[28] A.M. Lowman and N.A. Peppas. Analysis of the complexation/decomplexation phenomena in graft copolymer networks. *Macromolecules*, 30:4959, 1997.
[29] N.A. Peppas, P. Bures, W. Leobandung, and H. Ichikawa. Hydrogels in pharmaceutical formulations. *Eur. J. Pharm. Biopharm.*, 50:27, 2000.
[30] R.A. Scott and N.A. Peppas. Compositional effects on network structure of highly cross-linked copolymers of PEG-containing multiacrylates with-acrylic acid. *Macromolecules*, 32:6139, 1999.
[31] K. Podual, F.J. Doyle III, and N.A. Peppas. Glucose-sensitivity of glucose oxidase-containing cationic copolymer hydrogels having poly(ethylene glycol) grafts. *J. Control Rel.*, 67:9, 2000.
[32] A.J. Lee and K. Park. Synthesis and characterization of sol-gel phase-reversible hydrogels sensitive to glucose. *J. Mol. Recognit.*, 9:549, 1996.
[33] C. Gao and D. Yan. Hyperbranched polymers: From synthesis to applications. *Progress in Polymer Science*, 29:183, 2004.
[34] J. Jansen, E. van den Berg, and E.W. Meijer. Encapsulation of guest molecules into a dendritic box. *Science*, 266:1226, 1994.
[35] L.G. Griffith and S. Lopina. Microdistribution of substratum-bound ligands affects cell function: Hepatocyte spreading on PEO-tethered galactose. *Biomaterials*, 19:979, 1998.
[36] S.C. Zimmerman, M.S. Wendland, N.A. Rakow, I. Zharov, and K.S. Suslick. Synthetic hosts by monomolecular imprinting inside dendrimers. *Nature*, 418:399, 2002.

[37] E. Oral and N.A. Peppas. Responsive and recognitive hydrogels using star polymers. *J. Biomed. Mater. Res.*, 68A:439, 2004.
[38] Y. Ito. Photolithographic synthesis of intelligent microgels. *J. Intell. Mater. Syst. Struct.* 10:541, 1999.
[39] G. Chen, Y. Imanishi, and Y. Ito. Photolithographic synthesis of hydrogels. *Macromolecules*, 31:4379, 1998.
[40] G. Chen, Y. Ito, and Y. Imanishi. Micropattern immobilization of a pH-sensitive polymer. *Macromolecules*, 30:7001, 1997.
[41] Y. Nakayama, J.M. Anderson, and T. Matsuda. Laboratory-scale mass production of a multi-micropatterned grafted surface with different polymer regions. *J. Biomed. Mater. Res. (Appl. Biomater.)*, 53:584, 2000.
[42] K.M. DeFife, E. Colton, Y. Nakayama, T. Matsuda, and J.M. Anderson. Spatial regulation and surface chemistry control of monocyte/macrophage adhesion and foreign body giant cell formation by photochemically micropatterned surfaces. *J. Biomed. Mater. Res.*, 45:148, 1999.
[43] J. Higashi, Y. Nakayama, R.E. Marchant, and T. Matsuda. High-spatioresolved microarchitectural surface prepared by photograft copolymerization using dithiocarbamate: Surface preparation and cellular responses. *Langmuir*, 15:2080, 1999.
[44] Y. Nakayama, K. Nakamata, Y. Hirano, K. Goto, and T. Matsuda. Surface hydrogelation of thiolated water-soluble copolymers on gold. *Langmuir*, 14:3909, 1998.
[45] Y. Nakayama and T. Matsuda. Surface macromolecular architectural designs using photo-graft copolymerization based on photochemistry of benzyl N,N-diethyldithiocarbamate. *Macromolecules*, 29:8622, 1996.
[46] J.H. Ward, R. Bashir, and N.A. Peppas. Micropatterning of biomedical polymer surfaces by novel UV polymerization techniques. *J. Biomed. Mater. Res.*, 56:351, 2001.
[47] D.J. Beebe, J.S. Moore, J.M. Bauer, Q. Yu, R.H. Liu, C. Devadoss, and B. Jo. Functional hydrogel structures for autonomous flow control inside microfluidic channels. *Nature*, 404:588, 2000.
[48] D.J. Beebe, J.S. Moore, Q. Yu, R.H. Liu, M.L. Kraft, B. Jo, and C. Devadoss. Microfluidic tectonics: A comprehensive construction platform for microfluidic systems. *Proc. Natl. Acad. Sci. U.S.A.*, 97:13488, 2000.
[49] B. Zhao and J.S. Moore. Fast pH- and ionic strength-responsive hydrogels in microchannels. *Langmuir*, 17:4758, 2001.
[50] Q. Yu, J.M. Bauer, J.S. Moore, and D.J. Beebe. Responsive biomimetic hydrogel valve for microfluidics. *Appl. Phys. Lett.*, 78:2589, 2001.
[51] L. Low, S. Seetharaman, K. He, and M.J. Madou. Microactuators toward microwaves for responsive controlled drug delivery. *Sens. Actu. B*, 67:149, 2000.
[52] K. Sirkar and M.V. Pishko. Amperometric biosensors based on oxidoreductases immobilized in photopolymerized poly(ethylene glycol) redox polymer hydrogels. *Anal. Chem.*, 70:2888, 1998.
[53] A. Munoz, R. Mas, C.A. Galan-Vidal, C. Dominiguez, J. Garcia-Raurich, and S. Alegret. Thin-film microelectrodes for biosensing. *Quimica. Analytica.* 18:155, 1999.
[54] G. Jobst, I. Moser, M. Varahram, P. Svasek, E. Aschauer, Z. Trajanoski, P. Wach, P. Kotanko, F. Skrabal, and G. Urban. Thin-film microbiosensors for glucose-lactate monitoring. *Anal. Chem.*, 68:3173, 1996.
[55] C. Jimenez, J. Bartrol, N.F. de Rooij, and M. Koudelka. Use of photopolymerizable membranes based on polyacrylamide hydrogels for enzymatic microsensor construction. *Anal. Chim. Acta.*, 351:169, 1997.
[56] B.S. Ebarvia, C.A. Binag, and F. Sevilla III. Surface and potentiometric properties of a SO_2 sensor based on a hydrogel coated pH-FET. *Sens. Actu. B*, 76:644, 2001.
[57] N.F. Sheppard, Jr., M.J. Lesho, P. McNally, and A.S. Francomacaro. Microfabricated conductimetric pH sensor. *Sens. Actu. B*, 28:95, 1995.
[58] M.J. Lesho and N.F. Sheppard, Jr., Adhesion of polymer films to oxidized silicon and its effect on performance of a conductometric pH sensor. *Sens. Actu. B*, 37:61, 1996.
[59] N.F. Sheppard Jr., R.C. Tucker, and S. Salehi-Had. Design of a conductimetric pH microsensor based on reversibly swelling hydrogels. *Sens. Actu. B*, 10:73, 1993.
[60] S. Brahim, A.M. Wilson, D. Narinesingh, E. Iwuoha, and A. Guiseppi-Elie. Chemical and biological sensors based on electrochemical detection using conducting electroactive polymers. *Microchim. Acta*, 143:123, 2003.
[61] Y. Zheng, K.M. Gattas-Asfura, C. Li, F.M. Andreopoulos, S.M. Pham, and R.M. Leblanc. Design of a membrane fluorescent sensor based on photo-cross-linked PEG hydroge. *J. Phys. Chem. B*, 107:483, 2003.
[62] A. Revzin, R.J. Russell, V.K. Yadavalli, W. Koh, C. Deister, D.D. Hile, M.B. Mellott, and M.V. Pishko. Fabrication of poly(ethylene glycol) hydrogel microstructures using photolithography. *Langmuir*, 17:5440, 2001.

[63] F.J. Arregui, Z. Ciaurriz, M. Oneca, and I.R. Matias. An experimental study about hydrogels for the fabrication of optical fiber humidity sensors. *Sens. Actu. B*, 96:165, 2003.
[64] K. Koh, A. Revzin, and M.V. Pishko. Poly(ethylene glycol) hydrogel microstructures encapsulating living cells. *Langmuir*, 18:2459, 2002.
[65] C. Ruan, K.G. Ong, C. Mungle, M. Paulose, N.J. Nickl, and C.A. Grimes. A wireless pH sensor based on the use of salt-independent micro-scale polymer spheres. *Sens. Actu. B*, 96:61, 2003.
[66] C. Ruan, K. Zeng, and C.A. Grimes. A mass-sensitive pH sensor based on a stimuli-responsive polymer. *Anal. Chim. Acta*, 497:123, 2003.
[67] I. Han, M. Han, J. Kim, S. Lew, Y.J. Lee, F. Horkay, and J.J. Magda. Constant-volume hydrogel osmometer: A new device concept for miniature biosensors. *Biomacromolecules*, 3:1271, 2002.
[68] L. Zhang and W.R. Seitz. A pH sensor based on force generated by pH-dependent polymer swelling. *Anal. Bioanal. Chem.*, 373:555, 2002.
[69] S. Herber, W. Olthuis, and P. Bergveld. A swelling hydrogel-based P-CO_2 sensor. *Sens. Actu. B*, 91:378, 2003.
[70] R. Bashir, J.Z. Hilt, A. Gupta, O. Elibol, and N.A. Peppas. Micromechanical cantilever as an ultrasensitive pH microsensor. *Appl. Phys. Lett.*, 81:3091, 2002.
[71] J.Z. Hilt, A.K. Gupta, R. Bashir, and N.A. Peppas. Ultrasensitive biomems sensors based on microcantilevers patterned with environmentally responsive hydrogels. *Biomed. Microdev.*, 5:177, 2003.
[72] Y. Zhang, H. Ji, G.M. Brown, and T. Thundat. Detection of CrO42- using a hydrogel swelling microcantilever sensor. *Anal. Chem.*, 75:4773, 2003.
[73] K. Liu and H. Ji. Detection of Pb^{2+} using a hydrogel swelling microcantilever sensor. *Anal. Sci.*, 20:9, 2004.

II

Processing and Integrated Systems

7

A Multi-Functional Micro Total Analysis System (μTAS) Platform for Transport and Sensing of Biological Fluids using Microchannel Parallel Electrodes

Abraham P. Lee[a,b], John Collins[a], and Asuncion V. Lemoff[c]

[a] *Department of Biomedical Engineering*
[b] *Department of Mechanical & Aerospace Engineering, University of California at Irvine, 204 Rockwell Engineering Center, Irvine, CA 92697-2715, U.S.A.*
[c] *Biotechnology Consultant, Union City, CA 94587, U.S.A.*

7.1. INTRODUCTION

The field of micro total analysis systems (μTAS) is developing technologies to integrate sample acquisition, sample separation, target purification, and cellular/molecular level detection schemes on microscale common platforms. The foundation of μTAS is a fluid transport platform that enables the manipulation of small volumes of fluids in microscale channels and chambers. Ideally, it would function much like integrated circuits (IC), except that it would be dealing with fluids instead of electrons. For μTAS, in lieu of the voltage sources would be micropumps and for transistors the microvalves and microfluidic switches. In ICs, batch fabrication processes such as CMOS (complementary metal-oxide semiconductors) have enabled a low cost (per chip), multi-functional design platform that integrates logic and control elements on the same chip. Analogously, the same impact can be made on μTAS if an integrated batch fabrication process is developed for high density complex fluidic routing. However, integrated microfluidics for μTAS faces much tougher challenges for the following reasons. First, the fluid being pumped is far from being uniform in properties or homogeneous in its constituents. The fluids routed not only are different depending on the application, but their properties change over time as they are being processed. Second, it is much more important to consider the three-dimensional aspects of the flow channels versus the virtually one-dimensional thin film wires in ICs. Third, microfluidic channels function not only as resistors, due to fluid viscosity and interfacial forces between fluid and the solid channel walls, but they also store mechanical energy in a way analogous to capacitors and inductors. Microfluidic devices with parallel electrodes can address many of these challenges since they have the following desirable features: large scale integration of flow control elements, compatible with biological samples (low ac voltage), generates continuous volume flow instead of pulsatile surface flow, and establishes a common microfabrication platform for multi-functional elements such as flow rate sensors, viscosity sensors, impedance sensors, micro mixers, and droplet generators. In addition, it can manipulate unprocessed biofluids with wide ranges of properties as long as the solution is slightly conductive.

In the 1960s MHD was investigated as a method to generate quiet propulsion of marine vehicles (ships and submarines) using the conducting characteristics of seawater [19, 23, 29]. However, the propulsion efficiency was low at the large scales necessary for practical applications, requiring superconducting magnets and heavy weight structural materials. Other applications of large-scale MHD instruments include generators, heat extractors (using liquid metals in nuclear reactors), high temperature plasma controllers, spacecraft propulsion, and metallurgy for casting [4]. Recently, the advent of MEMS and microfluidics has enabled the implementation and application of MHD in the micro-scale.

This chapter introduces the various components to manipulate biological fluids enabled by parallel electrodes in microchannels. Components include MHD microfluidic pumps (AC micropump [16] and microfluidic switch [17], mixers, droplet generators), impedance based sensors, and biomolecular separation/extraction devices. In addition, various applications of these integrated microfluidics platforms are introduced to motivate those interesting in further developing this exciting technology.

7.2. MHD MICROPUMP FOR SAMPLE TRANSPORT USING MICROCHANNEL PARALLEL ELECTRODES

7.2.1. Principle of Operation

The pumping mechanism for a magnetohydrodynamic pump results from the Lorentz force produced when an electrical current is applied across a channel filled with conducting solution in the presence of a perpendicular magnetic field (Fig. 7.1). The Lorentz force is both perpendicular to the current in the channel and the magnetic field, and is given by

$$\frac{F}{V_e} = J \times B \qquad (7.1a)$$

$$P = J \times Bl_e \qquad (7.1b)$$

where F is the MHD propulsion force in the channel, J the electrical current density across the channel, B the magnetic field (or magnetic flux density in webers/m^2), V_e the fluid

A MULTI-FUNCTIONAL MICRO TOTAL ANALYSIS SYSTEM (µTAS) PLATFORM

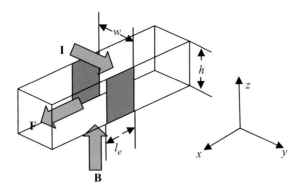

FIGURE 7.1. Vector diagram of MHD pump. I is the current between the blue electrodes, B is the magnetic field perpendicular to the substrate, and F is the Lorentz force generated in the microchannel.

volume in between the electrodes, P the MHD generated pressure drop in the channel, and l_e the length of the electrodes. This geometry is shown in Fig. 7.1. As Eq. 7.1 shows, the MHD force scales poorly with fluid volume (length cubed) but the pressure drop scales more favorably with only one length unit, l_e. Since the total current across the electrodes, $I = JLh$, Eq. 7.1b becomes:

$$P = IB/h \qquad (7.2)$$

where h is the height of the electrode. For flow analysis the Navier-Stokes partial differential equation can be written as [22, 25, 31]:

$$\rho \frac{DU}{Dt} = -\nabla P + \mu \nabla^2 U + J \times B \qquad (7.3)$$

where ρ is the fluid density, U the velocity as a function of x and t, and τ the viscosity of the fluid. This equation is provided for those interested in solving more complex MHD flow problems. In this chapter a simplified analysis is provided.

In microchannels, assuming laminar, Newtonian flow then Poiseuille's law governs that the pressure drop P is:

$$P = QR \qquad (7.4)$$

where Q is the volumetric flow rate and R is the fluidic resistance which is dependent on the geometry of the channels [31]. For rectangular channels, the fluidic resistance is given by

$$R = \frac{8\mu L(w+h)^2}{w^3 h^3} \qquad (7.5)$$

where w is the distance between the electrodes, L the total length of the channel and µ is the viscosity of the fluid. Substituting Eq. 7.5 into Eq. 7.4 and then equating with Eq. 7.2 gives the flow rate as

$$Q = \frac{IBw^3 h^2}{8\mu L(w+h)^2} \qquad (7.6)$$

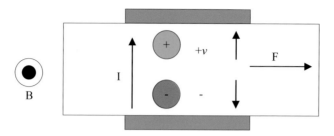

FIGURE 7.2. Top-view of MHD pump with magnetic field coming out of the page. Positive and negative charges are pumped in the same direction.

An equivalent expression for the Lorentz force is given by

$$\mathbf{F} = q\mathbf{v} \times B \tag{7.7}$$

where q is the charge, v the velocity and B the magnetic field. Unlike electrophoresis, both the positive and negative charges are pumped in the same direction (since qv is always same sign). This is shown in Fig. 7.2.

In our micropump design, an AC electrical current is applied in a perpendicular, synchronous AC magnetic field from an electromagnet. When an AC current of sufficiently high frequency is passed through an electrolytic solution, the chemical reactions are reversed rapidly such that there isn't sufficient electrochemical ionic exchange to form bubbles and cause electrode degradation. In this case, the time-averaged Lorentz force not only depends on the amplitudes of the electrical current or the magnetic field but also depends on the phase of the magnetic field, relative to the electrode current, and is given by

$$F = IBw \int_0^{2\pi} \sin \omega t \, \sin(\omega t + \phi) d\omega t \tag{7.8}$$

The ability to control the phase allows for controlling not only the flow speeds but also flow direction. The integrand can have a value between $-1/2$ to $1/2$. At $0°$, the integrand is positive and corresponds to a flow in one direction. At $180°$, the integrand is negative and correponds to flow in the opposite direction and at $90°$, the integrand is zero corresponding to no flow.

7.2.2. Fabrication of Silicon MHD Microfluidic Pumps

The microchannels were fabricated by etching a v-groove through a silicon wafer 380 μm thick (see Fig. 7.3). A thin oxide was grown for electrical insulation before metal electrodes (200A° Ti / 2000A° Pt) were deposited on to the side-walls of the channel using a shadow mask. The silicon wafer was then anodically bonded between two glasses. The top glass has holes that were ultrasonically drilled for fluidic input and electrical contact. Figure 7.3 shows a photograph of a fluidic chip in which liquid is pumped around a square loop.

The electromagnet is positioned underneath the fluidic chip as shown in Fig. 7.3. The magnetic field strength is measured underneath the device and above the device. The

A MULTI-FUNCTIONAL MICRO TOTAL ANALYSIS SYSTEM (μTAS) PLATFORM

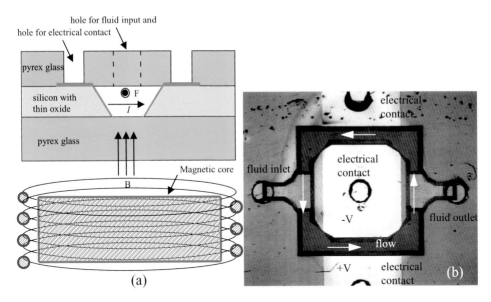

FIGURE 7.3. (a) Cross-section of AC MHD micropump set-up. The Lorentz force produced is directed into the paper. (b) Photograph of top view of circular pump. Channel depth of 380 μm and top width of 800 μm. Electrode width of 4 mm.

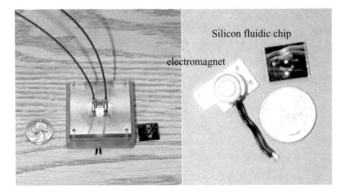

Photos of the MHD system. *Left* Actual MHD package. *Right* The mini-electromagnet and the silicon fluidic chip compared to a US quarter dollar coin.

electromagnet used is commercially available through Edmund Scientific and is shown in 0 along with the MHD packaging used. The core measures 1/4" in diameter and 1/4" in height. The actual core material is not known.

7.2.3. Measurement Setup and Results

The measurement set-up is shown in Fig. 7.4. The MHD components (microchannel electrodes and electromagnet) have the same basic electrical circuit. The electrodes and the electromagnet are driven separately by a function generator which is connected to an

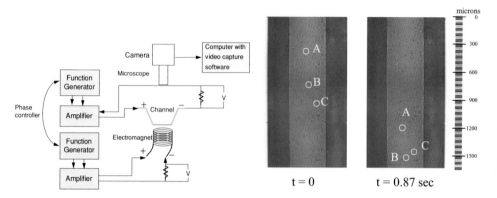

FIGURE 7.4. (a) Left: Measurement set-up for MHD micropump; (b) Right: Video capture of 5 µm polystyrene beads flowing through a microchannel 800 µm wide. Flow velocities for the three particles measured in mm/s are: A = 1.06, B = 1.02, C = 0.67

amplifier. A resistor is used in series with the MHD components to measure the current and phase going through the device. The two function generators are locked to enable phase-control. A microscope is positioned above the fluidic chip for viewing. This microscope has a CCD camera which is hooked up to a computer with video capture software.

The solutions used are mixed with 5 µm polystyrene beads. Flow measurements are done by recording a 3–5 second movie using video capture software. This allows us track the beads frame by frame. The distance a particle has travelled can be measured within a given time which enables flow speeds to be deduced. Fig. 7.4 shows the evolution of three beads between two frames 0.87 seconds apart. The resulting velocity profile is consistent with pressure-driven flow.

Measurements were done on varying concentrations of NaCl solution to determine the maximum current allowed in the microchannels before bubble formation is observed. This is shown in Fig. 7.5. Bubbles are observed at lower currents in solutions of lower concentrations. One possible explanation is that solutions with lower concentrations have much longer Debye length, and hence larger volumes near the electrodes where hydrolysis can occur [22]. Increasing the frequency allows higher currents to be achieved without bubble formation. All measurements were done with a top channel width of 800 µm and an electrode area of 4 mm × 380 µm. For channels of different width but the same electrode area, the same bubble current threshold is observed. For smaller widths, there is a lower voltage drop and for larger widths, there is a higher voltage drop across the solution, since the resistance depends upon the length of the current path in the solution. Another consequence of hydrolysis is a change in pH very near to the electrodes. Neither the magnitude nor the spatial extent of this pH variation was measured in our system. Since biological specimens are quite sensitive to pH, it will be necessary to consider this aspect when designing practical systems.

7.2.3.1. Electromagnet Field Strength Since there were no technical specifications available on the electromagnet, the magnetic field strength of the electromagnet was measured as a function of frequency given the same driving voltage using a gauss meter.

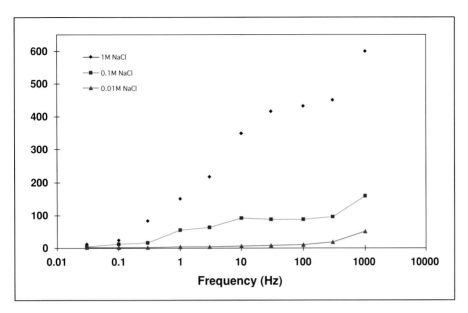

FIGURE 7.5. Bubble current threshold for varying concentrations of NaCl solution as a function of frequency.

Beyond 1 kHz, no magnetic field was observed. For the same driving voltage, the magnetic field strength has a maximum at 60 Hz, where commercially available electromagnets are optimized. Because higher frequencies allow for higher currents in the channel, 1 kHz was chosen to be the operating frequency for our micropump. However, because the electromagnet is not optimized for operation at 1 kHz, the electromagnet consumes high power, up to 2W amplitude. All measurements are done using the Edmund Scientific mini-electromagnet.

Other electromagnetic cores are being investigated to allow for operation at higher frequencies with minimal power. At present, the best candidates for an electromagnetic core are ferrite materials or metallic glass (METGLAS). In addition, the magnetic field strength in the channel can be dramatically increased if a second electromagnet is situated above the channel. This configuration was not used in this experiment since the second electromagnet would obscure the view of the channel necessary for flow measurement using the video-capture method.

7.2.3.2. Measurements in AC Magnetohydrodynamic Micropump There are two ways to vary the flow speed in the AC MHD micropump using the maximum current possible in the channel. One is by varying the magnetic field and another is by varying the relative phase between the channel current and the magnetic field. The maximum speed with opposite direction is seen at 0° and 180° relative phase and no flow is observed at 90° relative phase. These are characteristic of any conducting solution with the only difference being the amplitude of the maximum flow speeds. Various concentrations of electrolytic solutions were also tried using the same AC MHD micropump, including solutions at near neutral pH and DNA solutions. These results are summarized in Table 7.1.

TABLE 7.1. *Flow velocity and calculated flow rates for other conducting solutions with the same magnetic field of 187 gauss underneath the device and 74 gauss above the device.*

Solution	Flow velocity (mm/s)	Channel current (mA)	Calculated Flow Rate (μL/min)
1 M NaCl	1.51	140	18.3
0.1 M NaCl	0.51	100	6.1
0.01 M NaCl	0.34	36	4.1
0.01 M NaOH	0.30	24	3.6
PBS ph = 7.2	0.16	12	1.9
Lambda DNA in 5 mM NaCl	0.11	10	1.3

7.2.4. MHD Microfluidic Switch

The MHD [17] microfluidic switch is a basic microfluidic logic element that can be implemented by a combination of 2 independently controlled AC MHD micropumps. One microfluidic switch configuration is shown in Fig. 7.6. In this configuration, there are three arms arranged in a "Y" pattern. Arms 1 and 2, the "branches", each have an identical electrode pair for pumping. Arm 3, the "trunk", can be switched to either arm 1, arm 2, or some combination. Flow can be in either direction, depending upon the application. When only the pump in arm 1 is actuated (for example in the direction towards arm 3), the flow that is produced divides into both arm 2 and arm 3. In this case, the flow in arm 2 will be in the direction opposite to the flow in arm 1, but with a lower flow rate (since the flow in arm 1 is divided between arms 2 and 3). In order to stop the flow in arm 2, while flow continues from arm 1 to arm 3, the pump in arm 2 must be actuated to produce the necessary pressure to cancel the pressures in that arm. This pressure will in general be smaller than the pumping pressure in arm 1. If the same electromagnet is used to actuate both pumps, then one can switch flow by either tuning the relative electrode current between arms 1 and 2 or by adjusting the phase differences. The latter method was chosen since change in current amplitude may have undesirable effects on the temperature and pH of the solution, and therefore the current amplitude in each branch was fixed throughout the phase switching process.

A photograph of a fluidic circuit in which liquid is switched between two flow loops is shown in Fig. 7.7. Although there are additional MHD electrode pairs patterned in arm 3, the upper loop and the lower loop, only the two inner MHD electrode pairs in arms 1 and 2,

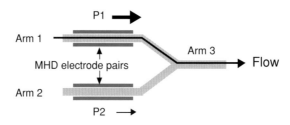

FIGURE 7.6. Conceptual diagram of an MHD microfluidic switch.

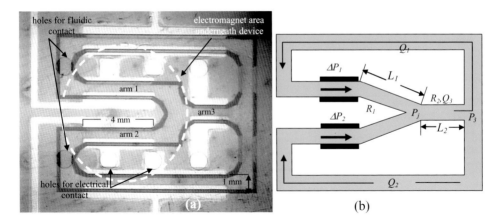

FIGURE 7.7. (a) Photograph of fluidic switch connecting two flow loops. Only two MHD electrode pairs (arms 1 and 2) are actuated. Channels are 1 mm wide and 300 μm deep and electrode lengths in arms 1 and 2 are 4 mm. (b) Diagram illustrating the microfluidic switch circuitry.

which are enclosed in the diameter of the electromagnet core, were actuated and switched. The other MHD electrode pairs could be actuated with a larger diameter electromagnet core to cover the entire fluidic circuit layout. As depicted in Fig. 7.7b, the simplified microfluidic circuitry of Fig. 7.7 can be modelled as a linear network. As MHD pump 1 is turned on to exert a pressure ΔP_1, flow will be induced in both the upper channel (Q1) and the lower channel (Q2). However, if a counter pressure ΔP_2 from MHD pump 2 is gradually increased from zero, it will eventually reach equilibrium resulting in $Q_2 = 0$. The linear fluidic network can be solved using Amphere's Law and Kirchoff's Law:

$$\cos \phi_2 = \frac{R_2}{R_{uloop}} \quad (7.9)$$

where ϕ_2 is the phase difference between pumps 2 and 1, R_2 the resistance between point j and point 3, and R_{uloop} the total resistance in the upper loop.

The experimental set up is similar to the AC MHD micropump except that two function generators are phase-locked to drive the two MHD micropumps. Ideally, with the same electromagnet and the same applied electrode current, both arms should produce the same flow. In reality, the flow can differ due to variations in magnetic field strength between the two arms. To compensate for this, a potentiometer was added in series with one of the arms. The potentiometer can be adjusted to ensure that identical flow is produced in each arm for a given input voltage.

Arms 1 and 2 of the microfluidic switch were actuated with the same electrode current of 189 mA and the same magnetic field of 0.024 Tesla underneath the microfluidic switch. Flow could be switched between the two flow arms at a velocity of 0.3mm/sec. While arm 1 was kept at 0° relative phase with respect to the electromagnet, the phase of arm 2 was varied to determine the phase required to cancel the flow in that arm. This was found to be approximately 45° relative phase. With this phase, flow in arm 2 is completely blocked. This is far off from the calculated 78° based on Eq. 7.9, which assumes a linear, sequential resistance circuit diagram. The discrepancy is largely due to the pressure drops along the

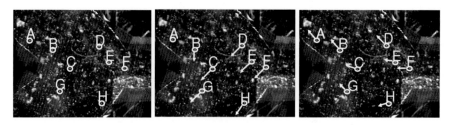

FIGURE 7.8. Tracking 8 particles in 3 consecutive frames 0.033 seconds apart. Arrows show direction of displacement from the previous frame.

cross-sections of the junction that results in viscous swirling. There was also observable leakage at the junctions that may further contribute to the unaccounted pressure drops in channel L_2 of Fig. 7.7.

Within the resolution of the measurement system, the switching of flow between the two arms is instantaneous. Fig. 7.8 shows three consecutive video frames, captured at the moment of switching, each separated by 0.033 seconds. Between the first two frames, each of the tracked particles moves in the direction of the second arm, as indicated by the arrows in the diagram. Between the second and third frames, all of the particles have moved in the direction of the first arm, indicating that flow switching has already occurred.

7.2.5. Other MHD Micropumps and Future Work

The Lorentz force can be produced using either a DC or an AC set-up. In a DC configuration, a DC current is applied across the channel in the presence of a uniform magnetic field from a permanent magnet. There have been several DC MHD micropumps presented in the literature. Professor Haim Bau and his group at the University of Pennsylvania have been developing DC MHD micropumps fabricated by low temperature co-fired ceramic tapes [5, 11]. The electrodes pairs were patterned on the bottom of the fluidic channels to drive both mercury and saline solutions. Professor Seung Lee and his group at the Pohang University of Technology in Korea have demonstrated a DC MHD pump in silicon where the electrodes faced each other in the vertical direction while the magnetic field was parallel to the substrate (across the channels) [12]. Practical issues of a DC set-up were reported due to gas bubbles generated by electrolysis at the electrodes that impede fluid [12] flow and cause electrode degradation. Recently, both Manz' group at the Imperial College in London [9] and van den Berg's group at the University of Twente [26] have presented μTAS devices using the MHD pumping principle.

Many other microfluidic devices can be easily implemented on an integrated platform once a microfabrication process is established. Examples of other MHD devices include microfluidic droplet generators, droplet mixers, sample loaders, and various versions of combinatorial mixers. Researchers at the University of Pennsylvania have developed an innovative microfluidic mixer by patterning electrode stripes on the bottom of the flow channel (perpendicular to flow direction) to induce cellular convection [6] (see Fig. 7.9).

There are several challenges that need to be addressed to make MHD microfluidics a viable microTAS platform. First of all, it is critical to understand whether heat generation

A MULTI-FUNCTIONAL MICRO TOTAL ANALYSIS SYSTEM (μTAS) PLATFORM

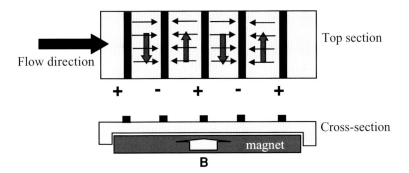

FIGURE 7.9. Microfluidic mixer in [6] where electrode stripes (black arrows) on the bottom of the flow channel induce flow patterns (red arrows) that split the flow perpendicular to flow direction.

by the electrical current would be detrimental to biological samples. Initial tests on DNA show no effect on the biological viability since subsequent amplification by PCR verified the original sequence. However, heat may degenerate proteins, cells, and other biological constituents. Another challenge is to develop a microfabrication process that can integrate high density MHD microfluidic components on chip-scale platforms. As different length scales are required for different applications, it is important to carry out a parametric study to understand the scaling effects of different parameters (channel dimensions, electrode sizes). Our group is also looking to develop devices with local flow rate feedback control with the integration of MHD micropumps and flow sensors. For AC MHD devices, it is critical to identify high frequency AC magnets at a reasonable cost.

7.3. MICROCHANNEL PARALLEL ELECTRODES FOR SENSING BIOLOGICAL FLUIDS

7.3.1. MHD Based Flow Sensing

Moving charges in the presence of a perpendicular magnetic field are subjected to a force that is both perpendicular to the direction of motion and the magnetic field. The movement of charges due to the magnetic field results in a charge separation as shown in Fig. 7.10.

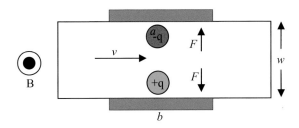

FIGURE 7.10. Top view of flow meter. Magnetic field is pointed out of the page. Positive and negative are deflected away from each other due to the Lorentz force.

The work done in moving a charge from point a to point b is simply the force multiplied by the distance w given by:

$$W_{ab} = Fw. \quad (7.10)$$

Substituting the Lorentz force into Eq. 7.10 results in:

$$W_{ab} = qvBw, \quad (7.11)$$

where q is the charge of the particle, v is the velocity of particle and B the magnetic field. The voltage difference is defined as the work per unit charge which is given by:

$$V_{ab} = vBw. \quad (7.12)$$

The flow velocity is related to the volumetric flow rate Q by:

$$v = \frac{Q}{wh}, \quad (7.13)$$

where wh is the cross-sectional area of the channel. Substituting Eq. 7.13 into 7.12 gives:

$$V_{ab} = \frac{QB}{h}. \quad (7.14)$$

Thus, measuring the voltage difference across the channels gives us the flow rate.

7.3.2. MHD Based Viscosity meter

A viscosity meter is simply a pump and a flow meter in series with one another. The first set of electrodes is used as the pump and the second set of electrodes as a flow meter. Measuring the voltage difference from the flow meter allows us to determine the flow rate. For the MHD pump, the flow rate for a rectangular cross-section channel is given by Eq. 7.6. Since I, B, w, h, and L are known, from the pump, and Q is measured by the flow-meter, the viscosity μ, can be deduced. Viscosity measurements are important in µTAS since biological particles in a solution affect the viscosity of the solution.

7.3.3. Impedance Sensors with MicroChannel Parallel Electrodes

7.3.3.1. Electrical Double Layer In a microchannel with a pair of electrodes interfacing the liquid the electrostatic charges on the electrode surface will attract the counter ions in the liquid. The rearrangement of the charges on the electrode surface balances the charges in the liquid. This gives rise to electrical double layer [10] (EDL). Because of the electrostatic interaction, the counter ion concentration near the electrode surface is higher than that in the bulk liquid far away from the solid surface. Immediately next to the solid surface, there is a layer of ions that are strongly attracted to the solid surface and are immobile. This compact layer, normally less than 1 nm thick is called Helmholtz layer, labeled as inner (IHP) and outer (OHP) Helmholtz plane in Fig. 7.11. From the compact layer to the uniform bulk liquid, the counter ion concentration gradually reduces to that of bulk liquid. Ions in this region are affected less by the electrostatic interaction and are mobile. This layer

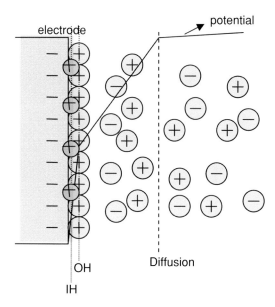

FIGURE 7.11. Electrical Double Layer.

is called the diffuse layer of the EDL as proposed by Gouy and Chapman. The thickness of the diffuse layer depends on the bulk ionic concentration and electrical properties of the liquid, ranging from a few nanometers for high ionic concentration solutions up to 1 mm for distilled water and pure organic liquids. Stern put together compact layer and diffusion layer while Grahame proposed the possibility of some ionic or uncharged species to penetrate in to the zone closest to the electrodes. The potential rises rapid when it moves from Helmholtz plane to diffusion layer and goes to saturation in the middle of the channel. Thus, the flow of liquids along a pair of electrodes reorients the ionic distribution in the channel.

7.3.3.2. Fabrication of Channel Electrodes and Microfluidic Channel The impedance sensors were generally fabricated with thin surface metallic electrodes, fabricated [32] on a glass substrate (Fig. 7.12) and the microfluidic channel being

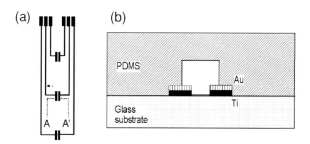

FIGURE 7.12. (a) Layout design of fabricated electrodes and the wiring for measuring the current flow. (b) Cross-section of fabricated flow sensor along the electrodes at AA' in (a).

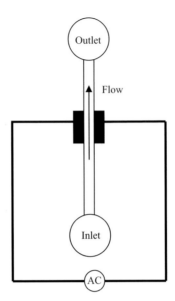

FIGURE 7.13. Measurements using Channel Electrodes.

made on polydimethylsiloxane (PDMS) using a SU8 mold [8]. Electrodes have also been electroplated for 3-d electrodes which results in uniform electric fields for higher sensitivity and accuracy [20, 21].

7.3.3.3. Flow Sensing When a liquid is forced through a microchannel under an applied hydrostatic pressure, the counter ions in the diffuse layer of the EDL are carried toward the downstream end, resulting in an electrical current in the pressure-driven flow direction. This current is called the streaming current. However, this current is measured with two electrodes on the line of direction of flow through the channel. On the other hand, in order to use the channel electrodes for measuring flow impedance or admittance measurements across the channel electrodes is considered. An application of an ac signal across the electrodes results in an increase in electrical admittance across the electrodes. This increase of admittance increases with flow rate of the liquid flow through the channel. Thus the channel electrodes act as a flow sensor in the channel with the measurement of flow induced admittance [7] (see Fig. 7.13).

In hydrodynamic conditions, forced convection dominates the transport of ions to the electrodes within the flow channels. When the width of the microfluidic channel is very small compared to the length of the channel, the lateral diffusion of the ions is significant under laminar flow. Under an ac electrical signal applied across the channel, the equivalent circuit [28] of the microsystem is shown in Fig. 7.14a. The electrical double layer [10] formed across the channel is formed from two capacitances namely diffuse layer capacitance (C_s) and the outer Helmholtz plane capacitance (C_e). The former is due to ion excess or depletion in the channel, and the latter is due to the free electrons at the electrodes and is independent of the electrolyte concentration. The smaller of these capacitances dominates the admittance since these two capacitances are in series. The frequency of the applied ac voltage,

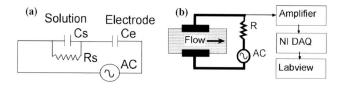

FIGURE 7.14. (a) The equivalent circuit for the channel and electrodes flow sensor cell. The solution in the channel offers a parallel resistive (Rs) and capacitive (Cs) impedance while the electrodes by themselves offer serial capacitive (Ce) impedance with the solution. (b) Experimental Setup for measuring current increase due to flow of electrolytes (Standard Resistance R = 1kΩ, AC is the ac signal source, NI DAQ is PCI 6024E).

flow rate and conductivity of the fluid are the factors affecting the admittance of the fluidic system and our flow sensing principle is based on the optimization of these parameters.

$$D_A \frac{\partial^2 [A]}{\partial y^2} - v_x \frac{\partial [A]}{\partial x} = 0 \qquad (7.15)$$

$$\frac{\partial A}{\partial t} = 0 \qquad (7.16)$$

$$i_L = 0.925 n F [A]_{bulk} D_A^{2/3} Q^{1/3} w \cdot 3\sqrt{x_e^2/h^2 d} \qquad (7.17)$$

For an electrochemical oxidation of a species A to A$^+$ in a microchannel, the convective-diffusive equation for mass transport under steady state condition is given by equation (1), where [A] is the concentration of the species, D_A is the diffusion coefficient and v_x is the velocity in the direction of flow. The first term is the lateral diffusion in the microchannel and the second term is the transport along the length of the channel. Under steady state flow condition the boundary condition is given by equation (2). The solution of this equation predicts the mass transport limited current, i_L [18] as a function of flow rate, Q as given by equation (3), where n is the number of electrons transferred, F, the Faraday constant, x_e is the electrode length, h, the cell half-height, d, the width of the cell and w, the electrode width. It is to be noted that the current due to flow of electrolyte is directly proportional to the cube root of volume flow rate of the fluid. AC voltage signal is considered rather than dc voltage since the application of an ac voltage in the flow sensor does not promote any electrode reaction.

In order to measure the flow induced admittance an ac voltage is applied across the channel electrodes in series with a standard resistor and the current flowing across the pair of electrodes is calculated. The voltage across the resistor is fed to a data acquisition card through an amplifier as shown in Fig. 14b. The rms values of voltage across the standard resistor are measured using a programmable interface to the computer.

Microfluidic flow of an electrolyte (Eg. NaOH) is maintained at a constant flow rate in the channel using a syringe pump. An ac signal of low voltage (Eg. 0.05 V) is applied in the circuit by a signal generator. The current exponentially grows when the flow is switched on and then stays constant. After switching off the flow, the current again decays exponentially until it reaches a constant value. The difference between two constant values of current gives the current increase due to flow and this current is a key figure in the measurements.

The flow sensor is optimized for the electrical parameters (f = 500 Hz, V = 0.4 V and concentration = 0.2 M) and the flow sensor is capable of measuring very low values

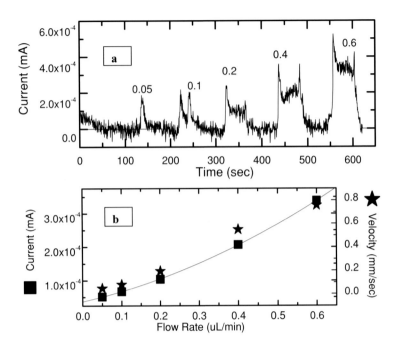

FIGURE 7.15. Flow sensor calibration (a) Time versus current across the sensor electrodes when flow of fluid is turned on for 1 minute and turned off for 1 minutes at the flow rates 0.05 µL/min to 0.6 µL/min (b) Flow induced currents calculated from (a) and bead velocities at different flow rates.

of flow rate starting at 0.05 µL/min (< 1 nL/sec). The current measured is proportional to the flow rate as shown in Fig. 7.15. In another experiment, fluorescent beads of diameter 2.5 µm are mixed with NaOH and sent through the channel. The motion of the beads at the flow rates .05, 0.1, 0.2, 0.4, and 0.6 µL/min are recorded using optical video microscopy at 30, 60, 120, 250, 250 frames/sec respectively, and the beads at a particular stream are analyzed and averaged to predict the velocity response at very low flow rates. The sensor results are compared with the velocity of beads (the symbol '*' in Fig. 7.15b) and shows similar response. Thus the calibration of the flow sensor is accomplished using the velocity measurements with beads.

7.3.3.4. Measurement of Solution Properties Microfabricated impedance sensors [3] have demonstrated the ability to sense variations in solution temperatures, ionic concentrations, and even antigen-antibody binding (immunosensors) in microchannels. Traditional admittance spectra of solution represent different ionic dispersions at broad frequency range. It is very difficult to quantitatively analyse the spectra of different solutions. On the other hand, the flow of a solution along the impedance sensing channel electrodes in the flow induced admittance measurement configuration gives more information on the solution. The flow induced admittance depends on the frequency of the ac signal applied across the electrodes for the admittance measurement. The flow induced admittance frequency spectra of different fluids flowing across the channel is characteristic of the molecules or constituents of the fluid.

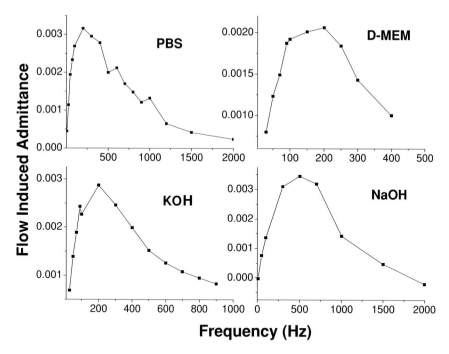

FIGURE 7.16. Flow induced admittance spectra of some analytical solutions.

Fig. 7.16 shows the flow induced admittance spectra for cell culture reagents PBS and D-MEM and ionic solutions KOH and NaOH. These spectra for different solutions not only differ in magnitude but also show clear shift in frequency and width of the peak. The peak magnitude, frequency and band width of the spectra depend upon the ionic properties like ionic strength, valency of the ions, concentration of the ions, etc, present in the solutions.

7.3.3.5. Measurement of Particles in Solution Measurement electrical properties of tiny particles suspended in solution or any liquid have been measured in their bulk form using microelectrodes are electrodes arrays built in a well. With the advent of microfluidics, a single particle is pumped across electrical sensors based on capacitance or impedance measurement and are sensed. Thus a solution containing non-identical particles can be sensed continuously with the pair of electrodes in the channel. These sensors, not only can count the particles based on their electrical response, but also can detect the nature of the particles based on the relationship between the electrical parameters and the nature of the particle.

Particle sensing has been carried out traditionally using fluorescence or radioactive tagging. Measurements based on such methods are very sensitive but that require optical staining or radioactive labelling and other manipulations. Electrical measurements are better than such techniques in the sense that they do not require sophisticated sample preparation techniques.

Depending on the electrical behaviour of the liquids and the particles, capacitance or impedance measurements are employed. Generally, capacitance measurements are sensitive

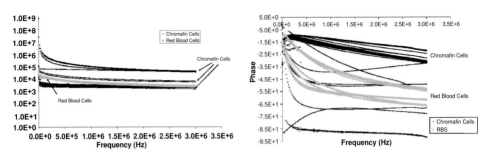

FIGURE 7.17. Single cell impedance spectroscopy of bovine chromaffin and red blood cells [21].

at low frequencies and impedance measurements are sensitive at high frequencies. If the liquid (pH buffer, saline solution) where the particles are suspended is more conducting than the particles dielectric of the particles are predominant and so capacitance measurements are more sensitive. On the other hand if the liquid is a non-conductor (eg. DI water, oil, solvent) impedance measurements can detect the particles suspended in the liquid. In a typical microfluidic sensor, a pair of electrodes is built across the channel where the particles are flowed in the channel. The length of the electrodes is comparable to the size of the particles to be measured. In order to make sure that the particles are flowing one by one, the width of the channel is less than twice the size of the particles.

Single cell characterization [21] of bovine chromaffin cells and red blood cells has been conducted using electrical impedance spectroscopy over a frequency range of 40 Hz to 3 MHz. In order to trap the cell in between opposing electrodes, vacuum or pressure and dielectrophoresis techniques have been utilized. The impedance measurements are done for the cell media in order to provide baseline for the impedance data recorded for the cells along with the media. At lower frequencies of the impedance spectra, large difference in cell impedance were observed whereas a characteristic impedance value develops at higher frequencies due to the elimination of the membrane capacitive component. The phase data is very sensitive for different cells types than the magnitude spectrum of the cells as shown in Fig. 7.17.

7.3.3.6. Particle Cytometry using Capacitance and Impedance Measurements In cytometry, particles are sensed one by one continuously so that monitoring of every particle is possible. Micro Coulter particle counter principle is utilized in most of the cytometries. Capacitance cytometry [27] is based on ac capacitance measurement by probing the polarization response of the particles in an external electric field. Capacitance measurements have been used to assay cell-cycle progression [1], differentiate normal and malignant white blood cells [24], DNA content within the nucleus of the cell [27], cell growth etc. Capacitance cytometry of cells is carried using the channel electrodes by measuring the capacitance of the cells when they flow through the channel.

The integrated microfluidic device for the cell cytometry consists of a pair of electrodes where the cells are sensed, and the PDMS microfluidic channel with inlet and outlet holes for fluid.

A MULTI-FUNCTIONAL MICRO TOTAL ANALYSIS SYSTEM (µTAS) PLATFORM

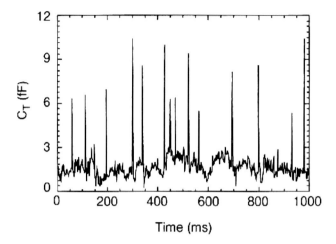

FIGURE 7.18. Capacitance Cytometry of mouse cells correlated to the DNA contents of the cells under different metabolism cycles [27].

The electrodes are made of gold and are 50 µm wide. The distance, d, separating the electrodes is 30 µm. The width of the PDMS microfluidic channel is also d, the length, L, is 5 mm, and the height, h, is either 30 µm or 40 µm. Fluid delivery is accomplished with a syringe pump at nonpulsatile rates ranging from 1 to 300 µl/hr. By electrically shielding the device and controlling the temperature precisely (to within $\pm 0.05°C$), noise levels of ≈ 5 aF when the microfluidic channel is dry and 0.1-2 fF when wet, are achieved.

Figure 7.18 shows the device response over a course of 1,000 ms to fixed mouse myeloma SP2/0 cells suspended in 75% ethanol/25% PBS solution at 10°C. Distinct peaks present in the data correspond to a single cell flowing past the electrodes. The channel height of the device was 30 µm. The peak values of the capacitance are correlated to the DNA content of the different cells and also to the metabolism phase cycle of the cells.

7.4. PARALLEL MICROCHANNEL ELECTRODES FOR SAMPLE PREPARATION

7.4.1. A Microfluidic Electrostatic DNA Extractor

7.4.1.1. DNA Extractor Principle DNA is captured and concentrated electrostatically using a microfluidic device that utilizes the inherent negative charge on a DNA molecule for its capture. Due to the advances in molecular biology, techniques for reading the genome (DNA sequencing) and for identifying the existence of a known sequence (DNA detection) have been developed. However, extracting the DNA from a raw sample, such as a blood or urine sample, can be labor-intensive and time consuming. For blood samples, DNA extraction involves several steps by a trained technician. These steps involve measuring the sample volume, cell sorting using a centrifuge, lysing the cell (breaking down the cell membrane), and filtering out the DNA for detection. Each of the steps mentioned involves a carefully controlled set of procedures in order to be done correctly. DNA extractor chip has been developed by Cepheid using silicon pillars [14]. The silicon pillars are oxidized which allow for DNA to bind to the surface as biological samples are pumped through. Once the DNA has been concentrated, a wash solution followed by a buffer solution is flowed through the microstructure to release the DNA.

We present another method to extract DNA from a sample using the inherent net negative charge on the DNA. The DNA Extractor described uses the H-filter™ design developed by Micronics Technologies combined with electrostatic forces. The principle of the H-filter™ relies on the absence of turbulent mixing in a microfluidic channel. Thus, two flow streams can flow next to one another without mixing. Movement of particles from one flow stream to another occurs due to diffusion coefficient, particle size, viscosity of the solution and temperature.

7.4.1.2. DNA Extractor Design and Experiment DNA molecules have a net charge of 2 negative charges per base pair. Thus, in the presence of a DC electric field, DNA molecules are attracted to the positive electrode. Using the H-filter™ design, DNA extraction can be achieved by patterning electrodes along the bar of the H-filter™ and applying a DC voltage across the channel. Instead of relying on diffusion to extract DNA from one flow stream to another, electrostatic forces are used to transport DNA from one stream to another.

The DNA extractor can be used to remove the DNA from a lysed spore or cell for example. The lysing solution breaks down the spore coating or cell membrane. This allows the DNA to be transported from the lysing solution stream to a buffer solution. The extraction not only allows DNA concentration but also allows the DNA to be separated from other debris that could inhibit DNA detection techniques, such as PCR (polymerase chain reaction). This is illustrated in Fig. 7.19.

The DNA extractor has two sample inputs. The first sample input is the sample volume containing the DNA. The test sample volume consists of 250 μl of DNA (48 kbp Lambda DNA from Sigma) in concentrations of 500 μg/ml of deionized water, stained with 25 μl of 1 mM YOYO-1 dye (from Molecular Probes) diluted 100:1 in distilled water and 25 μl

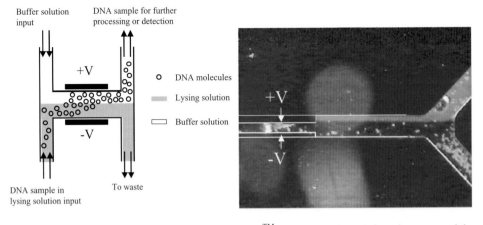

FIGURE 7.19. (a) DNA extractor chip based on the H-filter™. DNA in a lysing solution migrates toward the positive electrode. (b) The light band along the upper half of the device is fluorescence, which indicates the presence of DNA only in the upper output channel [15].

of 0.05% Tween 20, a surfactant to prevent sticking of the DNA to the glass/electrode surface.

The second sample input consists of 25 mM NaCl solution. An external syringe pump was used to provide flow. The infusion flow rate was set at 0.6 ml/hr. An epi-fluorescent microscope with a CCD camera is positioned above the device to view DNA migration. DNA transport is observed when the fluorescence moves from one flow stream to another.

7.4.1.3. DNA Extractor Demonstration Measurements were done to determine what DC voltage would result in electrolysis. For our channel width and exposed electrode area, 1 V was used. Any voltage above 1 V resulted in bubbles which impeded flow for both fluid streams. When the two input solutions are flowed through the device with no electrode voltage, there is sufficient diffusion over the length of the electrode, that at the output end of the electrode, DNA is present throughout the full width of the channel. Consequently, fluorescence is observed in both of the output channels, indicating the presence of DNA flowing to both outputs. When the experiment is repeated with the electrode voltage turned on, fluorescence is only observed in the output channel corresponding to the +V electrode. This is shown quite clearly in Fig. 7.19b, which clearly demonstrates the utility of this device for DNA concentration.

As can be seen in Fig. 7.19b, when the device is operating with voltage on, the DNA is concentrated sufficiently close to the +V electrode so as to be completely obscured by the electrode until it reaches the output. This is perfectly fine, but it makes it difficult to view the DNA in the electrode region. To illustrate the speed of the DNA migration, we first flow the DNA through the channel with the voltage off. Then, while viewing the fluorescence in the channel between the electrodes, the voltage is turned on, and rapid migration occurs toward the +V electrode.

7.4.2. Channel Electrodes for Isoelectric Focusing Combined with Field Flow Fractionation

Isoelectric focusing is an electrophoretic separation based on isoelectric point of proteins. The separation is done in a non-sieving medium (sucrose density gradient, agarose, or polyacrylamide gel) in the presence of carrier ampholytes, which establish a pH gradient increasing from the anode to the cathode during the electrophoreis. As the protein migrates into an acidic region of the gel, it will gain positive charge via protonation of the carboxylic and amino groups. At some point, the overall positive charge will cause the protein to migrate away from the anode (+) to a more basic region of the gel. As the protein enters a more basic environment, it will lose positive charge and gain negative charge, via ammonium and carboxylic acid group deprotonation, and consequently, will migrate away from the cathode (−). Eventually, the protein reaches a position in th pH gradient where its net charge is zero (defined as its pI or isoelectric point). At that point, the electrophoretic mobility is zero and is said to be focused.

Field-flow fractionation (FFF) is an elution chromatographic method for separating, concentrating, and collecting complex macromolecules, colloidal suspensions, emulsions, viruses, bacteria, cells, subcellular components, and surface-modified particles. In EFFF, an electric field, E, is applied across the channel and particles are subjected to the applied field according to their electrophoretic mobility. Fractionation occurs as different particles migrate at different rates in the applied field, reaching different positions in the parabolic flow profile. Continuous separation of analytes performed by free-flow electrophoresis (FFE) where a mixture of charged particles is continuously injected into the carrier stream flowing between two electrode plates. When an electric field is applied, the particles are deflected from the direction of flow according to their electrophoretic mobility or pI.

In a microfluidic transverse isoelectric focusing device [13] two walls of the channels are formed by gold or palladium electrodes. The electrodes were in direct contact with the solution, so that the acid and base generated as a result of water electrolysis; OH^- at the cathode, and H^+ at the anode, can be exploited to form the pH gradient. The partial pressures of oxygen and hydrogen gases also produced by electrolysis could be kept below the threshold of bubble formation by keeping the voltages low, so no venting is required. Both acid-base indicators and protein conjugated with fluorescent dyes with experimentally determined pI values are used to monitor the formation of the pH gradients in the presence of pressure-driven flow. The micro IEF technique is utilized to separate and concentrate subcellular organelles (eg. Nuclei, peroxisomes, and mitochondria) from crude cell lysate [20].

7.5. SUMMARY

There are two types of integrated microfluidic devices for micro total analysis systems (µTAS): passive flow through devices and active programmable devices. The former (passive) devices are designed specifically for certain fixed biological and chemical assays where process steps are sequential in flow-through configurations. One example is described in [2] where mixing, reaction, separation, and self-calibration of immunoassays are performed on a microchip. The advantages of passive devices include reduced need for valves, simplified design, and likely higher manufacturing yield. On the other hand, passive devices are limited to fixed, predetermined assays, and are more difficult to design for multi-analyte detection. Active programmable microfluidic devices are more difficult to design and fabricate since they require integrated microvalves and micropumps. However,

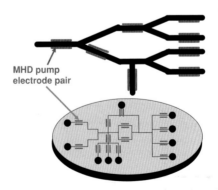

FIGURE 7.20. Illustration of complex fluidic routing on an integrated chip-scale platform using microchannel parallel electrodes enabling truly integrated, programmable "lab-on-a-chip".

active devices such as those implemented by the microchannel parallel electrodes platform are programmable, reconfigurable, and have the potential to become universal modules for biochemical processes.

Applications of MHD microfluidics are abundant. MHD-based microfluidic switching can route samples/reagents into different detection systems. Programmable combinatorial chemistry and biology assays can be implemented by a generic MHD microfluidic control platform. Local pumps can enable the integration of diffusion-based assays [30], flow cytometry, micro titration, or sample extractors. Ultimately the most powerful implementation of MHD-based microfluidics will be a general platform as shown in Fig. 7.20 for the design of new chemical and biological assays with few limits on complexity.

REFERENCES

[1] K. Asami, E. Gheorghiu, and T. Yonezawa. Real-time monitoring of yeast cell division by dielectric spectroscopy. *Biophys. J.*, 76:3345–3348, 1999.

[2] S. Attiya, X.C. Qiu, G. Ocvirk, N. Chiem, W.E. Lee, and D.J. Harrison. Integrated microsystem for sample introduction, mixing, reaction, separation and self calibration. In D.J.H. a. A. v. d. Berg, (ed.) *Micro Total Analysis Systems '98*, Kluwer Academic Publishers, Dordrecht,Banff, Canada, pp. 231–234, 1998.

[3] H.E. Ayliffe, A.B. Frazier, and R.D. Rabitt. Electrical impedance spectroscopy using microchannels with integrated metal electrodes. *IEEE J. MEMS*, 8:50–57, 1999.

[4] R.S. Baker and M.J. Tessier. *Handbook of Electromagnetic Pump Technology*. Elsevier, New York, 1987.

[5] H.H. Bau. A case for magneto-hydrodynamics (MHD). In J.A. Shaw and J. Main, (eds.), *International Mechanical Engineering Congress and Exposition*, American Society of Mechanical Engineers, New York City, 2001.

[6] H.H. Bau, J. Zhong, and M. Yi. A minute magneto hydro dynamic (MHD) mixer. *Sens. Actu. B, Chem.*, 79:207, 2001.

[7] J. Collins and A.P. Lee. Microfluidic flow transducer based on the measurement of electrical admittance. *Lab. Chip.*, 4:7–10, 2004.

[8] D.C. Duffy, J.C. McDonald, O.J.A. Schueller, and G.M. Whitesides. Rapid prototyping of microfluidic systems in poly(dimethylsiloxane), *Anal. Chem.*, 70:4974–4984, 1998.

[9] J.C.T. Eijkel, C. Dalton, C.J. Hayden, J.A. Drysdale, Y.C. Kwok, and A. Manz. Development of a micro system for circular chromatography using wavelet transform detection. In J.M.R. et. al. *Micro Total Analysis Systems 2001*, Kluwer Academic Publishers, Monterey, CA, pp. 541–542, 2001.

[10] A.C. Fisher. *Electrode Dynamics*, Oxford University Press, Oxford, 1996.

[11] M.Y.J. Zhong and H.H. Bau. Magneto hydrodynamic (MHD) pump fabricated with ceramic tapes. *Sens. Actu. A, Phys.*, 96:59–66, 2002.

[12] J. Jang and S.S. Lee. Theoretical and experimental study of MHD (magnetohydrodynamic) micropump. *Sens. Actu. A*, 80:84, 2000.

[13] K. Macounova, C.R. Cabrera, and P. Yager. *Anal. Chem.*, Concentration and separation of proteins in microfluidic channels on the basis of transverse IEF 73:1627–1633, 2001.

[14] K.P.L.A. Christel, W. McMillan, and M.A. Northrup. Rapid, automated nucleic acid probe assays using silicon microstructures for nucleic acid concentration. *J. Biomed. Eng.*, 121:22–27, 1999.

[15] A.V. Lemoff. Flow Driven Microfluidic Actuators for Micro Total Analysis Systems: Magnetohydrodynamic Micropump and Microfluidic Switch, Electrostatic DNA Extractor, Dielectrophoretic DNA Sorter, PhD Thesis, Dept. of Applied Science, UC Davis, 2000.

[16] A.V. Lemoff and A.P. Lee. An AC magnetohydrodynamic micropump. *Sens. Actu. B*, 63:178, 2000.

[17] A.V. Lemoff and A.P. Lee. *Biomed. Microdev.*, 5(1); 55–60, 2003.

[18] V.G. Levich. *Physicochemical Hydrodynamics*, Prentice Hall, Englewood Cliffs, NJ, 1962.

[19] T.F. Lin and J.B. Gilbert. Analyses of magnetohydrodynamic propulsion with seawater for underwater vehicles, *J. Propulsion*, 7:1081–1083, 1991.

[20] H. Lu, S. Gaudet, P.K. Sorger, M.A. Schmidt, and K.F. Jensen. Micro isoelectric free flow separation of subcellular materials. In M.A. Northrup, K.F. Jensen and D.J. Harrison, (eds.), *7th International Conference on Miniaturized Chemical and Biochemical Analysis Systems Squaw Valley*, California, USA, pp. 915–918, 2003.

[21] S.K. Mohanty, S.K. Ravula, K. Engisch, and A.B. Frazier. Single cell analysis of bovine chromaffin cells using micro electrical impedance spectroscopy. In Y. Baba, A. Berg, S. Shoji, (eds.), *6th International Conference on Miniaturized Chemical and Biochemical Analysis Systems, Micro Total Analysis Systems*, Nara, Japan, November 2002, Dordrecht, Kluwer Academic, 2002.

[22] W.J. Moore. *Phys. Chem.*, Prentice-Hall, 1972.

[23] O.M. Phillips. The prospects for magnetohydrodynamic ship propulsion. *J. Ship Res.*, 43:43–51, 1962.

[24] Y. Polevaya, I. Ermolina, M. Schlesinger, B.Z. Ginzburg, and Y. Feldman. Time domain dielectric spectroscopy study of human cells: II. Normal and malignant white blood cells. *Biochim. Biophys. Acta*, 15:257–271, 1999.

[25] J.I. Ramos and N.S. Winowich. Magnetohydrodynamic channel flow study. *Phys. Fluids*, 29:992–997, 1986.

[26] R.B.M. Schasfoort, R. Lüttge, and A. van den Berg. Magneto-hydrodynamically (MHD) directed flow in microfluidic networks. In J.M. Ramsey. *Micro Total Analysis Systems 2001*. Kluwer Academic Publishers, Monterey, CA, pp. 577–578, 2001.

[27] L.L. Sohn, O.A. Saleh, G.R. Facer, A.J. Beavis, R.S. Allan, and D.A. Notterman. *PNAS*, 97:10687–10690, 2000.

[28] K.Y. Tam, J.P. Larsen, B.A. Coles, and R.G. Compton. *J. Electroanal. Chem.*, 407:23–35, 1996.

[29] S. Way and C. Delvin. *AIAA*, 67:432, 1967.

[30] B.H. Weigl and P. Yager. *Science*, 15:346–347, 1999.

[31] F.M. White. *Fluid Mech.*, McGraw-Hill, 1979.

[32] Y. Xia, E. Kim, and G.M. Whitesides. *Chem. Mater.*, 8:1558–1567, 1996.

8

Dielectrophoretic Traps for Cell Manipulation

Joel Voldman

Department of Electrical Engineering, Room 36-824, Massachusetts Institute of Technology Cambridge, MA 02139

8.1. INTRODUCTION

One of the goals of biology for the next fifty years is to understand how cells work. This fundamentally requires a diverse set of approaches for performing measurements on cells in order to extract information from them. Manipulating the physical location and organization of cells or other biologically important particles is an important part in this endeavor. Apart from the fact that cell function is tied to their three-dimensional organization, one would like ways to grab onto and position cells. This lets us build up controlled multicellular aggregates, investigate the mechanical properties of cells, the binding properties of their surface proteins, and additionally provides a way to move cells around. In short, it provides physical access to cells that our fingers cannot grasp.

Many techniques exist to physically manipulate cells, including optical tweezers [78], acoustic forces [94], surface modification [52], etc. Electrical forces, and in particular dielectrophoresis (DEP), are an increasingly common modality for enacting these manipulations. Although DEP has been used successfully for many years to separate different cell types (see reviews in [20, 38]), in this chapter I focus on the use of DEP as "electrical tweezers" for manipulating individual cells. In this implementation DEP forces are used to trap or spatially confine cells, and thus the chapter will focus on creating such traps using these forces. While it is quite easy to generate forces on cells with DEP, it is another thing altogether to obtain predetermined quantitative performance. The goal for this chapter is to help others develop an approach to designing these types of systems. The focus will be on trapping cells—which at times are generalized to "particles"—and specifically mammalian cells, since these are more fragile than yeast or bacteria and thus are in some ways more challenging to work with.

I will start with a short discussion on what trapping entails and then focus on the forces relevant in these systems. Then I will discuss the constraints when working with cells, such as temperature rise and electric field exposure. The last two sections will describe existing trapping structures as well as different approaches taken to measure the performance of those structures. The hope is that this overview will give an appreciation for the forces in these systems, what are the relevant design issues, what existing structures exist, and how one might go about validating a design. I will not discuss the myriad other uses of dielectrophoresis; these are adequately covered in other texts [39, 45, 60] and reviews.

8.2. TRAPPING PHYSICS

8.2.1. Fundamentals of Trap Design

The process of positioning and physically manipulating particles—cells in this case—is a trapping process. A trap uses a set of confining forces to hold a particle against a set of destabilizing forces. In this review, the predominant confining force will be dielectrophoresis, while the predominant destabilizing forces will be fluid drag and gravity. The fundamental requirement for any deterministic trap is that it creates a region where the net force on the particle is zero. Additionally, the particle must be at a stable zero, in that the particle must do work on the force field in order to move from that zero [3]. This is all codified in the requirement that $\mathbf{F}_{net} = 0$, $\mathbf{F}_{net} \cdot d\mathbf{r} < 0$ at the trapping point, where $\mathbf{F}_{net}$ is the net force and $d\mathbf{r}$ is an increment in any direction.

The design goal is in general to create a particle trap that meets specific requirements. These requirements might take the form of a desired trap strength or maximum flowrate that trapped particles can withstand, perhaps to meet an overall system throughput specification. For instance, one may require a minimum flowrate to replenish the nutrients around trapped cells, and thus a minimum flowrate against which the cells must be trapped. When dealing with biological cells, temperature and electric-field constraints are necessary to prevent adverse effects on cells. Other constraints might be on minimum chamber height or width—to prevent particle clogging—or maximum chamber dimensions—to allow for proximate optical access. In short, *predictive quantitative* trap design. Under the desired operating conditions, the trap must create a stable zero, and the design thus reduces to ensuring that stable zeros exist under the operating conditions, and additionally determining under what conditions those stable zeros disappear (i.e., the trap releases the particle).

Occasionally, it is possible to analytically determine the conditions for stable trapping. When the electric fields are analytically tractable and there is enough symmetry in the problem to make it one-dimensional, this can be the best approach. For example, one can derive an analytical expression balancing gravity against an exponentially decaying electric field, as is done for field-flow fractionation [37]. In general, however, the fields and forces are too complicated spatially for this approach to work. In these cases, one can numerically calculate the fields and forces everywhere in space and find the net force ($\mathbf{F}_{net}$) at each point, then find the zeros.

A slightly simpler approach exists when the relevant forces are conservative. In this case one can define scalar potential energy functions U whose gradient gives each force (i.e., $\mathbf{F} = -\nabla U$). The process of determining whether a trap is successfully confining the particle then reduces to determining whether any spatial minima exist within the trap. This approach is nice because energy is a scalar function and thus easy to manipulate by hand and on the computer.

In general, a potential energy approach will have limited applicability because dissipation is usually present. In this case, the energy in the system depends on the specifics of the particle motion—one cannot find a U that will uniquely define $\mathbf{F}$. In systems with liquid flow, for example, an energy-based design strategy cannot be used because viscous fluid flow is dissipative. In this case, one must use the vector force-fields and find stable zeroes.

In our lab, most modeling incorporates a range of approaches spanning analytical, numerical, and finite-element modeling. In general, we find it most expedient to perform finite-element modeling only when absolutely necessary, and spend most of the design combining those results with analytical results in a mixed-numerical framework run on a program such as Matlab© (Mathworks, Natick, MA). Luckily, one can run one or two finite-element simulations and then use simple scaling laws to scale the resulting data appropriately. For instance, the linearity of Laplace's equation means that after solving for the electric fields at one voltage, the results can be linearly scaled to other voltages. Thus, FEA only has to be repeated when the geometry scales, if at all.

To find the trapping point (and whether it exists), we use MATLAB to compute the three isosurfaces where each component of the net force (F_x, F_y, F_z) is zero. This process is shown in Figure 8.1 for a planar quadrupole electrode structure. Each isosurface—the three-dimensional analog of a contour line—shows where in space a single component of the force is zero. The intersection of all three isosurfaces thus represents points where all three force components, and thus the net force, is zero. In the example shown in Figure 8.1,B–D, increasing the flowrate changes the intersection point of the isosurfaces, until at some threshold flowrate (Figure 8.1D), the three isosurfaces cease to intersect, and the particle is no longer held; the strength of the trap has been exceeded [83]. In this fashion we can determine the operating characteristics (e.g., what voltage is needed to hold a particular cell against a particular flow) and then whether those characteristics meet the system requirements (exposure of cells to electric fields, for instance).

A few caveats must be stated regarding this modeling approach. First, the problem as formulated is one of determining under what conditions an already trapped particle will remain trapped; I have said nothing about how to get particles in traps. Luckily this is not a tremendous extension. Particle inertia is usually insignificant in microfluidic systems, meaning that particles will follow the streamlines of the force field. Thus, with numerical representations of the net force, one can determine, given a starting point, where that particle will end up. Matlab in fact has several commands to do this (e.g., streamline). By placing test particles in different initial spots, it is possible to determine the region from within which particles will be drawn to the trap.

Another implicit assumption is that only one particle will be in any trap, and thus that particle-particle interactions do not have to be dealt with. In actuality, designing a trap that will only hold one particle is quite challenging. To properly model this, one must account for the force perturbations created when the first particle is trapped; the second particle sees a force field modified by the first particle. While multiple-particle modeling is still largely

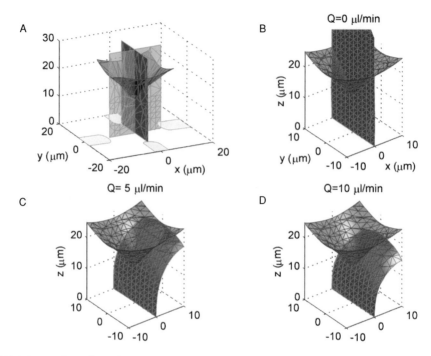

FIGURE 8.1. Surfaces of zero force describe a trap. (**A**) Shown are the locations of planar quadrupole electrodes along with the three isosurfaces where one component of the force on a particle is zero. The net force on the particle is zero where the three surfaces intersect. (**B–D**) As flow increases from left to right, the intersection point moves. The third isosurface is not shown, though it is a vertical sheet perpendicular to the $F_x = 0$ isosurface. (**D**) At some critical flow rate, the three isosurfaces no longer intersect and the particle is no longer trapped.

unresolved, the single-particle approach presented here is quite useful because one can, by manipulating experimental conditions, create conditions favorable for single-particle trapping, where the current analysis holds.

Finally, we have constrained ourselves to deterministic particle trapping. While appropriate for biological cells, this assumption starts to break down as the particle size decreases past ∼1μm because Brownian motion makes trapping a probabilistic event. Luckily, as nanoparticle manipulation has become more prevalent, theory and modeling approaches have been determined. The interested reader is referred to the monographs by Morgan and Green [60] and Hughes [39].

8.2.2. Dielectrophoresis

The confining force that creates the traps is dielectrophoresis. Dielectrophoresis (DEP) refers to the action of a body in a non-uniform electric field when the body and the surrounding medium have different polarizabilities. DEP is easiest illustrated with reference to Figure 8.2. On the left side of Figure 8.2, a charged body and a neutral body (with different permittivity than the medium) are placed in a uniform electric field. The charged body feels a force, but the neutral body, while experiencing an induced dipole, does not feel a net

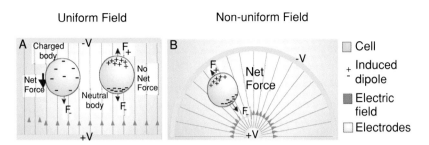

FIGURE 8.2. Dielectrophoresis. The left panel (A) shows the behavior of particles in uniform electric fields, while the right panel shows the net force experienced in a non-uniform electric field (B).

force. This is because each half of the induced dipole feels opposite and equal forces, which cancel. On the right side of Figure 8.2, this same body is placed in a non-uniform electric field. Now the two halves of the induced dipole experience a different force magnitude and thus a net force is produced. This is the dielectrophoretic force.

The force in Figure 8.2, where an induced dipole is acted on by a non-uniform electric field, is given by [45]

$$\mathbf{F}_{dep} = 2\pi \varepsilon_m R^3 \text{Re}[\underline{CM}(\omega)] \cdot \nabla |\mathbf{E}(\mathbf{r})|^2 \tag{8.1}$$

where ε_m is the permittivity of the medium surrounding the particle, R is the radius of the particle, ω is the radian frequency of the applied field, $\mathbf{r}$ refers to the vector spatial coordinate, and $\mathbf{E}$ is the applied vector electric field. The Clausius-Mossotti factor ($\underline{CM}$)—CM factor— gives the frequency (ω) dependence of the force, and its sign determines whether the particle experiences positive or negative DEP. Importantly, the above relation is limited to instances where the field is spatially invariant, in contrast to traveling-wave DEP or electrorotation (see [39, 45]).

Depending on the relative polarizabilities of the particle and the medium, the body will feel a force that propels it toward field maxima (termed positive DEP or p-DEP) or field minima (negative DEP or n-DEP). In addition, the direction of the force is independent of the polarity of the applied voltage; switching the polarity of the voltage does not change the direction of the force—it is still toward the field maximum in Figure 8.2. Thus DEP works equally well with both DC and AC fields.

DEP should be contrasted with electrophoresis, where one manipulates charged particles with electric fields [30], as there are several important differences. First, DEP does not require the particle to be charged in order to manipulate it; the particle must only differ electrically from the medium that it is in. Second, DEP works with AC fields, whereas no net electrophoretic movement occurs in such a field. Thus, with DEP one can use AC excitation to avoid problems such as electrode polarization effects [74] and electrolysis at electrodes. Even more importantly, the use of AC fields reduces membrane charging of biological cells, as explained below. Third, electrophoretic systems cannot create stable non-contact traps, as opposed to DEP—one needs electromagnetic fields to trap charges (electrophoresis can, though, be used to trap charges at electrodes [63]). Finally, DEP forces increase with the square of the electric field (described below), whereas electrophoretic forces increase linearly with the electric field.

This is not to say that electrophoresis is without applicability. It is excellent for transporting charged particles across large distances, which is difficult with DEP (though traveling-wave versions exist [17]). Second, many molecules are charged and are thus movable with this technique. Third, when coupled with electroosmosis, electrophoresis makes a powerful separation system, and has been used to great effect [30].

8.2.2.1. The Clausius-Mossotti Factor The properties of the particle and medium within which it resides are captured in the form of the Clausius-Mossotti factor (*CM*)—CM factor. The Clausius-Mossotti factor arises naturally during the course of solving Laplace's equation and matching the boundary conditions for the electric field at the surface of the particle (for example, see [45]). For a homogeneous spherical particle, the CM factor is given by

$$CM = \frac{\underline{\varepsilon}_p - \underline{\varepsilon}_m}{\underline{\varepsilon}_p + 2\underline{\varepsilon}_m} \quad (8.2)$$

where $\underline{\varepsilon}_m$ and $\underline{\varepsilon}_p$ are the complex permittivities of the medium and the particle, respectively, and are each given by $\underline{\varepsilon} = \varepsilon + \sigma/(j\omega)$, where ε is the permittivity of the medium or particle, σ is the conductivity of the medium or particle, and j is $\sqrt{-1}$.

Many properties lie within this simple relation. First, one sees that competition between the medium ($\underline{\varepsilon}_m$) and particle ($\underline{\varepsilon}_p$) polarizabilities will determine the sign of CM factor, which will in turn determine the sign—and thus direction—of the DEP force. For instance, for purely dielectric particles in a non-conducting liquid ($\sigma_p = \sigma_m = 0$), the CM factor is purely real and will be positive if the particle has a higher permittivity than the medium, and negative otherwise.

Second, the real part of the CM factor can only vary between +1 ($\underline{\varepsilon}_p \gg \underline{\varepsilon}_m$, e.g., the particle is much *more* polarizable than the medium) and −0.5 ($\underline{\varepsilon}_p \ll \underline{\varepsilon}_m$, e.g., the particle is much *less* polarizable than the medium). Thus n-DEP can only be half as strong as p-DEP. Third, by taking the appropriate limits, one finds that at low frequency the CM factor (Eqn. (8.2)) reduces to

$$CM_{\omega \to 0} = \frac{\sigma_p - \sigma_m}{\sigma_p + 2\sigma_m} \quad (8.3)$$

while at high frequency it is

$$CM_{\omega \to \infty} = \frac{\varepsilon_p - \varepsilon_m}{\varepsilon_p + 2\varepsilon_m} \quad (8.4)$$

Thus, similar to many electroquasistatic systems, the CM factor will be dominated by relative permittivities at high frequency and conductivities at low frequencies; the induced dipole varies between a free charge dipole and a polarization dipole. The relaxation time separating the two regimes is

$$\tau_{MW} = \frac{\varepsilon_p + 2\varepsilon_m}{\sigma_p + 2\sigma_m} \quad (8.5)$$

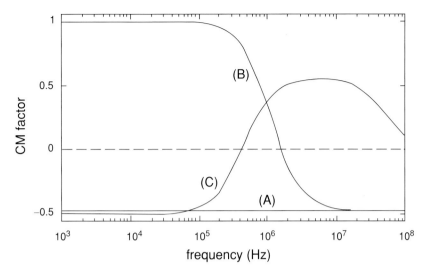

FIGURE 8.3. CM factor for three situations. (**A**) A non-conducting uniform sphere with $\varepsilon_p = 2.4$ in non-conducting water ($\varepsilon_m = 80$). The water is much more polarizable than the sphere, and thus the CM factor is ~ -0.5. (**B**) The same sphere, but with a conductivity $\sigma_p = 0.01$ S/m in non-conducting water. Now there is one dispersion—at low frequencies the bead is much more conducting than the water & hence there is p-DEP, while at high frequencies the situation is as in (A). (**C**) A spherical shell (approximating a mammalian cell), with ($\varepsilon_{cyto} = 75$, $c_m = 1$ μF/cm^2, $\sigma_{cyto} = 0.5$ S/m, $g_m = 5$ mS/cm^2) in a 0.1 S/m salt solution, calculated using results from [45]. Now there are two interfaces and thus two dispersions. Depending on the frequency, the shell can experience n-DEP or p-DEP.

and is denoted τ_{MW} to indicate that the physical origin is Maxwell-Wagner interfacial polarization [73].

This Maxwell-Wagner interfacial polarization causes the frequency variations in the CM factor. It is due to the competition between the charging processes in the particle and medium, resulting in charge buildup at the particle/medium interface. If the particle and medium are both non-conducting, then there is no charge buildup and the CM factor will be constant with no frequency dependence (Figure 8.3A). Adding conductivity to the system results in a frequency dispersion in the CM factor due to the differing rates of interfacial polarization at the sphere surface (Figure 8.3B).

While the uniform sphere model is a good approximation for plastic microspheres, it is possible to extend this expression to deal with more complicated particles such as biological cells, including non-spherical ones.

Multi-Shelled Particulate Models Because we are interested in creating traps that use DEP to manipulate cells, we need to understand the forces on cells in these systems. Luckily, the differences between a uniform sphere and a spherical cell can be completely encompassed in the CM factor; the task is to create an electrical model of the cell and then solve Laplace's equation to derive its CM factor (a good review of electrical properties of cells is found in Markx and Davey [57]). The process is straightforward, though tedious, and has been covered in detail elsewhere [39, 43, 45]. Essentially, one starts by adding a thin shell to the uniform sphere and matches boundary conditions at now two interfaces, deriving a CM factor very similar to Eqn (8.2) but with an effective complex permittivity $\underline{\varepsilon}'_p$ that subsumes the effects of the complicated interior (see §5.3 of Hughes [39]). This process can be repeated multiple times to model general multi-shelled particles.

Membrane-Covered Spheres: Mammalian Cells, Protoplasts Adding a thin shell to a uniform sphere makes a decent electrical model for mammalian cells and protoplasts. The thin membrane represents the insulating cell membrane while the sphere represents the cytoplasm. The nucleus is not modeled is this approximation. For this model the effective complex permittivity can be represented by:

$$\underline{\varepsilon}'_p = \frac{\underline{c}_m R \cdot \underline{\varepsilon}_{cyto}}{\underline{c}_m R + \underline{\varepsilon}_{cyto}} \qquad (8.6)$$

where $\underline{\varepsilon}_{cyto}$ is the complex permittivity of the cytoplasmic compartment and $\underline{c}_m$ refers to complex membrane capacitance per unit area and is given by

$$\underline{c}_m = c_m + g_m/(j\omega) \qquad (8.7)$$

where c_m and g_m are the membrane capacitance and conductance per unit area (F/m^2 and S/m^2) and can be related to the membrane permittivity and conductivity by $c_m = \varepsilon_m/t$ and $g_m = \sigma_m/t$, where t is the membrane thickness. The membrane conductance of intact cells is often small and can be neglected. Because cell membranes are comprised of phospholipid bilayers whose thickness and permittivity varies little across cell types, the membrane capacitance per unit area is fairly fixed at $c_m \sim 0.5 - 1 \mu F/cm^2$ [64].

Plotting a typical CM factor for a mammalian cell shows that it is more complicated than for a uniform sphere. Specifically, since it has two interfaces, there are two dispersions in its CM factor, as shown in Figure 8.3C. In low-conductivity buffers, the cell will experience a region of p-DEP, while in saline or cell-culture media the cells will only experience n-DEP. This last point has profound implications for trap design. If one wishes to use cells in physiological buffers, one is restricted to n-DEP excitation, irrespective of applied frequency. Only by moving low-conductivity solutions can one create p-DEP forces in cells. While, as we discuss below, p-DEP traps are often easier to implement, one must then deal with possible artifacts due to the artificial media.

One challenge for the designer in applying different models for the CM factor is getting accurate values for the different layers. In Table 8.1 we list properties culled from the literature for several types of particles, along with the appropriate literature references. Care must be taken in applying these, as some of the properties may be dependent on the cell type, cell physiology, and suspending medium, as well as limited by the method in which they were measured. Besides the values listed below, there are also values on Jurkat cells [67] and other white blood cells [21].

Sphere with Two Shells: Bacteria and Yeast Bacteria and yeast have a cell wall in addition to a cell membrane. Iterating on the multi-shell model can be used to derive a CM factor these types of particles [35, 76, 95]. Griffith *et al.* also used a double-shell model, this time to include the nucleus of a mammalian cells, in this case the human neutrophil [29].

TABLE 8.1. Parameters for the electrical models of different cells and for saline.

Particle type	Radius (μm)	Inner compartment		Membrane			Wall		
		ε	σ (S/m)	ε	σ (S/m)	thickness (nm)	ε	σ(S/m)	thickess (nm)
Latex microspheres	nm–μm	2.5	2e-4	—	—	—	—	—	—
Yeast [96, 97]	4.8	60	0.2	6	250e-9	8	60	0.014	~200
E. coli [76]	1	60	0.1	10	50e-9	5	60	0.5	20
HSV-1 virus [40]	0.25	70	8e-3	10	$\sigma_p = 3.5$ nS	—	—	—	—
HL-60 [37]	6.25	75	0.75	1.6 μF/cm²	0.22 S/cm²	1	—	—	—
PBS	—	78–80	1.5	—	—	—	—	—	—

Surface Conduction: Virus and Other Nanoparticles Models for smaller particles must also accommodate surface currents around the perimeter of the particle. As particles get smaller, this current path becomes more important and affects the CM factor (by affect the boundary conditions when solving Laplace's equation). In this case, the conductivity of the particle can be approximated by [39]

$$\sigma_p + \frac{2K_s}{R} \qquad (8.8)$$

where K_s represents the surface conductivity (in Siemens). One sees that this augments the bulk conductivity of the particle (σ_p) with a surface-conductance term inversely proportional to the particle radius.

Non-Spherical Cells Many cells are not spherical, such as some bacteria (e.g., E. coli) and red blood cells. The CM factor can be extended to include these effects by introducing a depolarizing factor, described in detail in Jones' text [45].

8.2.2.2. Multipolar Effects The force expression given in Eqn (8.1) is the most commonly used expression for the DEP force applied to biological particles, and indeed accurately captures most relevant physics. However, it is not strictly complete, in that the force calculated using that expression assumes that only a dipole is induced in the particle. In fact, arbitrary multipoles can be induced in the particle, depending on the spatial variation of the field that it is immersed in. Specifically, the dipole approximation will become invalid when the field non-uniformities become great enough to induce significant higher-order multipoles in the particle. This can easily happen in microfabricated electrode arrays, where the size of the particle can become equal to characteristic field dimensions. In addition, in some electrode geometries there exists field nulls. Since the induced dipole is proportional to the electric field, the dipole approximation to the DEP force is zero there. Thus at least the quadrupole moment must be taken into account to correctly model the DEP forces in such configurations.

In the mid-90's Jones and Washizu extended their very successful effective-moment approach to calculate all the induced moments and the resultant forces on them [50, 51, 91, 92]. Gascoyne's group, meanwhile, used an approach involving the Maxwell's stress tensor to arrive at the same result [87]. Thus, it is now possible to calculate the DEP forces in arbitrarily polarized non-uniform electric fields. A compact tensor representation of the final result in is

$$\mathbf{F}_{dep}^{(n)} = \frac{\overline{\overline{\overline{p}}}^{(n)}[\cdot]^n(\nabla)^n \mathbf{E}}{n!} \tag{8.9}$$

where n refers to the force order ($n = 1$ is the dipole, $n = 2$ is the quadropole, etc.), $\overline{\overline{\overline{p}}}^{(n)}$ is the multipolar induced-moment tensor, and $[\cdot]^n$ and $(\nabla)^n$ represent n dot products and gradient operations. Thus one sees that the n-th force order is given by the interaction of the n-th-order multipolar moment with the n-th gradient of the electric field. For $n = 1$ the result reverts to the force on a dipole (Eqn (8.1)).

A more explicit version of this expression for the time-averaged force in the i-th direction (for sinusoidal excitation) is

$$\langle F_i^{(1)} \rangle = 2\pi \varepsilon_m R^3 Re \left[\underline{CM}^{(1)} \underline{E}_m \frac{\partial}{\partial x_m} \underline{E}_i^* \right]$$

$$\langle F_i^{(2)} \rangle = \frac{2}{3}\pi \varepsilon_m R^5 Re \left[\underline{CM}^{(2)} \frac{\partial}{\partial x_m} \underline{E}_n \frac{\partial^2}{\partial x_n \partial x_m} \underline{E}_i^* \right] \tag{8.10}$$

$$\vdots$$

for the dipole ($n = 1$) and quadrupole ($n = 2$) force orders [51]. The Einstein summation convention has been applied in Eqn. (8.10). While this approach may seem much more difficult to calculate than Eqn (8.1), compact algorithms have been developed for calculating arbitrary multiples [83]. The multipolar CM factor for a uniform lossy dielectric sphere is given by

$$\underline{CM}^{(n)} = \frac{\underline{\varepsilon}_p - \underline{\varepsilon}_m}{n\underline{\varepsilon}_p + (n+1)\underline{\varepsilon}_m} \tag{8.11}$$

It has the same form as the dipolar CM factor (Eqn. (2)) but has smaller limits. The quadrupolar CM factor ($n = 2$), for example, can only vary between $+1/2$ and $-1/3$.

8.2.2.3. Scaling Although the force on a dipole in a non-uniform field has been recognized for decades, the advent of microfabrication has really served as the launching point for DEP-based systems. With the force now defined, I will now investigate why downscaling has enabled these systems.

Most importantly, reducing the characteristic size of the system reduces the applied voltage needed to generate a given field gradient, and for a fixed voltage increases that field gradient. A recent article on scaling in DEP-based systems [46] illustrates many of the relevant scaling laws. Introducing the length scale L into Eqn (8.1) and appropriately approximating derivatives, one gets that the DEP force (dipole term) scales as

$$F_{dep} \sim R^3 \frac{V^2}{L^3} \tag{8.12}$$

illustrating the dependency. This scaling law has two enabling implications. First, generating the forces required to manipulate micron-sized bioparticles (∼pN) requires either large voltages (100's–1000's V) for macroscopic systems (1–100 cm) or small voltages (1–10 V) for microscopic systems (1–100 µm). Large voltages are extremely impractical to generate at the frequencies required to avoid electrochemical effects (kHz–MHz). Slew rate limitations in existing instrumentation make it extremely difficult to generate more than 10 V_{pp} at frequencies above 1 MHz. Once voltages are decreased, however, one approaches the specifications of commercial single-chip video amplifiers, commodity products that can be purchased for a few dollars.

The other strong argument for scaling down is temperature. Biological systems can only withstand certain temperature excursions before their function is altered. Electric fields in conducting liquids will dissipate power, heating the liquid. Although even pure water has a finite conductivity (∼5 µS/m), the problem is more acute as the conductivity of the water increases. For example, electrolytes typically used to culture cells are extremely conductive (∼1 S/m). While the exact steady-state temperature rise is determined by the details of electrode geometry and operating characteristics, the temperature rise, as demonstrated by Jones [46], scales as

$$\Delta T \sim \sigma \cdot V^2 \cdot L^3 \qquad (8.13)$$

where σ is the conductivity of the medium. It is extremely difficult to limit these rises by using convective heat transfer (e.g., flowing the media at a high rate); in these microsystems conduction is the dominant heat-transfer mechanism unless the flowrate is dramatically increased. Thus, one sees the strong ($\sim L^3$) argument for scaling down; it can enable operation in physiological buffers without significant concomitant temperature rises.

Temperature rise has other consequences besides directly affecting cell physiology. The non-uniform temperature distribution creates gradients in the electrical properties of the medium (because permittivity and conductivity are temperature-dependent). These gradients in turn lead to free charge in the system, which, when acted upon by the electric field, drag fluid and create (usually) destabilizing fluid flows. These electrothermal effects are covered in §2.3.3.

Thus, creating large forces is limited by either the voltages that one can generate or the temperature rises (and gradients) that one creates, and is always enhanced by decreasing the characteristic length of the system. All of these factors point to microfabrication as an enabling fabrication technology for DEP-based systems.

8.2.3. Other Forces

DEP interacts with other forces to produce a particle trap. The forces can either be destabilizing (e.g., fluid drag, gravity) or stabilizing (e.g., gravity).

8.2.3.1. Gravity The magnitude of the gravitational force is given by

$$F_{grav} = \frac{4}{3}\pi R^3 (\rho_p - \rho_m) g \qquad (8.14)$$

where ρ_m and ρ_p refer to the densities of the medium and the particle, respectively, and g is the gravitational acceleration constant. Cells and beads are denser than the aqueous media and thus have a net downward force.

8.2.3.2. Hydrodynamic Drag Forces Fluid flow past an object creates a drag force on that object. In most systems, this drag force is the predominant destabilizing force. The fluid flow can be intentional, such as that created by pumping liquid past a trap, or unintentional, such as electrothermal flows.

The universal scaling parameter in fluid flow is the Reynolds number, which gives an indication of the relative strengths of inertial forces to viscous forces in the fluid. At the small length scales found in microfluidics, viscosity dominates and liquid flow is laminar. A further approximation assumes that inertia is negligible, simplifying the Navier-Stokes equations even further into a linear form. This flow regime is called creeping flow or Stokes flow and is the common approximation taken for liquid microfluidic flows.

In creeping flow, a sphere in a uniform flow field will experience a drag force—called the Stokes' drag—with magnitude

$$F_{drag} = 6\pi \eta R v \qquad (8.15)$$

where η is the viscosity of the liquid and v is the far-field relative velocity of the liquid with respect to the sphere. As an example, a 1-μm-diameter particle in a 1-mm/s water flow will experience ∼10 pN of drag force.

Unfortunately, it is difficult to create a uniform flow field, and thus one must broaden the drag force expression to include typically encountered flows. The most common flow pattern in microfluidics is the flow in a rectangular channel. When the channel is much wider than it is high, this flow can be approximated as the one-dimensional flow between to parallel plates, or plane Poiseuille flow. This flow profile is characterized by a parabolic velocity distribution where the centerline velocity is 1.5× the average linear flow velocity

$$v(z) = 1.5 \frac{Q}{wh} \left(1 - \left(\frac{z - h/2}{h/2} \right)^2 \right) \qquad (8.16)$$

where Q is the volume flowrate, w and h are the width and height of the channel, respectively, and z is the height above the substrate at which the velocity is evaluated. The expression in Eqn (8.15) can then be refined by using the fluid velocity at the height of the particle center.

Close to the channel wall ($z \ll h$) the quadratic term in Eqn (8.16) can be linearly approximated, resulting a velocity profile known as plane shear or plane Couette flow

$$v(z) = 1.5 \frac{Q}{wh} \left(4 \frac{z}{h} \right) = 6 \frac{Q}{wh} \frac{z}{h} \qquad (8.17)$$

The error between the two flow profiles increases linearly with z for $z \ll h/2$; the error when $z = 0.1 \cdot h$ is ∼10%.

Using Eqn (8.15) with the modified fluid velocities is a sufficient approximation for the drag force in many applications, and is especially useful in non-analytical flow profiles

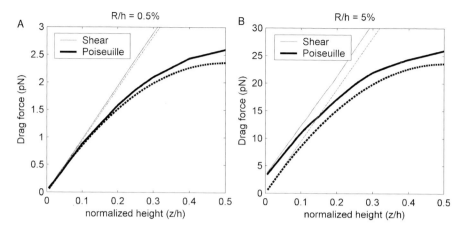

FIGURE 8.4. Drag force using different approximations for a particle that is 0.5% (**A**) and 5% (**B**) of the chamber height. For the smaller particle (A), all approaches give the same result near the surface. For larger particles (B), the exact formulations (—) give better results than approximate approaches (- - -).

derived by numerical modeling. In that case one can compute the Stokes' drag at each point by multiplying Eqn. (8.15) with the computed 3-D velocity field. To get a more exact result, especially for particles that are near walls, one can turn to solved examples in the fluid mechanics literature. Of special interest to trapping particles, the drag force on both a stationary and moving sphere near a wall in both plane Poiseuille [19] and shear flow [26] has been solved. The calculated drag forces have the same form as Eqn (8.15) but include a non-dimensional multiplying factor that accounts for the presence of the wall.

In Figure 8.4 I compare drag forces on 1 μm and 10 μm-diameter spheres using the different formulations. In both cases, the channel height is fixed at 100 μm. One sees two very different behaviors. When the sphere size is small compared to the channel height ($R = 0.5\%$ of h), all four formulations give similar results near the chamber wall (Figure 8.4A), with the anticipated divergence of the shear and Poiseuille drag profiles away from the wall. However, as the sphere becomes larger compared to the chamber height (Figure 8.4B), the different formulations diverge. Both the shear and parabolic profiles calculated using a single approach converge to identical values at the wall, but the two approaches yield distinctly different results. In this regime the drag force calculated using Eqn. (8.15) consistently underestimates the drag force, in this case by about 2 pN. This has a profound effect near the wall, where the actual drag force is 50% higher than that estimated by the simple approximation. Thus, for small particles ($R \ll h$) away from walls ($z \gg R$), the simple approximation is fine to within better than 10%, while in other cases one should use the exact formulations.

While spheres approximate most unattached mammalian cells as well as yeast and many bacteria, other cells (e.g., *E. coli*, erythrocytes) are aspherical. For these particles, drag forces have the same form as Eqn. (8.15) except that term $6\pi\eta R$ is replaced by different "friction" factors, nicely catalogued by Morgan and Green [60].

8.2.3.3. Electrothermal Forces The spatially non-uniform temperature distribution created by the power dissipated by the electric field can lead to flows induced by electrothermal effects. These effects are covered in great detail by Morgan and Green [60]. Briefly, because the medium permittivity and conductivity are functions of temperature, temperature gradients directly lead to gradients in ε and σ. These gradients in turn generate free charge which can be acted upon by an electric field to move and drag fluid along with it, creating fluid flow. This fluid flow creates a drag force on an immersed body just as it does for conventional Stokes' drag (Eqn. (8.15)). In general, derivations of the electrothermal force density, the resulting liquid flow, and the drag require numerical modeling because the details of the geometry profoundly impact the results. Castellanos *et al.* have derived solutions for one simple geometry, and have used it to great effect to derive some scaling laws [6].

8.3. DESIGN FOR USE WITH CELLS

Since dielectrophoretic cell manipulation exposes cells to strong electric fields, one needs to know how these electric fields might affect cell physiology. Ideally, one would like to determine the conditions under which the trapping will not affect the cells and use those conditions to constrain the design. Of course, cells are poorly understood complex systems and thus it is impossible to know for certain that one is not perturbing the cell. However, all biological manipulations—cell culture, microscopy, flow cytometer, etc.—alter cell physiology. What's most important is to minimize known influences on cell behavior and then use proper controls to account for the unknown influences. In short, good experimental design.

The known influences of electric fields on cells can be split into the effects due to current flow, which causes heating, and direct interactions of the fields with the cell. We'll consider each of these in turn.

8.3.1. Current-Induced Heating

Electric fields in a conductive medium will cause power dissipation in the form of Joule heating. The induced temperature changes can have many effects on cell physiology. As mentioned previously, microscale DEP is advantageous in that it minimizes temperature rises due to dissipated power. However, because cells can be very sensitive to temperature changes, it is not assured that any temperature rises will be inconsequential.

Temperature is a potent affecter of cell physiology [4, 11, 55, 75]. Very high temperatures ($>4\,°C$ above physiological) are known to lead to rapid mammalian cell death, and research has focused on determining how to use such knowledge to selectively kill cancer cells [81]. Less-extreme temperature excursions also have physiological effects, possibly due to the exponential temperature dependence of kinetic processes in the cell [93]. One well-studied response is the induction of the heat-shock proteins [4, 5]. These proteins are molecular chaperones, one of their roles being to prevent other proteins from denaturing when under environmental stresses.

While it is still unclear as to the minimum temperature excursion needed to induce responses in the cell, one must try to minimize any such excursions. A common rule of thumb for mammalian cells is to keep variations to $<1\,°C$, which is the approximate daily variation in body temperature [93]. The best way we have found to do this is to numerically solve for the steady-state temperature rise in the system due to the local heat sources given by σE^2. Convection and radiation can usually be ignored, and thus the problem reduces to

8.3.2. Direct Electric-Field Interactions

Electric fields can also directly affect the cells. The simple membrane-covered sphere model for mammalian cells can be used to determine where the fields exist in the cell as the frequency is varied. From this one can determine likely pathways by which the fields could impact physiology [31, 73]. Performing the analysis indicates that the imposed fields can exist across the cell membrane or the cytoplasm. A qualitative electrical model of the cell views the membrane as a parallel RC circuit connected in-between RC pairs for the cytoplasm and the media. At low frequencies (<MHz) the circuit looks like three resistors in series and because the membrane resistance is large the voltage is primarily dropped across it. This voltage is distinct from the endogenous transmembrane potential that exists in the cell. Rather, it represents the voltage derived from the externally applied field. The total potential difference across the cell membrane would be given by the sum of the imposed and endogenous potentials. At higher frequencies the impedance of the membrane capacitor decreases sufficiently that the voltage across the membrane starts to decrease. Finally, at very high frequencies (100's MHz) the model looks like three capacitors in series and the membrane voltage saturates.

Quantitatively, the imposed transmembrane voltage can be derived as [73]

$$|V_{tm}| = \frac{1.5|\mathbf{E}|R}{\sqrt{1+(\omega\tau)^2}} \tag{8.18}$$

where ω is the radian frequency of the applied field and τ is the time constant given by

$$\tau = \frac{Rc_m(\rho_{cyto} + 1/2\rho_{med})}{1 + Rg_m(\rho_{cyto} + 1/2\rho_{med})} \tag{8.19}$$

where ρ_{cyto} and ρ_{med} med are the cytoplasmic and medium resistivities (Ω-m). At low frequencies $|V_m|$ is constant at $1.5|\mathbf{E}|R$ but decreases above the characteristic frequency $(1/\tau)$. This model does not take into account the high-frequency saturation of the voltage, when the equivalent circuit is a capacitive divider.

At the frequencies used in DEP—10's kHz to 10's MHz—the most probably route of interaction between the electric fields and the cell is at the membrane [79]. There are several reasons for this. First, electric fields already exist at the cell membrane, leading to transmembrane voltages in the 10's of millivolts. Changes in these voltages could affect voltage-sensitive proteins, such as voltage-gated ion channels [7]. Second, the electric field across the membrane is greatly amplified over that in solution. From Eqn. (6.18) one gets that at low frequencies

$$|V_{tm}| = \frac{1.5|\mathbf{E}|R}{\sqrt{1+(\omega\tau)^2}} \approx 1.5|\mathbf{E}|R$$
$$|E_{tm}| \approx |V_m|/t = (1.5R/t) \cdot |\mathbf{E}| \tag{8.20}$$

and thus at the membrane the imposed field is multiplied by a factor of 1.5 R/t (~1000), which can lead to quite large membrane fields (E_{tm}). This does not preclude effects due to cytoplasmic electric fields. However, these effects have not been as intensively studied, perhaps because 1) those fields will induce current flow and thus heating, which is not a direct interaction, 2) the fields are not localized to an area (e.g., the membrane) that is likely to have field-dependent proteins, and 3) unlike the membrane fields, the cytoplasmic fields are not amplified.

Several studies have investigated possible direct links between electric fields and cells. At low frequencies, much investigation has focused on 60-Hz electromagnetic fields and their possible effects, although the studies thus far are inconclusive [54]. DC fields have also been investigated, and have been shown to affect cell growth [44] as well as reorganization of membrane components [68]. At high frequencies, research has focused on the biological effects of microwave radiation, again inconclusively [65].

In the frequency ranges involved in DEP, there has been much less research. Tsong has provided evidence that some membrane-bound ATPases respond to fields in the kHz–MHz range, providing at least one avenue for interaction [79]. Electroporation and electrofusion are other obvious, although more violent, electric field-membrane coupling mechanisms [98].

Still other research has been concerned specifically with the effects of DEP on cells, and has investigated several different indicators of cell physiology to try to elucidate any effects. One of the first studies was by the Fuhr et al., who investigated viability, anchorage time, motility, cell growth rates, and lag times after subjecting L929 and 3T3 fibroblast cells in saline to short and long (up to 3 days) exposure to 30–60 kV/m fields at 10–40 MHz near planar quadrupoles [16]. They estimated that the transmembrane load was <20 mV. The fields had no discernable effect.

Another study investigated changes in cell growth rate, glucose uptake, lactate and monoclonal antibody production in CHO & HFN 7.1 cells on top of interdigitated electrodes excited at 10 MHz with ~10^5 V/m in DMEM (for the HFN 7.1 cells) or serum-free medium (for the CHO cells) [12]. Under these conditions they observed no differences in the measured properties between the cells and control populations.

Glasser and Fuhr attempted to differentiate between heating and electric-field effects on L929 mouse fibroblast cells in RPMI to the fields from planar quadrupoles [24]. They imposed ~40 kV/m fields of between 100 kHZ and 15 MHz for 3 days and observed monolayers of cells near the electrodes with a video microscopy setup, similar to their previous study [16]. They indirectly determined that fields of ~40 kV/m caused an ~2 °C temperature increase in the cells, but did not affect cell-division rates. They found that as they increased field frequency (from 500 kHZ to 15 MHz) the maximum tolerable field strength (before cell-division rates were altered) increased. This is consistent with a decrease in the transmembrane load with increasing frequency.

Wang et al. studied DS19 murine erythroleukemia cells exposed to fields (~10^5 V/m) of 1 kHz–10 MHz in low-conductivity solutions for up to 40-min [90]. They found no effects due to fields above 10 kHz. They determined that hydrogen peroxide produced by reactions at the electrode interfaces for 1 kHz fields caused changes in cell growth lag phase, and that removal of the peroxide restored normal cell growth.

On the p-DEP side, Archer et al. subjected fibroblast-like BHK 21 C13 cells to p-DEP forces produced by planar electrodes arranged in a sawtooth configuration [1]. They used low-conductivity (10 mS/m) isoosmotic solutions and applied fields of $\sim 10^5$ V/m at 5 MHz. They monitored cell morphology, cell doubling time, oxidative respiration (mitochondrial stress assay), alterations in expression of the immediate-early protein fos, and non-specific gene transcription directly after a 15 minute exposure and after a 30-min time delay. They observed 20–30% upregulation of fos expression and a upregulation of a few unknown genes (determined via mRNA analysis). Measured steady-state temperatures near the cells were <1 °C above normal, and their calculated transmembrane voltage under their conditions was <100 µV, which should be easily tolerable. The mechanism—thermal or electrical—of the increased gene expression was left unclear. It is possible that artifacts from p-DEP attraction of the cells to the electrodes led to observed changes. Either way, this study certainly demonstrates the possibility that DEP forces could affect cell physiology.

Finally, Gray *et al.* exposed bovine endothelial cells in sucrose media (with serum) to different voltages—and thus fields—for 30-min in order to trap them and allow them to adhere to their substrates. They measured viability and growth of the trapped cells and found that cell behavior was the same as controls for the small voltages but that large voltages caused significant cell death [27]. This study thus demonstrates the p-DEP operation in artificial media *under the proper conditions* does not grossly affect cell physiology.

In summary, studies specifically interested in the effects of kHz–MHz electroquasistatic fields on cells thus far demonstrate that choosing conditions under which the transmembrane loads and cell heating are small—e.g., >MHz frequencies, and fields in ~ 10's kV/m range—can obviate any gross effects. Subtler effects, such as upregulation of certain genetic pathways or activation of membrane-bound components could still occur, and thus DEP, as with any other assay technique, must be used with care.

8.4. TRAP GEOMETRIES

The electric field, which creates the DEP force, is in turn created by electrodes. In this section I will examine some of the electrode structures used in this field and their applicability to trapping cells and other microparticles. The reader is also encouraged to read the relevant chapters in Hughes' [39] and Morgan and Green's texts [60], which contain descriptions of some field geometries.

One can create traps using either p-DEP or n-DEP. Using n-DEP a zero-force point is created away from electrodes at a field minimum and the particle is trapped by pushing at it from all sides. In p-DEP the zero-force point is at a field maximum, typically at the electrode surface or at field constrictions. Both approaches have distinct advantages and disadvantages, as outlined in Table 8.2. For each application, the designer must balance these to select the best approach.

8.4.1. n-DEP Trap Geometries

Although an infinite variety of electrode geometries can be created, the majority of research has focused on those that are easily modeled or easily created.

TABLE 8.2. Comparison of advantages and disadvantages of p-DEP and n-DEP approaches to trapping cells.

p-DEP	n-DEP
Must use low conductivity *artificial* media (−)	Can use saline or other high-salt buffers (+)
CM factor can go to +1 (+)	CM factor can go to −0.5 (−)
Less heating (+)	More heating (−)
Typically easier to trap by pulling (+)	Typically harder to trap by pushing (−)
Traps usually get stronger as V increases (+)	Traps often do not get stronger with increasing V (−)
Cells stick to or can be damaged by electrodes (−)	Cells are physically removed from electrodes (+)
Cells go to maximum electric field (−)	Cells go to minimum electric field (+)

8.4.1.1. Interdigitated Electrodes Numerous approximate and exact analytical solutions exist for the interdigitated electrode geometry (Figure 8.5A), using techniques as varied as conformal mapping [23, 82], Green's function [10, 86], and Fourier series [33, 61]. Recently, an elegant exact closed-form solution was derived [8]. Numerical solutions are also plentiful [28].

While the interdigitated electrode geometry has found much use in DEP separations, it does not make a good trap for a few reasons. First, the long extent of the electrodes in one direction creates an essentially 2-D field geometry and thus no trapping is possible along the length of the electrodes. Further, the spatial variations in the electric field—which create the DEP force—decrease exponentially away from the electrode surface. After about one electrode's worth of distance away from the susbtrate, the field is mostly uniform at a given height, and thus DEP trapping against fluid flows or other perpendicular forces cannot occur. Increasing the field to attempt to circumvent this only pushes the particle farther away from the electrodes, a self-defeating strategy; like the planar quadrupole [83], this trap is actually strongest at lower voltages, when the particle is on the substrate.

8.4.1.2. Quadrupole Electrodes Quadrupole electrodes are four electrodes with alternating voltage polarities applied to every other electrode (Figure 8.5B). The field for four point charges can be easily calculated by superposition, but relating the charge to voltage (via the capacitance) is difficult in general and must be done numerically.

Planar quadrupoles can create rudimentary particle traps (Figure 8.5B), and can trap single particles down to 100's of nm [40]. Using n-DEP, they provide in-plane particle confinement, and can provide three-dimensional confinement if the particle is denser than the suspending medium. As with interdigitated electrodes, however, these traps suffer from the drawback that increasing the field only pushes the particle farther out of the trap and does not necessarily increase confinement. We showed this in 2001 with measurements of the strength of these traps [83]. Unexpectedly, the traps are strongest at an intermediate voltage, just before the particle is about to be levitated (Figure 8.6).

A variant of the quadrupole electrodes is the polynomial electrode geometry (Figure 8.5C), introduced by Huang and Pethig in 1991 [36]. By placing the electrode edges at the equipotentials of the applied field, it is possible to analytically specify the field between the electrodes. One caveat of this approach is that it solves the 2-D Laplace equation, which is not strictly correct for the actual 3-D geometry; thus, the electric field is at best only truly specified right at the electrode surface, and not in all of space.

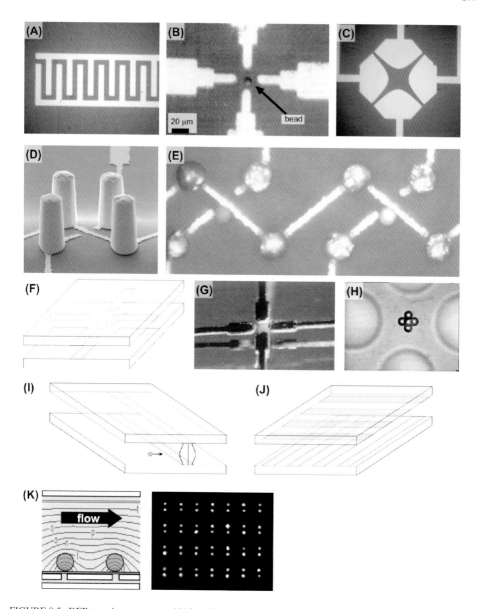

FIGURE 8.5. DEP trapping structures. (**A**) Interdigitated electrodes. (**B**) A planar quadrupole, showing a bead in the center. (**C**) Quadrupolar polynomial electrodes. (**D**) A 3-D view of an extruded quadrupole trap, showing the four gold post electrode electrodes and the gold wiring on the substrate. (**E**) A top-down image of two extruded quadrupole traps showing living trapped HL-60 cells in liquid. (**F–H**) Schematic (F), stereo image (G), and top-down view (H) of the oppose ocotpole, showing beads trapped at the center. (**I**) Schematic of the strip electrodes, showing the non-uniform electric field between them that creates an n-DEP force wall to incoming particles. (**J**) Schematic of the crossed-electrode p-DEP structure of Suehiro and Pethig [77]. (**K**) Side view schematic of Gray *et al.*'s p-DEP trap, showing the bottom point electrodes and the top plate, along with a top-down image of endothelial cells positioned at an array of traps.

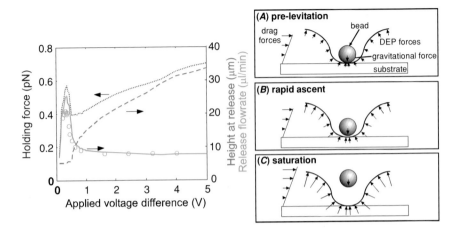

FIGURE 8.6. Behavior of planar quadrupole trap at different voltages, showing the measured (o) and simulated (—) release flowrate, the holding force (···), and the height of the particle when it is released (- - -). (A) Pre-levitation. At very low voltages, the z-directed DEP force cannot overcome the gravitation force, and the bead is not levitated; (B) Rapid ascent. At a certain voltage the bead will just become levitated and the holding characteristics will peak; (C) Saturation. At high voltages, the increase in holding force is balanced by the increased particle levitation height, resulting in a flat release flowrate profile.

One way to avoid this behavior is to extend the electrodes into the third-dimension, creating extruded quadrupole traps (Figure 8.5D–E, [84, 85]). These traps, while much more difficult to make, are orders of magnitude stronger than the planar quadrupole traps, and can successfully hold single cells against significant liquid flows. These electrode geometries are sufficiently complicated that only numerical simulation can derive the correct field solution.

8.4.1.3. Octopole Electrodes Another way to increase the strength of quadrupole electrode traps is to put another quadrupole on the chamber ceiling to provide further particle confinement (Figure 8.5F–H). These opposed octopole traps are significantly stronger than planar quadrupoles, and are routinely used for single-particle trapping [69, 71]. They are much simpler to fabricate than the extruded quadrupoles, but are more complex to align and package.

8.4.1.4. Strip Electrodes Strip electrodes are simply two electrodes opposed from one-another, with one on the substrate and one on the chamber ceiling (Figure 8.5I). Introduced by Fiedler et al. in 1998, these have been used to create n-DEP "barriers" to herd particles [14]. The solution to this geometry has been analytically solved using conformal mapping [72]. As with the interdigitated electrodes, strip electrodes are of limited use for particle trapping because they only provide one dimension of confinement.

8.4.1.5. Other Electrode Structures Several other microscale trapping structures have been introduced. Some, like the castellated electrodes [22, 59] or round electrodes [34], which have been successfully used for particle separation, are not well-suited for trapping particles because of their planar format; they suffer the same drawbacks as the interdigitated and planar quadrupoles.

Recently, a team in Europe has been developing an active n-DEP-based trapping array [56]. Essentially, their device consists of a two-dimensional array of square electrodes and a conductive lid. The key is that incorporating CMOS logic (analog switches and memory) allows each square electrode to be connected to in-phase or out-of-phase AC voltage in a programmable fashion. By putting a center square at +V and the surrounding squares at −V, they can create an in-plane trap. Further putting the chamber top at +V closes the cage, giving 3-D confinement. The incorporation of CMOS further means that very few leads are required to control an indefinite number of sites, creating a readily scalable technology. Using this trap geometry, they have successfully manipulated both beads and cells, although moving cells from one site to another is currently quite slow (∼sec).

8.4.2. p-DEP Trap Geometries

p-DEP traps, while easier to create, have seen less use, probably because the required low-conductivity media can perturb cell physiology (at least for mammalian cells) and because of concerns about electrode-cell interactions. As stated earlier, obtaining p-DEP with mammalian cells requires low-conductivity buffer, and this can create biological artifacts in the system. Nonetheless, several geometries do exist.

An early p-DEP-based trapping system was described by Suehiro and Pethig (Figure 8.5J, [77]). This used a set of parallel individually addressable electrodes on one substrate and another set of electrodes on the bottom substrate that were rotated 90°. By actuating one electrode on top and bottom, they could create a localized field maximum that could be moved around, allowing cell manipulation.

Another example is a concentric ring levitator that uses feedback-controlled p-DEP to actually trap particles away from electrodes [66]. In an air environment, they can levitate drops of water containing cells by pulling up against gravity with an upper electrode, feeding back the vertical position of the droplet to maintain a constant height.

Recently Gray *et al.* created a geometry consisting of a uniform top plate and electrode points on the substrate to create the field concentrations (Figure 8.5K, [27]). They were able to pattern cells onto the stubs using p-DEP. Importantly, experiments showed that the low-conductivity buffer did not affect the gross physiology of the cells at reasonable voltages. Finally, Chou *et al.* used geometric constrictions in an insulator to create field maxima in a conductivity-dominated system [9]. These maxima were used to trap DNA.

8.4.3. Lessons for DEP Trap Design

The preceding discussion raises some important points for DEP trap design. First, the choice of whether to trap via p-DEP or n-DEP is a system-level partitioning problem. For instance, if one absolutely requires use in saline, then n-DEP must be used. If, however, minimizing temperature rises is most important, then p-DEP may be better, as the low-conductivity media will reduce temperature rises. The decision may also be affected by fabrication facilities, etc.

In general, p-DEP traps are easier to create than n-DEP traps, because it is easier to hold onto a particle by attracting it than repelling it. The tradeoff is that p-DEP requires artificial media for use with mammalian cells. Nonetheless, the key for effective p-DEP is the creation of isolated field maxima. Because the particles are pulled into the field, p-DEP traps always trap stronger at higher voltages.

Creating effective n-DEP traps is more difficult, and requires some sort of three-dimensional confinement. This is difficult (though not impossible) to do with planar electrode structures, because the +z-component of the DEP force scales with voltage just as much as the in-plane components. This fundamentally pushes the particle away from the trap when one increases the voltage, drastically limiting trap strength. Any planar electrode structure, including the planar quadrupoles and interdigitated electrodes described above fail this test and therefore make a poor n-DEP trap. The two extant structures that exhibit strong trapping create three-dimensional trapping by removing the net +z-directed DEP force. Both the extruded quadrupole and opposed octopole structures do this by creating a structure that cancels out z-directed DEP forces at the trap center, enabling one to increase voltage—and thus trap strength—without pushing the particle farther away.

8.5. QUANTITATING TRAP CHARACTERISTICS

In order to assess whether a quantitative design is successful, one needs some quantitative validation of the fields and forces in DEP traps. Given that the complete DEP theory is known and that the properties of at least some particles are known, it should be possible to quantitate trap parameters. Those that are of interest include trap strength, field strength, and the spatial extents of the trap.

Measuring traps requires a quantitative readout. This typically takes the form of a test particle (or particles), whose location or motion can be measured and then matched against some prediction. Quantitative matching gives confidence in the validity of a particular modeling technique, thus allowing predictive design of new traps.

Starting in the 1970's, Tom Jones and colleagues explored DEP levitation in macroscopic electrode systems [47–49, 53]. Using both stable n-DEP traps and p-DEP traps with feedback control, they could measure levitation heights of different particles under various conditions. Knowledge of the gravitation force on the particle could then be used to as a probe of the equally opposing DEP forces at equilibrium.

Levitation measurements have continued to the present day, but now applied to microfabricated electrode structures, such as levitation height measurements of beads in planar quadrupoles [15, 25, 32], or on top of interdigitated electrodes [37, 58]. In all these measurements, errors arise because of the finite depth of focus of the microscope objective and because it is difficult to consistently focus on the center of the particle. The boundary between levitation and the particle sitting on the ground is a "sharp" event and is usually easier to measure and correlate to predictions than absolute particle height [25].

Wonderful pioneering work in quantitating the shapes of the fields was reported by the group in Germany in the 1992 and 1993 when they introduced their planar quadrupole [15] and opposed octopole [69] trap geometries. In the latter paper, the authors trapped 10's of beads that were much smaller than the trap size. The beads packed themselves to minimize their overall energy, in the process creating surfaces that reflected the force distribution in the trap. By comparing the experimental and predicted surfaces, they could validate their modeling.

An early velocity-measurement approach was described X.-B. Wang *et al.*, who used spiral electrodes and measured radial velocity and levitation height of breast cancer cells as they varied frequency, particle radius, and medium conductivity [89]. They then matched the data to DEP theory, using fitting parameters to account for unknown material properties, and obtained good agreement. These researchers performed similar analyses using erythroleukemia cells in interdigitated electrode geometries, again obtaining good fits of the data to the theory [88].

Another approach that compares drag force to DEP force is described by Tsukahara *et al.*, where they measured the velocity as a particle moved toward or away from the minimum in a planar quadrupole polynomial electrode [80]. If the electric field and particle properties are known, it should be possible to relate the measured velocity to predictions, although, as described earlier, the use of Stokes drag introduces errors when the particle is near the wall and the forces they calculated for their polynomial electrodes are only valid at the electrode symmetry plane. This was reflected in the use of a fitting parameter to match predictions with experiment, although in principle absolute prediction should be possible.

The German team that initially introduced the idea of opposed electrodes on both the bottom and top of the chamber have continued their explorations into this geometry with great success. They have attempted to quantify the strength of their traps in two different ways. In the first approach, they measure the maximum flowrate against which a trap can hold a particle. Because of the symmetry of their traps, the particles are always along the midline of the flow, and by approximating the drag force on the particle with the Stokes drag (Eqn (8.15)) they can measure the strength of the trap in piconewtons [13, 62, 72]. Because they can calculate the electric fields and thus DEP forces, they have even been able to absolutely correlate predictions to experiment [72]. With such measurements they have determined that their opposed electrode devices can generate $\sim$20pN of force on 14.9-μm diameter beads [62].

The other approach that these researchers have taken to measuring trap strength is to combine DEP octopole traps with optical tweezers [2]. If the strength of one of the trapping techniques is known then it can be used to calibrate the other. In one approach, this was done by using optical tweezers to displace a bead from equilibrium in a DEP trap, then measuring the voltage needed to make that bead move back to center [18]. They used this approach to measure the strength of the optical tweezers by determining the DEP force on the particle at that position at the escape voltage. In principle, one could use this to calibrate the trap if the optical tweezer force constant was known.

In the other approach, at a given voltage and optical power, they measured the maximum that the bead could be displaced from the DEP minimum before springing back [70]. This is very similar to the prior approach, although it also allows one to generate a force-displacement characteristic for the DEP trap, mapping out the potential energy well.

A clever and conceptually similar approach was tried by Hughes and Morgan with a planar quadrupole [41], although in this case the unknown was the thrust exerted by *E. coli* bacteria. By measuring the maximum point that the bacteria could be displaced from the DEP trap minimum, they could back out the bacterial thrust if the DEP force characteristic in the trap was known. They achieved good agreement between predictions and modeling, at least at lower voltages.

For much smaller particles, where statistics are important, Chou *et al.* captured DNA in electrodeless p-DEP traps. They used the spatial distribution of the bacteria to measure the strength of the traps [9]. They measured the width of the fluorescence intensity distribution of labeled DNA in the trap, and assuming that the fluorescence intensity was linearly related to concentration, could extract the force of the trap by equating the "Brownian" diffusive force to the DEP force. The only unknown in this approach, besides the assumptions of linearity, was the temperature, which could easily be measured.

In our lab we have been interested in novel trap geometries to enable novel trapping functionalities. One significant aim has been to create DEP traps for single cells that are strong enough to hold against significant liquid flows, such that cells and reagents can be transported on and off the chips within reasonable time periods (~min). Our approach to measuring trap strength is similar to the one described above, where the fluid velocity necessary to break through a barrier is correlated to a barrier force [13, 62, 72]. This approach is also similar to those undertaken by the optical tweezer community, who calibrate their tweezers by measuring the escape velocity of trapped particles at various laser powers.

We have chosen to generalize this approach to allow for particles that may be near surfaces where Stokes drag is not strictly correct, where multipolar DEP forces may be important, and where electrode geometries may be complex [83]. In our initial validation of this approach, we were able to make absolute prediction of trap strength, as measured by the minimum volumetric flowrate needed for the particle to escape the trap. This volumetric flowrate can be related to a linear flowrate and then to a drag force using the analytical solutions for the drag on a stationary particle near a wall.

Our validation explained the non-intuitive trapping behavior of planar quadrupole traps (Figure 8.6), giving absolute agreement—to within 30%—between modeling and experiment with no fitting parameters [83]. We then extended this modeling to design a new, high-force trap created from extruded electrodes that could hold 13.2-μm beads with 95 pN of force at 2 V, and HL-60 cells with ~60 pN of force at the same voltage [84, 85]. Again, we could make absolute predictions and verify them with experiments. We continue to extend this approach to design traps for different applications.

8.6. CONCLUSIONS

In conclusion, DEP traps, when properly confined, can be used to confine cells, acting as electrical tweezers. In this fashion cells can be positioned and manipulated in ways not achievable using other techniques, due to the dynamic nature of electric fields and the ability to shape the electrodes that create them.

Achieving a useful DEP system for manipulating cells requires an understanding of the forces present in these systems and an ability to model their interactions so as to predict the operating system conditions and whether they are compatible with cell health, etc. I have presented one approach to achieving these goals that employs quantitative modeling of these systems, along with examples of others who have sought to quantitate the performance of their systems.

8.7. ACKNOWLEDGEMENTS

The author wishes to thank Tom Jones for useful discussions and Thomas Schnelle for the some of the images in Figure 8.4. The author also wishes to acknowledge support from NIH, NSF, Draper Laboratories, and MIT for this work.

REFERENCES

[1] S. Archer, T.T. Li, A.T. Evans, S.T. Britland, and H. Morgan. Cell reactions to dielectrophoretic manipulation. *Biochem. Biophys. Res. Commun.*, 257:687–698, 1999.
[2] A. Ashkin. Optical trapping and manipulation of neutral particles using lasers. *Proc. Natl. Acad. Sci. U.S.A.*, 94:4853–4860, 1997.
[3] A. Blake. *Handbook of Mechanics, Materials, and Structures*. Wiley, New York, pp. 710, 1985.
[4] R.H. Burdon. Heat-shock and the heat-shock proteins. *Biochem. J.*, 240:313–324, 1986.
[5] S.W. Carper, J.J. Duffy, and E.W. Gerner. Heat-shock proteins in thermotolerance and other cellular processes. *Cancer Res.*, 47:5249–5255, 1987.
[6] A. Castellanos, A. Ramos, A. Gonzalez, F. Morgan, and N. Green. AC Electric-Field-Induced Fluid Flow in Microelectrode Structures: Scaling Laws. Presented at Proceedings of 14th International Conference on Dielectric Liquids, Graz, Austria, 7–12 July 2002.
[7] W.A. Catterall. Structure and function of voltage-gated ion channels. *Ann. Rev. Biochem.*, 64:493–531, 1995.
[8] D.E. Chang, S. Loire, and I. Mezic. Closed-form solutions in the electrical field analysis for dielectrophoretic and travelling wave inter-digitated electrode arrays. *J. Phys. D: Appl. Phys.*, 36:3073–3078, 2003.
[9] C.F. Chou, J.O. Tegenfeldt, O. Bakajin, S.S. Chan, E.C. Cox, N. Darnton, T. Duke, and R.H. Austin. Electrodeless dielectrophoresis of single- and double-stranded DNA. *Biophys. J.*, 83:2170–2179, 2002.
[10] D.S. Clague and E.K. Wheeler. Dielectrophoretic manipulation of macromolecules: The electric field. *Phys. Rev. E*, 64:026605–8, 2001.
[11] E.A. Craig. The heat shock response. *CRC Crit. Rev. Biochem.*, 18:239–280, 1985.
[12] A. Docoslis, N. Kalogerakis, and L.A. Behie. Dielectrophoretic forces can be safely used to retain viable cells in perfusion cultures of animal cells. *Cytotechnology*, 30:133–142, 1999.
[13] M. Durr, J. Kentsch, T. Muller, T. Schnelle, and M. Stelzle. Microdevices for manipulation and accumulation of micro- and nanoparticles by dielectrophoresis. *Electrophoresis*, 24:722–731, 2003.
[14] S. Fiedler, S.G. Shirley, T. Schnelle, and G. Fuhr. Dielectrophoretic sorting of particles and cells in a microsystem. *Anal. Chem.*, 70:1909–1915, 1998.
[15] G. Fuhr, W.M. Arnold, R. Hagedorn, T. Muller, W. Benecke, B. Wagner, and U. Zimmermann. Levitation, holding, and rotation of cells within traps made by high-frequency fields. *Biochimi. Et Biophys. Acta*, 1108:215–223, 1992a.
[16] G. Fuhr, H. Glasser, T. Muller, and T. Schnelle. Cell manipulation and cultivation under AC electric-field influence in highly conductive culture media. *Biochimi. Et Biophys. Acta-Gen. Sub.*, 1201:353–360, 1994.
[17] G. Fuhr, R. Hagedorn, T. Muller, W. Benecke, and B.Wagner. Microfabricated electrohydrodynamic (EHD) pumps for liquids of higher conductivity. *J. Microelectromech. Sys.*, 1:141–146, 1992b.
[18] G. Fuhr, T. Schnelle, T. Muller, H. Hitzler, S. Monajembashi, and K.O. Greulich. Force measurements of optical tweezers in electro-optical cages. *Appl. Phys. A-Mater. Sci. Process.*, 67:385–390, 1998.
[19] P. Ganatos, R. Pfeffer, and S. Weinbaum. A strong interaction theory for the creeping motion of a sphere between plane parallel boundaries. Part 2. Parallel motion. *J. Fluid Mech.*, 99:755–783, 1980.
[20] P.R.C. Gascoyne and J. Vykoukal. Particle separation by dielectrophoresis. *Electrophoresis*, 23:1973–1983, 2002.
[21] P.R.C. Gascoyne, X.-B. Wang, Y. Huang, and F.F. Becker. Dielectrophoretic separation of cancer cells from blood. *IEEE Trans. Ind. Appl.*, 33:670–678, 1997.
[22] P.R.C. Gascoyne, H. Ying, R. Pethig, J. Vykoukal, and F.F. Becker. Dielectrophoretic separation of mammalian cells studied by computerized image analysis. *Measure. Sci. Technol.*, 3:439–445, 1992.
[23] W.J. Gibbs. *Conformal Transformations in Electrical Engineering*. Chapman & Hall, London, pp. 219, 1958.
[24] H. Glasser and G. Fuhr. Cultivation of cells under strong ac-electric field - differentiation between heating and trans-membrane potential effects. *Bioelectrochem. Bioenerget.*, 47:301–310, 1998.
[25] H. Glasser, T. Schnelle, T. Muller, and G. Fuhr. Electric field calibration in micro-electrode chambers by temperature measurements. *Thermochimi. Acta*, 333:183–190, 1999.
[26] A.J. Goldman, R.G. Cox, and H. Brenner. Slow viscous motion of a sphere parallel to a plane wall - II Couette flow. *Chem. Eng. Sci.*, 22:653–660, 1967.
[27] D.S. Gray, J.L. Tan, J. Voldman, and C.S. Chen. Dielectrophoretic registration of living cells to a microelectrode array. *Biosens. Bioelectron.*, 19:1765–1774, 2004.

[28] N.G. Green, A. Ramos, and H. Morgan. Numerical solution of the dielectrophoretic and travelling wave forces for interdigitated electrode arrays using the finite element method. *J. Electrostat.*, 56:235–254, 2002.
[29] A.W. Griffith and J.M. Cooper. Single-cell measurements of human neutrophil activation using electrorotation. *Anal. Chem.*, 70:2607–2612, 1998.
[30] A.J. Grodzinsky and M.L. Yarmush. Electrokinetic separations. In G. Stephanopoulus (ed.), *Bioprocessing*, Weinheim, Germany; New York, VCH, pp. 680–693, 1991.
[31] C. Grosse and H.P. Schwan. Cellular membrane potentials induced by alternating fields. *Biophys. J.*, 63:1632–1642, 1992.
[32] L.F. Hartley, K. Kaler, and R. Paul. Quadrupole levitation of microscopic dielectric particles. *J. Electrostat.*, 46:233–246, 1999.
[33] Z. He. Potential distribution within semiconductor detectors using coplanar electrodes. *Nuclear Instruments & Methods in Physics Research, Section A (Accelerators, Spectrometers, Detectors and Associated Equipment)* 365:572–575, 1995.
[34] Y. Huang, K.L. Ewalt, M. Tirado, T.R. Haigis, A. Forster, D. Ackley, M.J. Heller, J.P. O'Connell, and M. Krihak. Electric manipulation of bioparticles and macromolecules on microfabricated electrodes. *Anal. Chem.*, 73:1549–1559, 2001.
[35] Y. Huang, R. Holzel, R. Pethig, and X.-B. Wang. Differences in the AC electrodynamics of viable and non-viable yeast cells determined through combined dielectrophoresis and electrorotation studies. *Phys. Med. Biol.*, 37:1499–1517, 1992.
[36] Y. Huang and R. Pethig. Electrode design for negative dielectrophoresis. *Measure. Sci. Technol.*, 2:1142–1146, 1991.
[37] Y. Huang, X.-B. Wang, F.F. Becker, and P.R.C. Gascoyne. Introducing dielectrophoresis as a new force field for field-flow fractionation. *Biophys. J.*, 73:1118–1129, 1997.
[38] M.P. Hughes. Strategies for dielectrophoretic separation in laboratory-on-a-chip systems. *Electrophoresis*, 23:2569–2582, 2002.
[39] M.P. Hughes. *Nanoelectromechanics in Engineering and Biology*, CRC Press, Boca Raton, Fla., pp. 322, 2003.
[40] M.P. Hughes and H. Morgan. Dielectrophoretic trapping of single sub-micrometre scale bioparticles. *J. Phys. D-Appl. Phys.*, 31:2205–2210, 1998.
[41] M.P. Hughes and H. Morgan. Measurement of bacterial flagellar thrust by negative dielectrophoresis. *Biotechnol. Prog.*, 15:245–249, 1999.
[42] M.P. Hughes, H. Morgan, F.J. Rixon, J.P. Burt, and R. Pethig. Manipulation of herpes simplex virus type 1 by dielectrophoresis. *Biochim. Biophys. Acta*, 1425:119–126, 1998.
[43] A. Irimajiri, T. Hanai, and A. Inouye. A dielectric theory of "multi-stratified shell" model with its application to a lymphoma cell. *J. Theoret. Biol.*, 78:251–269, 1979.
[44] L.F. Jaffe and M.M. Poo. Neurites grow faster towards the cathode than the anode in a steady field. *J. Exp. Zool.*, 209:115–128, 1979.
[45] T.B. Jones. *Electromechanics of Particles*. Cambridge University Press, Cambridge, pp. 265, 1995.
[46] T.B. Jones. Influence of scale on electrostatic forces and torques in AC particulate electrokinetics. *IEEE Proc.-Nanobiotechnnol.*, 150:39–46, 2003.
[47] T.B. Jones and G.W. Bliss. Bubble dielectrophoresis. *J. Appl. Phys.*, 48:1412–1417, 1977.
[48] T.B. Jones, G.A. Kallio, and C.O. Collins. Dielectrophoretic levitation of spheres and shells. *J. Electrostat.*, 6:207–224, 1979.
[49] T.B. Jones and J.P. Kraybill. Active feedback-controlled dielectrophoretic levitation. *J. Appl. Phys.*, 60:1247–1252, 1986.
[50] T.B. Jones and M. Washizu. Equilibria and dynamics of DEP-levitated particles: Multipolar theory. *J. Electrostat.*, 33:199–212, 1994.
[51] T.B. Jones and M. Washizu. . Multipolar dielectrophoretic and electrorotation theory. *J. Electrostat.*, 37:121–134, 1996.
[52] D.R. Jung, R. Kapur, T. Adams, K.A. Giuliano, M. Mrksich, H.G. Craighead, and D.L. Taylor. Topographical and physicochemical modification of material surface to enable patterning of living cells. *Crit. Rev. Biotechnol.*, 21:111–154, 2001.
[53] K.V.I.S. Kaler and T.B. Jones. Dielectrophoretic spectra of single cells determined by feedback-controlled levitation. *Biophys. J.*, 57:173–182, 1990.

[54] A. Lacy-Hulbert, J.C. Metcalfe, and R. Hesketh. Biological responses to electromagnetic fields. *FASEB J.*, 12:395–420, 1998.
[55] S. Lindquist. The heat-shock response. *Annu. Rev. Biochem.*, 55:1151–1191, 1986.
[56] N. Manaresi, A. Romani, G. Medoro, L. Altomare, A. Leonardi, M. Tartagni, and R. Guerrieri. A CMOS chip for individual cell manipulation and detection. *IEEE J. Solid-State Circ.*, 38:2297–2305, 2003.
[57] G.H. Markx and C.L. Davey. The dielectric properties of biological cells at radiofrequencies: Applications in biotechnology. *Enzyme Microb. Technol.*, 25:161–171, 1999.
[58] G.H. Markx, R. Pethig, and J. Rousselet. The dielectrophoretic levitation of latex beads, with reference to field-flow fractionation. *J. Phys. D (Applied Physics)*, 30:2470–2477, 1997.
[59] G.H. Markx, M.S. Talary, and R. Pethig. Separation of viable and non-viable yeast using dielectrophoresis. *J. Biotechnol.*, 32:29–37, 1994.
[60] H. Morgan and N.G. Green. *AC Electrokinetics: Colloids and Nanoparticles*. Research Studies Press, Baldock, Hertfordshire, England, 2003.
[61] H. Morgan, A.G. Izquierdo, D. Bakewell, N.G. Green, and A. Ramos. The dielectrophoretic and travelling wave forces generated by interdigitated electrode arrays: Analytical solution using Fourier series. *J. Phys. D-Appl. Phys.*, 34:1553–1561, 2001.
[62] T. Muller, G. Gradl, S. Howitz, S. Shirley, T. Schnelle, and G. Fuhr. A 3-D microelectrode system for handling and caging single cells and particles. *Biosens. Bioelectron.*, 14:247–256, 1999.
[63] M. Ozkan, T. Pisanic, J. Scheel, C. Barlow, S. Esener, and S.N. Bhatia. Electro-optical platform for the manipulation of live cells. *Langmuir*, 19:1532–1538, 2003.
[64] R. Pethig and D.B. Kell. The passive electrical-properties of biological-systems - their significance in physiology, biophysics and biotechnology. *Phys. Med. Biol.*, 32:933–970, 1987.
[65] C. Polk and E. Postow. *Handbook of Biological Effects of Electromagnetic Fields*. CRC Press, Boca Raton, FL, pp. 618, 1996.
[66] L. Qian, M. Scott, K.V.I.S. Kaler, and R. Paul. Integrated planar concentric ring dielectrophoretic (DEP) levitator. *J. Electrostat.*, 55:65–79, 2002.
[67] C. Reichle, T. Schnelle, T. Muller, T. Leya, and G. Fuhr. A new microsystem for automated electrorotation measurements using laser tweezers. *Biochim. Et. Biophys. Acta-Bioenerg.*, 1459:218–229, 2000.
[68] T.A. Ryan, J. Myers, D. Holowka, B. Baird, and W.W. Webb. Molecular crowding on the cell surface. *Science*, 239:61–64, 1988.
[69] T. Schnelle, R. Hagedorn, G. Fuhr, S. Fiedler, and T. Muller. 3-Dimensional electric-field traps for manipulation of cells - calculation and experimental verification. *Biochim. Et Biophys. Acta*, 1157:127–140, 1993.
[70] T. Schnelle, T. Muller, and G. Fuhr. The influence of higher moments on particle behaviour in dielectrophoretic field cages. *J. Electrostat.*, 46:13, 1999a.
[71] T. Schnelle, T. Muller, and G. Fuhr. Trapping in AC octode field cages. *J. Electrostat.*, 50:17–29, 2000.
[72] T. Schnelle, T. Muller, G. Gradl, S.G. Shirley, and G. Fuhr. Paired microelectrode system: Dielectrophoretic particle sorting and force calibration. *J. Electrostat.*, 47:121–132, 1999b.
[73] H.P. Schwan. Dielectrophoresis and rotation of cells. In E. Neumann, A.E. Sowers, and C.A. Jordan, (eds.) *Electroporation and Electrofusion in Cell Biology*. New York, Plenum Press, pp. 3–21, 1989.
[74] H.P. Schwan. Linear and nonlinear electrode polarization and biological materials. *Ann. Biomed. Eng.*, 20:269–288, 1992.
[75] J.R. Subjeck and T.T. Shyy. Stress protein systems of mammalian-cells. *Am. J. Physiol.*, 250:C1–C17, 1986.
[76] J. Suehiro, R. Hamada, D. Noutomi, M. Shutou, and M. Hara. Selective detection of viable bacteria using dielectrophoretic impedance measurement method. *J. Electrostat.*, 57:157–168, 2003.
[77] J. Suehiro and R. Pethig. The dielectrophoretic movement and positioning of a biological cell using a three-dimensional grid electrode system. *J. Phys. D-Appl. Phys.*, 31:3298–3305, 1998.
[78] K. Svoboda and S.M. Block. Biological applications of optical forces. *Ann. Rev. Biophys. Biomol. Struc.*, 23:247–285, 1994.
[79] T.Y. Tsong. Molecular recognition and processing of periodic signals in cells study of activation of membrane ATPases by alternating electric fields. *Biochim. et Biophys. Acta*, 1113:53–70, 1992.
[80] S. Tsukahara, T. Sakamoto, and H. Watarai. Positive dielectrophoretic mobilities of single microparticles enhanced by the dynamic diffusion cloud of ions. *Langmuir*, 16:3866–3872, 2000.
[81] M. Urano and E.B. Douple. *Thermal effects on cells and tissues*. VSP, Utrecht, The Netherlands, pp. 80, 1988.

[82] P. Van Gerwen, W. Laureyn, W. Laureys, G. Huyberechts, M.O. De Beeck, K. Baert, J. Suls, W. Sansen, P. Jacobs, L. Hermans, and R. Mertens. Nanoscaled interdigitated electrode arrays for biochemical sensors. *Sens. Actu. B-Chem.*, 49:73–80, 1998.

[83] J. Voldman, R.A. Braff, M. Toner, M.L. Gray, and M.A. Schmidt. Holding forces of single-particle dielectrophoretic traps. *Biophys. J.*, 80:531–541, 2001.

[84] J. Voldman, M. Toner, M.L. Gray, and M.A. Schmidt. A microfabrication-based dynamic array cytometer. *Anal. Chem.*, 74:3984–3990, 2002.

[85] J. Voldman, M. Toner, M.L. Gray, and M.A. Schmidt. Design and analysis of extruded quadrupolar dielectrophoretic traps. *J. Electrostat.* 57:69–90, 2003.

[86] X. Wang, X.-B. Wang, F.F. Becker, and P.R.C. Gascoyne. A theoretical method of electrical field analysis for dielectrophoretic electrode arrays using Green's theorem. *J. Phys. D (Applied Physics)*, 29:1649–1660, 1996.

[87] X. Wang, X.-B. Wang, and P.R.C. Gascoyne. General expressions for dielectrophoretic force and electrorotational torque derived using the Maxwell stress tensor method. *J. Electrostat.*, 39:277–295, 1997a.

[88] X.-B. Wang, Y. Huang, P.R.C. Gascoyne, and F.F. Becker. Dielectrophoretic manipulation of particles. *IEEE Trans. Ind. Appl.*, 33:660–669, 1997b.

[89] X.-B. Wang, Y. Huang, X. Wang, F.F. Becker, and P.R.C. Gascoyne. Dielectrophoretic manipulation of cells with spiral electrodes. *Biophys. J.*, 72:1887–1899, 1997c.

[90] X. J. Wang, J. Yang, and P.R.C. Gascoyne. Role of peroxide in AC electrical field exposure effects on Friend murine erythroleukemia cells during dielectrophoretic manipulations. *Biochimi. Et Biophys. Acta-Gen. Sub.*, 1426:53–68, 1999.

[91] M. Washizu and T.B. Jones. Multipolar dielectrophoretic force calculation. *J. Electrostat.*, 33:187–198, 1994.

[92] M. Washizu. and T. B. Jones. Generalized multipolar dielectrophoretic force and electrorotational torque calculation. *J. Electrostat.*, 38:199–211, 1996.

[93] J.C. Weaver, T.E. Vaughan, and G.T. Martin. Biological effects due to weak electric and magnetic fields: The temperature variation threshold. *Biophys. J.*, 76:3026–3030, 1999.

[94] J. Wu. Acoustical tweezers. *J. Acoust. Soc. Am.*, 89:2140–2143, 1991.

[95] G. Zhou, M. Imamura, J. Suehiro, and M. Hara. A dielectrophoretic filter for separation and collection of fine particles suspended in liquid. *Conference Record of the IEEE Industry Applications Conference*, 2:1404–1411, 2002.

[96] X.F. Zhou, J.P.H. Burt, and R. Pethig. Automatic cell electrorotation measurements: Studies of the biological effects of low-frequency magnetic fields and of heat shock. *Phys. Med. Biol.*, 43:1075–1090, 1998.

[97] X.F. Zhou, G.H. Markx, and R. Pethig. Effect of biocide concentration on electrorotation spectra of yeast cells. *Biochim. Et Biophys. Acta-Biomem.*, 1281:60–64, 1996.

[98] U. Zimmermann. Electrical Breakdown, Electropermeabilization and Electrofusion. *Rev. Phys. Biochem. Pharma.*, 105:175–256, 1986.

9
BioMEMS for Cellular Manipulation and Analysis

Haibo Li, Rafael Gómez-Sjöberg[a], and Rashid Bashir*
*Birck Nanotechnology Center and Bindley Biosciences Center, Discovery Park,
School of Electrical and Computer Engineering, Purdue University, Weldon School of
Biomedical Engineering, Purdue University, West Lafayette, IN. 47907, USA*
[a]*Now at Department of Bioengineering, Stanford Univ.
Stanford, CA. USA*

9.1. INTRODUCTION

Since the introduction of Micro-electro-mechanical systems in the early 70's, the significance of the biomedical applications of these miniature systems has been realized [54, 77]. BioMEMS, the abbreviation for Biomedical or Biological Micro-Electro-Mechanical-Systems, is now a heavily researched area with a wide variety of important biomedical applications. In general, BioMEMS, and its synonym BioChip, can be defined as "Devices or systems, constructed using techniques inspired from micro/ nanoscale fabrication, that are used for processing, delivery, manipulation, analysis, or construction of biological and chemical entities". A large number of BioMEMS devices and applications have been presented in [4, 34, 41, 57]. Technologies such as "lab-on-a-chip" and Micro-Total-Analysis-Systems (micro-TAS or µTAS), when used for biological applications, fall into the BioMEMS category. The use of these lab-on-a-chips for cellular analysis is justified by, (i) reducing the sensor element to the scale of size of cells and smaller and hence providing a higher sensitivity, (ii) reduced reagent volumes and associated costs, (iii) reduced time to result due to small volumes resulting in higher effective concentrations, (iv) amenability of portability and miniaturization of the entire system, and (v) ability to perform large numbers of assays or measurements in parallel.

Additionally, technologies that make possible the direct manipulation, probing and detection of individual cells (human, bacterial, from animals and plants) are of great interest

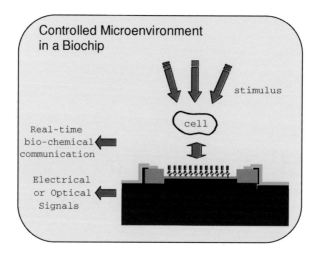

FIGURE 9.1. Micro-fluidic devices with controlled micro-environments for study of cells and the real time profiling of their proteins, mRNA, and other biochemicals (Reprinted with permission from Advanced Drug Delivery Reviews, vol. 56, 2004 and with kind permission from R. Bashir).

to biotechnology researchers and industries. These technologies can greatly facilitate the study of cellular physiology, structure and properties; development and testing of new drugs; rapid detection of pathogenic bacteria; etc. Because of the scales at which biochips operate, they are ideally suited for these applications. For example, until recently many electrical, mechanical and optical characteristics of cells had to be measured as averages over large numbers of cells, and sometimes the effects of extra-cellular materials could not be easily eliminated. Biochips provide an ideal platform for directly measuring electrical, mechanical, and optical properties of individual cells. Electrodes, channels, chambers, etc., in which cells can be manipulated and probed, are readily microfabricated with sizes similar to those of cells (~ 1 μm for bacterial cells, ~ 10 μm for mammalian cells).

BioMEMS holds a lot of promise for the analysis of single cells and the study of their function in real time. Micro and nano-scale systems and sensors could allow us to precisely measure the protein, mRNA, and chemical profiles of cells in real time, as a function of controlled stimulus and increase understanding of signaling pathways inside the cell. The development of micro-systems, as schematically shown in Fig. 9.1 [3], where cells can be precisely place, manipulated, lysed, and then analyzed using micro and nano-sensors in 'real-time', would have a significant impact on Systems Biology. Integration of sensors for detection of DNA, mRNA, proteins, and other parameters indicating cellular conditions such as oxygen, pH, etc. can be accomplished using BioMEMS platforms and nano-scale sensors. The following subsections will describe some microfabricated devices used for the manipulation, separation and detection of cells.

9.2. BIOMEMS FOR CELLULAR MANIPULATION AND SEPARATION

The ability to manipulate particles, especially living cells, in three-dimensional space is fundamental to many biological and medical applications, including isolation and detection

BIOMEMS FOR CELLULAR MANIPULATION AND ANALYSIS   189

of sparse cancer cells, concentration of cells from dilute suspensions, separation of cells according to specific properties, and trapping and positioning of individual cells for characterization. The current methods commonly used for manipulation, concentration and separation of biological particles employ several kinds of physical forces from mechanical, hydrodynamic, ultrasonic, optical, and electro-magnetic origins [73]. Among these methods, electrophoresis and dielectrophoresis (DEP) allow the non-invasive electrical manipulation and characterization of particles, including biological cells, by directly exploiting their electrical and dielectric properties.

9.2.1. *Electrophoresis*

Electrophoresis is the motion of a charged particle under the influence of an electric field, where the magnitude and direction of the force exerted on the particle are directly proportional to its charge. This effect is most commonly used in biological analysis for separation of DNA and proteins. Most cells have a net charge which will make them move when an electric field is applied. The direction and speed in which the cells move depend on the polarity and magnitude of the charge, respectively, so that cells with different polarities and amounts of charge can be separated by electrophoresis [49]. The electrophoretic mobility of a charged spherical particle can be approximated as:

$$\frac{\upsilon}{E} = \frac{q}{6\pi \eta a} \qquad (9.1)$$

where υ is the velocity of the particle, E is the electric field, q is the net charge on the particle, η is the viscosity of the solution and is the radius of the particle.

Poortinga et al. [58] used electrophoresis to deposit bacterial films onto electrodes. Chang et al. [9] used a micro-pore fabricated in an oxide-coated silicon diaphragm and placed between two chambers containing ionic buffer solutions to electrically characterize live and heat-inactivated *Listeria innocua* bacterial cells as shown in Fig. 9.2. Passage of the electrophoretically-driven cells through the pore was detected by the temporary decrease in the ionic current across the pore caused by the partial blockage of the pore by the traversing

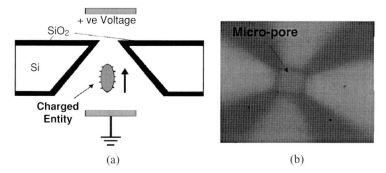

FIGURE 9.2. (a) Cross section of the micro-pore device along with the measurement concept, (b) an optical micrograph of the micro-pore, with pore size of ∼4um on a side (Reprinted with permission from Journal of Vacuum Science & Technology B, vol. 20, no. 5, 2002 and with kind permission from R. Bashir).

cells. Studies of the electrophoretic movement of live and heat-inactivated *Listeria innocua* cells on interdigitated fingered electrodes demonstrated that live cells have a net negative charge, while heat-inactivated ones are either neutral or positively charged.

9.2.2. Dielectrophoresis

Dielectrophoretic (DEP) forces occur when a non-uniform electric field interacts with a field-induced electrical dipole in a particle. The time-averaged dielectrophoretic force F for a dielectric sphere immersed in a medium is represented as [23, 55, 73]:

$$F = 2\pi \varepsilon_0 \varepsilon_m r^3 \text{Re}[f_{CM}] \nabla |E_{RMS}|^2 \quad (9.2)$$

where ε_0 is the vacuum dielectric constant, r is the particle radius, E_{RMS} is the root mean square of the electric field, and f_{CM} is known as the Clausius-Mossotti factor, defined as [23, 73]:

$$f_{CM} = \frac{\varepsilon_p^* - \varepsilon_m^*}{\varepsilon_p^* + 2\varepsilon_m^*} \quad (9.3)$$

where ε_p^* and ε_m^* are the relative complex permittivities of the particle and the medium respectively. These permittivities depend on the frequency of the applied field. When the permittivity of the particle is larger than that of the medium, $\text{Re}[f_{CM}] > 0$, the DEP is called positive and the particle moves towards the locations with the highest electric field gradient. When the permittivity of the particle is smaller than that of the medium, $\text{Re}[F_{CM}] < 0$, the DEP is called negative and the particle moves to the locations with the lowest electric field gradient. Early studies of cell dielectrophoresis employed electrode structures made from thin wires and foils, such as cone-plate electrodes in a levitation system [29, 31], simple pin-plate structures [46], and four-pole electrodes for characterization of cell plasma membranes [14]. More recently, MEMS technology has been used for the production of a number of devices with electrode arrays and chambers suitable for manipulation and measurement of biological objects. Integrated dielectrophoresis biochips provide the advantages of speed, flexibility, controllability, and ease of automation [23]. Examples include the fluid-integrated-circuit for single cell handling [74], polynomial electrode arrays for cell separation by trapping cells of different types at different locations using positive and negative DEP forces [25, 43, 72], and various three-dimensional arrays for cell positioning, trapping and levitation [52, 59, 68]. Among all of the above it is in trapping and cell separation that dielectrophoresis has its major applications. Significant examples include separations of viable and nonviable yeast cells [24, 45], leukemia [5] and breast-cancer [6] cells from blood, and the concentration of $CD34^+$ cells from peripheral-stem-cell harvests [61]. Particularly promising is the combination of dielectrophoretic forces with hydrodynamic forces, such as the combined dielectrophoretic-field flow fractionation (DEP-FFF) technique for continuous separation [44].

The heart of any dielectrophoresis system is formed by the electrodes from which the driving electric field is applied. The dependence of dielectrophoretic force on both the magnitude and the gradient of the applied electric field means that the electrode configuration needed for efficient dielectrophoretic collection should produce strong, highly non-uniform

FIGURE 9.3. Dielectrophoretic separation of viable and non-viable yeast cells using interdigitated, castellated electrodes. The viable yeast cells collect on the edges of and in pearl chains between the electrodes, whilst the non-viable cells collect in triangular aggregation between the electrodes and in diamond-shaped formations on top of the electrodes (Reprinted with permission from Journal of Biotechnology, vol. 32, 1994 and with kind permission from G. H. Markx).

fields. In this respect microfabricated electrodes present a very clear advantage over macro-scale ones because of the ability to create intense and highly non-uniform fields at low voltages (the voltage needed to produce a given field strength is directly proportional to the distance between electrodes). The required electrode configuration can be readily realized using planar micro-fabrication processes with a suitable choice of substrate, usually glass or silicon. The interdigitated electrode array is perhaps the simplest electrode structure that has been used in microbiological applications of dielectrophoresis [73]. From straightforward electrostatic considerations it is clear that the interdigitated electrode configuration has the electric field maximum in both its gradient and strength at the edges of the electrodes and minimum at the centers of electrodes and inter-electrode gaps. Separation of live and heat-treated *Listeria* bacteria was achieved on the microfabricated interdigitated electrodes reported by Li *et al.* [38]. Fig. 9.3 shows the separation of viable and non-viable yeast cells using interdigitated, castellated electrodes reported by Markx *et al.* [45].

A common electrode configuration used for negative dielectrophoresis research is a quadrupole arrangement where four electrodes point towards a central enclosed region [25] as shown in the schematic in Fig. 9.4 (left). By applying an AC field between adjacent electrodes with a phase angle of 180° particles will experience dielectrophoresis with a well-defined electric field minimum at the center of the electrode array and maxima at the electrode edges as shown in Fig. 9.4 (right) [26]. If the phase angle is reduced such that adjacent electrodes differ by 90°, particles will experience both electrorotation and dielectrophoresis and this technique can be used to trap single cells at specific sites.

More advanced and complex electrode structures have been developed for dielectrophoresis studies. Voldman *et al.* [67, 68, 70, 71] developed a 3-D extruded quadrupole

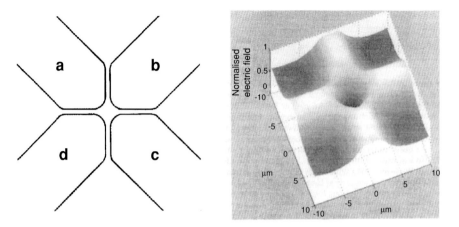

FIGURE 9.4. A schematic diagram of typical quadrupole electrode microstructure used in dielectrophoresis experiments (left). The gap between opposing electrodes in the center of the array is typically of the order 10–50 μm across, but can be as small as 500nm or as large as 1mm. The right figure shows a simulation of the electric field in the plane 5 μm above the electrode array shown in the left. The dark region at the center of the electrode array is the electric field minimum, surrounded by a ring of high electric field gradient. Particles experiencing negative dielectrophoresis are repelled into this minimum and become trapped. The electric field strength is high (white region) along the electrode edges, where particles experiencing positive dielectrophoresis will collect (Reprinted with permission from Nanotechnology, vol. 11, 2000 and with kind permission from M.P. Hughes).

structure as shown in Fig. 9.5 [70]. Each of the DEP traps in the array is electrically switchable and capable of holding a single cell, and provides better holding ability than a planar quadrupole trap.

T. Müller *et al.* [52] designed and constructed several 3-D microelectrode systems consisting of two metal layers with electrode structures acting as funnel, aligner, cage and

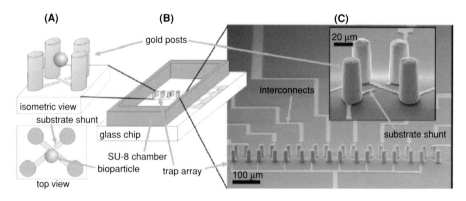

FIGURE 9.5. Schematic of the extruded quadrupole DEP trap array. (A) Two views of a single trap, illustrating the trapezoidal placement of the gold posts and a bioparticle suspended in the trap. (B) The trap is one of an array of traps. (C) SEMs of a 1 × 8 array of traps along with an exploded view of one trap (Reprinted with permission from Transducers, 2001 and with kind permission from J. Voldman).

BIOMEMS FOR CELLULAR MANIPULATION AND ANALYSIS

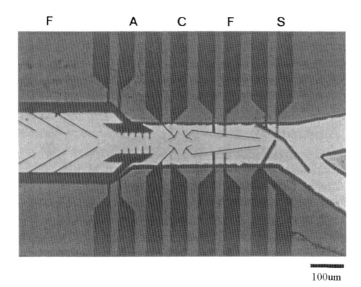

FIGURE 9.6. A 3-D microelectrode system. The electrode elements 'funnel' (F), 'aligner' (A), 'cage' (C), and 'switch' (S) are arranged over a small distance. Two identical layers of electrodes were overlayed and separated by a 40 μm polymer spacer forming channels (bright region) (Reprinted with permission from Biosensors and Bioelectronics, vol. 14, 1999 and with kind permission from T. Müller).

switch, separated by a 40μm thick polymer spacer forming a flow channel for handling and caging single cells and particles as shown in Fig. 9.6.

There have been a number of approaches to the application of dielectrophoretic techniques to the 'lab-on-a-chip' systems. Some researchers use dielectrophoresis as a method for isolating specific cells at a preliminary stage for further manipulation and/or analyses such as polymerase chain reaction (PCR), electroporation, or cell detection by biochemical methods. Others use dielectrophoresis to perform a range of functions, from separation to trapping and analysis [27]. 'Lab-on-a-chip' system for cell separation on microfabricated electrodes using dielectrophoretic/gravitational field-flow fractionation [79].

Recently Li *et al.* [39, 40] studied a microfabricated dielectrophoretic filter device with a thin chamber and interdigitated electrode array as a test bed for dielectrophoretic trapping of polystyrene beads and biological entities such as yeast cells, spores and bacteria. The device was characterized by both measurement and finite-element modeling of the holding forces against destabilizing flow-induced forces in both positive and negative dielectrophoretic traps. The combination of experiments and modeling has given insight into the DEP forces over the interdigitated electrodes and can be very useful in designing and operating a dielectrophoretic barrier or filter to sort and select biological particles for further analysis.

9.3. BIOMEMS FOR CELLULAR DETECTION

BioMEMS can combine a biologically sensitive element with a physical or chemical transducer to selectively and quantitatively detect the presence of specific compounds in

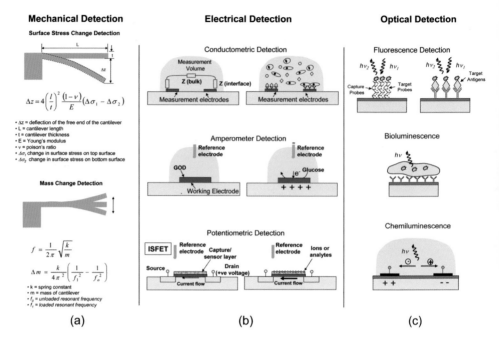

FIGURE 9.7. Some detection modalities used in BioMEMS and Biochip sensors (Reprinted with permission from Advanced Drug Delivery Reviews, vol. 56, 2004 and with kind permission from R. Bashir).

a given environment [67]. During the last decade, BioMEMS devices have been used as biosensors for sensitive, rapid, and real-time measurements [35, 75]. These BioMEMS sensors can be used to detect cells, proteins, DNA, or small molecules. Many demonstrations to date have been of single-sensor devices, but the biggest potential of BioMEMS lies on the ability to create massively parallel arrays of sensors. There are many detection methods used in BioMEMS sensors, including (i) Optical, (ii) Mechanical, (iii) and Electrical. Fig. 9.7 [3] shows a schematic of these key detection modalities as they are used in Biochips and BioMEMS sensors for detection of a wide variety of biological entities. In this review, however, we will focus on detection of cells.

9.3.1. Optical Detection

Optical detection techniques are perhaps the most common and prevalently used in biology and life-sciences. They can be based on fluorescence or chemiluminescence. Fluorescence detection techniques are based on fluorescent markers chemically linked to, for example, DNA or RNA strands or antibodies, which emit light at specific wavelengths in response to an external optical excitation. The presence, enhancement, or reduction in the fluorescence signal can indicate a binding reaction, as shown schematically in Fig. 9.7(c). Recent advances in fluorescence detection technology have enabled single molecule detection [50, 53, 67]. Fluorescence based detection in BioMEMS has been applied to detection of cells within micro-chips, using antibody-based assays (Enzyne-Linked ImmunoSorbent Assay, ELISA, type) as shown in Fig. 9.8 [50, 62].

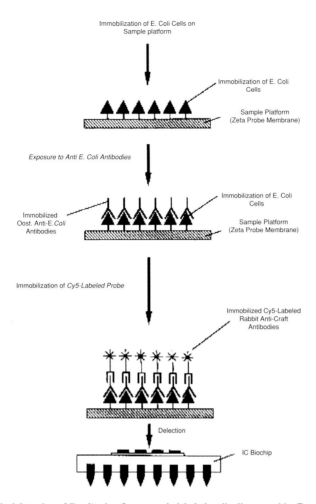

FIGURE 9.8. Optical detection of *E.coli* using fluorescently labeled antibodies on a chip (Reprinted with permission from Fresenius Journal of Analytical Chemistry, Vol. 369, 2001 and with kind permission from T. Vo-Dinh).

Chemiluminescence is the light generated by the release of energy as a result of a chemical reaction. Light emission from a living organism is commonly termed bioluminescence (sometimes called biological fluorescence), and light emission which takes place under excitation by an electrical current is designated electrochemiluminescence. One of the challenges for optical detection within biochips is the ability to integrate light sources and detectors in a miniaturized portable format. This integration requires fabrication of photo-diodes in silicon substrates [30] or heterogeneous integration of compound semiconductor LEDs and photodetectors within plastic or polymer platforms [10]. In the later study, microassembly of a hybrid fluorescence detection microsystem was demonstrated by heterogeneous integration of a CdS thin-film filter, an (In,Ga)N thin-film blue LED, and a disposable PDMS microfluidic device onto a Si PIN photodetector substrate.

McClain et al. [48] reported a microfluidic device that integrated cell handling, rapid cell lysis, and electrophoretic separation and detection of fluorescent cytosolic dyes. Cell analysis rates of 7–12 cells/min were demonstrated and are >100 times faster than those reported using standard bench-scale capillary electrophoresis. Hong et al. [21] recently developed microfluidic chips with parallel architectures for automated nucleic acid purification from small numbers of bacterial or mammalian cells. All processes, such as cell isolation, cell lysis, DNA and mRNA purification, and recovery, can be carried out on each single microfluidic chip in nanoliter volumes without any pre- or post-sample treatment.

9.3.2. Mechanical Detection

Mechanical detection of biochemical entities and reactions has more recently been realized through the use of micro and nano-scale cantilever sensors on a chip. As shown in Fig. 9.7(a), these cantilever sensors (diving board type structures) can be used in two modes, namely stress sensing and mass sensing. In stress sensing mode one side of the cantilever is usually coated with a Self-Assembled Monolayer (SAM) of biomolecules that bind to the analyte being detected. The binding of the analyte to the SAM produces a change in surface free energy, resulting in a change in surface stress, which in turn leads to a measurable bending of the cantilever. The bending can then be measured using optical means (laser reflection from the cantilever surface into a quad position detector, like in an Atomic Force Microscope) or electrical means (piezo-resistors incorporated near the fixed edge of the cantilever). To increase the stress sensitivity of the cantilever, the spring constant should be reduced, while the overall surface of the cantilever determines the number of molecules that should attach to the surface to cause a given stress change. In the mass sensing mode, the resonant frequency of the cantilever is constantly monitored as it vibrates due to an external driving force (i.e. a piezo-electric transducer) or in response to the background thermal noise. When the species being detected binds to the cantilever, it changes the cantilever mass and hence its resonant frequency. The mass of the detected species can be calculated from the change in resonant frequency. The resonant frequency can be measured using electrical or optical means, in the same way that bending is detected in stress sensing. To increase the mass sensitivity, in general, the mass of the cantilever should be made smaller, the quality factor should be increased, the resonant frequency should be chosen such that it is easily measured, and the detection system should be designed to measure as small a frequency shift as possible. The quality factor is decreased with increased damping, for example in a fluid, and hence the minimum detectable mass is much higher in damping mediums (liquids) as compared to low-damping mediums (air). Thus, the stress detection mode is inherently preferred in a fluid.

To have a significant change in surface stress, a large fraction of the cantilever area must be involved in the binding event that leads to detection, which precludes the use of the stress-based technique for detecting cells or large viruses. Even covering the whole cantilever surface (on one side) with cells bound to its surface *via* antibodies produces very small changes in surface stress because the effective binding area of each cell is just a small fraction of the total cantilever area. Detection of cells and microorganisms has been demonstrated using the mass detection method based on shifts in resonant frequency. Various examples of mass-based sensing are reported in the literature, for example, detection of the mass of E.coli O157:H7 using cantilevers [18, 28], detection of the mass of a single vaccinia

BIOMEMS FOR CELLULAR MANIPULATION AND ANALYSIS

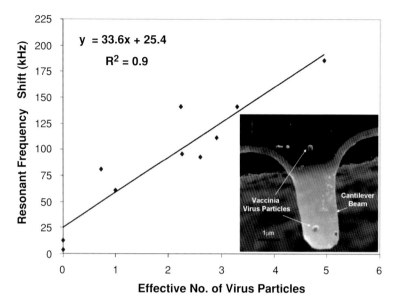

FIGURE 9.9. Shift (decrease) in resonant frequency with increasing number of virus particles. Inset shows an SEM of a nano-cantilever with a single Vaccinia virus particle (Reprinted with permission from Applied Physics Letters, vol. 84, no. 10, 2004 and with kind permission from R. Bashir).

virus particle, as shown in Fig. 9.9 [19], and mass change in a polymer upon absorption of a vapor [36].

9.3.3. Electrical Detection♦

Some of the earliest cell-related uses of micromachined devices were in the electrical probing of neurons by creating microscopic needles that could be inserted *in vitro* or *in vivo* in the neuron to stimulate it electrically and record the signals it produced [34]. Recent reports describe artificial structures where neurons are cultured and probed, while the configuration of interconnections between them is artificially patterned by microfabricated channels that guide axon growth [42, 47]. Devices for positioning and/or probing of other types of electrogenic cells, such as cardiac myocytes, were also reported in the literature [51, 65]. Most of these consist of arrays of electrodes, deposited either on a planar surface or at the bottom of cavities, on which the cells are located. In most cases the cells are placed on the electrodes manually, but Thielecke *et al.* [65] make use of an orifice at the center of each electrode, through which vacuum is created to move the cells towards the electrodes and hold them in place.

Monitoring the impedance of microfabricated electrodes over which adherent cells are cultured can reveal information about cellular motion, multiplication, metabolism, viability, etc. The signal produced in these devices arises from two main mechanisms: Cells growing

---

♦ Parts of this section are reprinted from: R. Gómez-Sjöberg, "Microfabricated device for impedance-based electronic detection of bacterial metabolism," Ph.D thesis, School of Electrical and Computer Engineering, Purdue University, West Lafayette, IN, December 2003, with kind permission from the author.

attached to the electrodes act as insulators, blocking current flow between electrodes; and the difference in dielectric constant between the cells and the growth medium modifies the capacitance of the electrodes. For example, Keese and Giaever [32] built a cell biosensor for environmental monitoring. The impedance of two electrodes in a cell growth chamber was modified by changes in the cell population produced by phenomena such as cell motion, multiplication, death, and metabolic activity. Borkholder et al. [7] used an array of 10 μm diameter electrodes to study the response of cells to certain toxins that block membrane channels. Similarly, Ehret et al. [13] used microfabricated interdigitated electrodes (fingers are 50 μm wide) on a sapphire substrate to monitor the behavior of mammalian cells, by measuring the capacitance of the electrodes at a frequency of 10 kHz. Cells were grown adherently over the surface of the electrodes. Very clear signals were observed by the authors when the cells were destroyed by adding the detergent Triton X-100. The response of cells to different concentrations of the toxic ion Cd^{2+} could also be monitored over time. Building upon the work of Ehret et al. [13], Wolf et al. [78] and Lehman et al. [37] developed the so-called "PhysioControl-Microsystem" and "Cell-Monitoring-System" that incorporate microfabricated temperature, pH, oxygen, and ion sensors, along with interdigitated electrodes, to gather detailed information on cellular metabolism. Ion sensitive field-effect transistors (ISFET) were used as pH, oxygen, and ion sensors. In these transistors the gate is covered with a film selective to the ions that are detected, so that adsorption of the ions into the film causes a shift in the gate potential with a concomitant change in the current flowing through the channel of the transistor.

A very interesting commercial system for the electrical monitoring of cellular metabolism is the "Cytosensor Microphysiometer" developed by Molecular Devices GmbH in Germany [8, 20]. This device measures pH changes in the cell growth medium, produced by excreted metabolites, using light-addressable potentiometric sensors (LAPS). The principle on which LAPS works is depicted in Fig. 9.10. A silicon substrate is coated with

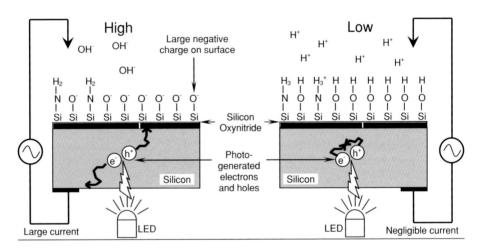

FIGURE 9.10. Operating principle for the light-addressable potentiometric sensor used to measure changes in the pH of cell cultures produced by cell metabolism (Adapted with permission from Biosensors & Bioelectronics, vol. 15, no. 3–4, 2000 and with kind permission from F. Hafner).

a silicon oxynitride film that will be in contact with the liquid medium being monitored. When hydrated, silanol (Si-OH) and silamine (Si-NH$_2$) groups will appear at the surface of the film. An ohmic contact is established with the silicon substrate and a reference electrode is immersed in the liquid medium, so that a voltage can be applied between the liquid and the silicon. Free electrons and holes are generated in the silicon by illuminating it with a pulsating LED. The silamine and silanol groups are ionized to different levels depending on the pH of the medium, affecting the surface charge on the film. When the pH of the medium is high, most of the silanol groups are ionized (they have donated a H$^+$ ion to the medium) so that a large negative charge exists on the surface. The electric field generated by this charge will separate the photo-generated holes and electrons, increasing their recombination time, and thus producing a large current across the electrodes. When the pH is low, most of the surface groups are neutral and little separation of electrons and holes occur, leading their rapid recombination and hence to a low current. The voltage between the silicon and the liquid is adjusted to have a constant photo-current, so that the required changes in voltage are proportional to changes in pH. This technique can detect changes as small as 5×10^{-4} pH units. Cells are kept in a chamber over the microfabricated LAP sensor, through which growth medium can flow. The flow of medium is stopped when the metabolic rate is being measured, so that metabolites accumulate in the chamber and change the pH. A very important feature of the LAPS technique is that local measurements of pH can be done by limiting the illumination to the area of interest. In this way, a pH map of the cell culture can be constructed by having an array of LEDs, each one illuminating a small section of the sensor. It is also worth mentioning the use of microcalorimetry for measuring metabolism, as exemplified by the device built by Verhaegen et al. [66]. This microcalorimeter uses aluminium/p$^+$-polysilicon junction thermopiles to measure the heat generated by metabolizing cells cultured in chambers microfabricated on a silicon substrate.

The electrical impedance of individual cells located between two microelectrodes formed on opposite sides of a microchannel can provide information such as the capacitance of the cell membrane and the dielectric constant and conductivity of the cytoplasm [1, 2, 57, 66]. Differences in impedance could be used to discriminate between different types of cells for sorting and counting. The same principle and geometry were used by Sohn et al. [60] to create a microfabricated cytometer for mammalian cells flowing one by one between electrodes in a microchannel. Since the capacitance is greatly affected by the DNA content of the cells, due to the large number of charges in DNA molecules, it can be used to track the multiplication phases of the cells. Suehiro et al. [63, 64] used DEP to trap E. coli cells and detect their presence over the DEP electrodes by monitoring the electrode impedance. The presence of cell bodies changes the impedance of the electrodes because their dielectric properties are different from those of the suspension medium, and the impedance can be correlated to the number of trapped cells. Koch et al. [33] fabricated a device to detect the passage of cells through fully enclosed microfluidic channels, based on the Coulter Counter principle. Two electrodes are placed across the channels, perpendicular to them, spaced 40 μm apart. When a cell passes between the electrodes the resistance measured across them changes (shown in Fig. 9.11) because cells are significantly less conductive than the liquid in which they are suspended (they can in fact be modeled as non-conductive particles).

Detecting viable bacteria is a very important goal in the development of novel biosensors, and microscale impedance-based monitoring of metabolic activity has the potential

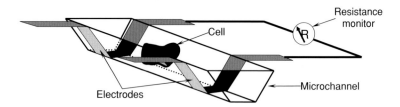

FIGURE 9.11. Microfabricated Coulter Counter used to detect the passage of cells through microchannels (Adapted with permission from Proceedings of Ninth Micromechanics Europe Workshop—MME'98, 1998 and with kind permission from M. Koch).

of realizing that goal in a simple and cost-effective way. Macroscale impedance-based detection is relatively slow when low numbers of cells are present. The lower the initial population of microorganisms, the longer it takes for the impedance to change by a measurable amount. This should be obvious, since it will take longer for a small number of organisms to produce enough ions to modify the impedance by a certain amount, than for a larger number of them. Eden & Eden [12] and Dupont et al. [11] showed that the detection time (incubation time needed to have the impedance change by a certain predefined value) decreases exponentially with increasing bacterial concentration. Consequently, the impedance method has the potential to detect bacterial contamination in a very short time if the bacterial concentration is somehow increased by several orders of magnitude. The number of bacterial cells in a given amount of food sample is an uncontrollable parameter, so the only other controllable parameter is the volume in which cells are placed for performing the measurement. The typical volumes in which impedance measurements are done range from 1ml to 100ml in the macro-scale impedance-based detection systems currently available commercially. With such large volumes, typical detection times are on the order of 8 hours or more for initial bacterial loads on the order of 10CFU/ml. By confining the same number of cells in a microfabricated volume, their effective concentration becomes very high and the detection time drops dramatically. Implementing the impedance monitoring method on a biochip, where detection chambers with volumes of even 0.1 nl can be readily fabricated, is an ideal way of exploring its potential for fast detection. Goméz et al. [15, 16, 17] have developed microfabricated devices (Fig. 9.12) with nanoliter-scale chambers used to study this bacterial detection method. A metabolic signal could be detected in off-chip incubated samples at cell concentrations equivalent to about 50 cells in a 5.27nl measuring volume in an initial biochip prototype [16, 17].

9.4. CONCLUSIONS AND FUTURE DIRECTIONS

Considerable progress has been made in the field of BioMEMS, especially in systems for cellular analysis as described above, and with the current drive towards nano-scale devices, micro- and nano-technologies are being combined with the emerging field of bionanotechnology [76]. BioMEMS are enabling us to probe, measure, and explore the micro- and nano-machinery in the biological world, including the inner workings of single cells. Such micro and nano-scale systems and sensors could allow us to precisely measure

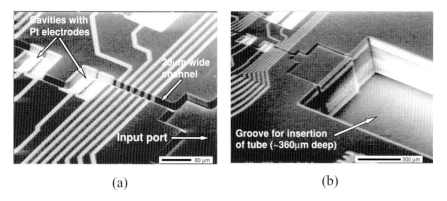

FIGURE 9.12. Scanning Electron Micrograph of a biochip used to explore the impedance-based detection of bacterial metabolism. (a) Chambers where bacterial cells are incubated for detection; (b) input/output port created by DRIE. (Reprinted with permission from Ph.D. thesis of R. Gómez-Sjöberg, 2004, School of Electrical and Computer Engineering, Purdue University, and with kind permission from R. Gómez-Sjöberg).

the protein, mRNA, and chemical profiles of cells in real time, as a function of controlled stimuli and increase understanding of signaling pathways inside the cell. These issues will also be the focus of the post-genomic era and also in the applications of systems theories to biology, also referred to as systems biology [22].

ACKNOWLEDGMENTS

The sponsorship of NIH (NIBIB), USDA Center for Food Safety Engineering at Purdue, NASA Institute of Nanoelectronics and Computing (INAC), and NSF Career Award is greatly appreciated. The authors would also like to thank all members of the Laboratory of Integrated Biomedical Micro/Nanotechnology and Applications (LIBNA) in the School of Electrical and Computer Engineering, Department of Biomedical Engineering, Purdue University, for providing the motivation for this review.

REFERENCES

[1] H. Ayliffe, A. Frazier, and R. Rabbitt. *Am. Soc. Mech. Eng., Bioeng. Div. (Publication) BED*, 35:485–486, 1997.
[2] H. Ayliffe, A. Frazier, and R. Rabbitt. *IEEE J. Microelectromech.*, 8:50–57, 1999.
[3] R. Bashir. *Advanc. Drug Del. Rev.*, 56:1565–1586, 2004.
[4] R. Bashir and S. Wereley (eds.). Biomolecular sensing, processing, and analysis. In *BioMEMS and Biomedical Nanotechnology*. Kluwer Academic Publishers, 2004.
[5] F.F. Becker, X.-B.Wang, Y. Huang, R. Pethig, J. Vykoukal, and P.R.C. Gascoyne. *J. Phys. D: Appl. Phys.*, 27:2659–2662, 1994.
[6] F.F. Becker, X.-B. Wang, Y. Huang, R. Pethig, J. Vykoukal, and P.R.C. Gascoyne. Separation of Human Breast Cancer Cells from Blood by Differential Dielectric Affinity. *Proceedings of National Academy of Science,* 1995.
[7] D. Borkholder, I. Opris, N. Maluf, and G. Kovacs. Planar Electrode Array Systems for Neural Recording and Impedance Measurements. *Proceedings of Annual International Conference of the IEEE Engineering in Medicine and Biology*, 1996.

[8] L. Bousse, R.J. Mcreynolds, G. Kirk, T. Dawes, P. Lam, W.R. Bemiss, and J.W. Farce. *Sens. Actuat. B-Chem.*, 20(2–3):145–150, 1994.
[9] H. Chang, A. Ikram, F. Kosari, G. Vasmatzis, A. Bhunia, and R. Bashir. *J. Vac. Sci. Technol. B.*, 20(5):2058–2064, 2002.
[10] J.A. Chediak, Z.S. Luo, J.G. Seo, N. Cheung, L.P. Lee, and T.D. Sands. *Sens. Actu. A-Phys.*, 111(1):1–7, 2004.
[11] J. Dupont, D. Menard, C. Herve, F. Chevalier, B. Beliaeff, and B. Minier. *J. Appl. Bacteriol.*, 80(1):81–90, 1996.
[12] R. Eden and G. Eden. *Impedance microbiology*. Research Studies Press Ltd., John Wiley & Sons Inc., 1984.
[13] R. Ehret, W. Baumann, M. Brischwein, A. Schwinde, K. Stegbauer, and B. Wolf. *Biosens. Bioelectron.*, 12(1):29–41, 1997.
[14] J. Gimsa, P. Marszalek, U. Loewe, and T.Y. Tsong. *Biophys. J.* 60:749–60, 1991.
[15] R. Gomez. Microfabricated Device for Impedance-based Electronic Detection of Bacterial Metabolism. Ph.D thesis, School of Electrical and Computer Engineering, West Lafayette, IN, Purdue University, 2003.
[16] R. Gomez, R. Bashir, and A.K. Bhunia. *Sens. Actu B: Chem.*, 86:198–208, 2002.
[17] R. Gomez, R. Bashir, A. Sarikaya, M. Ladisch, J. Sturgis, J. Robinson, T. Geng, A. Bhunia, H. Apple, and S. Wereley. *Biomed. Microdev.*, 3(3):201–9, 2001.
[18] A. Gupta, D. Akin, and R. Bashir. Resonant Mass Biosensor For Ultrasensitive Detection Of Bacterial Cells. *Microfluidics, Biomems, and Medical Microsystems Conference at Spie's Photonics West Micromachining and Microfabrication 2003 Symposium. Proceedings of SPIE—The International Society for Optical Engineering*, San Jose, CA, 2003.
[19] A. Gupta, D. Akin, and R. Bashir. *Appl. Phys. Lett.*, 84(11):1976–1978, 2004.
[20] F. Hafner. *Biosen. Bioelectron.*, 15(3–4):149–158, 2000.
[21] J.W. Hong, V. Studer, G. Hang, W.F. Anderson, and S.R. Quake. *Nat. Biotechnol.*, 22(4):435–439, 2004.
[22] L. Hood and D. Galas. *Nature*, 421(6921):444–448, 2003.
[23] Y. Huang, K.L. Ewalt, M. Tirado, R. Haigis, A. Forster, D. Ackley, M.J. Heller, J.P. O'Connell, and M. Krihak. *Anal. Chem.*, 73(7):1549–1559, 2001.
[24] Y. Huang, R. Holzel, R. Pethig, and X.-B. Wang. *Phys. Med. Biol.*, 37(7):1499–1517, 1992.
[25] Y. Huang and R. Pethig. *Measurement Sci. Technol.*, 2:1142–46, 1991.
[26] M. P. Hughes. *Nanotechnology*, 11:124–132, 2000.
[27] M.P. Hughes. *Electrophoresis*, 23:2569–82, 2002.
[28] B. Ilic, D. Czaplewski, H.G. Craighead, P. Neuzil, C. Campagnolo, and C. Batt. *Appl. Phys. Lett.*, 77(3):450–452, 2000.
[29] T.B. Jones and J.P. Kraybill. *J. Appl. Phys.*, 60:1247–52, 1986.
[30] A.M. Jorgensen, K.B. Mogensen, J.P. Kutter, and O. Geschke. *Sens. Actuat. B-Chem.*, 90(1–3):15–21, 2003.
[31] K.V.I.S. Kaler and T.B. Jones. *Biophys. J.*, 57:173–82, 1990.
[32] C. Keese and I. Giaever. A biosensor that monitors cell morphology with electrical fields. *IEEE Eng. Med. Biol. Mag.*, 3:402–408, 1994.
[33] M. Koch, A. Evans, and A. Brunnschweiler. Design and Fabrication of a Micromachined Coulter Counter. *Proceedings of Ninth Micromechanics Europe Workshop—MME'98*, 155–158, 1998.
[34] G.T.A. Kovacs. *Micromachined Transducers Sourcebook*, Boston, MA, WCB/Mcgraw-Hill, 1998.
[35] L.J. Kricka. *Clinica. Chimica. Acta*, 307(1):219–223, 2001.
[36] D. Lange, C. Hagleitner, A. Hierlemann, O. Brand, and H. Baltes. *Anal. Chem.*, 74(13):3084–3095, 2002.
[37] M. Lehmann, W. Baumann, M. Brischwein, H.J. Gahle, I. Freund, R. Ehret, S. Drechsler, H. Palzer, M. Kleintges, U. Sieben, and B. Wolf. *Biosens. Bioelectron.*, 16(3):195–203, 2001.
[38] H. Li and R. Bashir. *Sens. Actuat. B: Chem.*, 86:215–221, 2002.
[39] H. Li and R. Bashir. *Biomed. Microdev.*, 6(4):289–295, 2004.
[40] H. Li, Y. Zheng, D. Akin, and R. Bashir. *IEEE J. Microelectromech. Syst.*, 14(1):105–111 2005.
[41] M.J. Madou. *Fundamentals of Microfabrication: the Science of Miniaturization*. Boca Raton, FL, CRC Press, 2002.
[42] M.P. Maher, J. Pine, J. Wright, and Y.C. Tai. *J. Neurosci. Meth.*, 87(1):45–56, 1999.
[43] G.H. Markx, Y. Huang, X.-F. Zhou, and R. Pethig. *Microbiology*, 140:585–91, 1994a.
[44] G.H. Markx and R. Pethig. *Biotechnol. Bioeng.*, 45:337–343, 1995.
[45] G.H. Markx, M.S. Talary, and R. Pethig. *J. Biotechnol.*, 32:29–37, 1994b.
[46] P. Marszalek, J.J. Zielinski, and M. Fikus. *Biotechnol. Bioeng.*, 22:289–98, 1989.

[47] S. Martinoia, M. Bove, M. Tedesco, B. Margesin, and M. Grattarola. *J. Neurosci. Meth.*, 87(1):35–44, 1999.
[48] M.A. McClain, C.T. Culbertson, S.C. Jacobson, N.L. Allbritton, C.E. Sims, and J.M. Ramsey. *Anal. Chem.*, 75(21):5646–5655, 2003.
[49] J.N. Mehrishi and J. Bauer. *Electrophoresis*, 23(13):1984–1994, 2002.
[50] W.E. Moerner and M. Orrit. *Science*, 283(5408):1670–1676, 1999.
[51] A. Mohr, W. Finger, K.J. Fohr, W. Gopel, H. Hammerle, and W. Nisch. *Sens. Actuat. B-Chem.*, 34(1–3):265–269, 1996.
[52] T. Muller, G. Gradl, S. Howitz, S. Shirley, T. Schnelle, and G. Fuhr. *Biosens. Bioelectron.*, 14:247–256, 1999.
[53] S.M. Nie and R.N. Zare. *Ann. Rev. Biophys. Biomol. Struct.*, 26:567–596, 1997.
[54] K.E. Petersen. Silicon as a Mechanical Material. *Proceedings of the IEEE*, 1982.
[55] H.A. Pohl. *J. Appl. Phys.*, 22:869–871, 1951.
[56] D. Polla, P. Krulevitch, A. Wang, G. Smith, J. Diaz, S. Mantell, J. Zhou, S. Zurn, Y. Nam, L. Cao, J. Hamilton, C. Fuller, and P. Gascoyne. MEMS based Diagnostic Microsystems. *Proceedings of 1st Annual International IEEE-EMBS Special Topic Conference on Microtechnology in Medicine and Biology*, 2000.
[57] D.L. Polla. *Microdev. Med. Annual Rev. Biomed. Engin.*, 2:551–576, 2000.
[58] A.T. Poortinga, R. Bos, and H.J. Busscher. *Biotechnol. Bioengin.*, 67(1):117–120, 2000.
[59] T. Schnelle, T. Muller, and G. Fuhr. *J. Electrostat.*, 50:17–29, 2000.
[60] L. Sohn, O. Saleh, G. Facer, A. Beavis, and D. Notterman. Capacitance Cyotmetry: Measuring Biological Cells One by One. *Proceedings of the National Academy of Science,* 2000.
[61] M. Stephens, M.S. Talary, R. Pethig, A.K. Burnett, and K.I. Mills. *Bone Marrow Transplant*, 18:777–782, 1996.
[62] D.L. Stokes, G.D. Griffin, and V.D. Tuan. *Fresenius J. Analyt. Chem.*, 369(3-4):295–301, 2001.
[63] J. Suehiro, R. Hamada, D. Noutomi, M. Shutou, and M. Hara. *J. Electrostat.*, 57(2):157–168, 2003.
[64] J. Suehiro, R. Yatsunami, R. Hamada, and M. Hara. *J. Phys. D-Appl. Phys.*, 32(21):2814–2820, 1999.
[65] H. Thielecke, T. Stieglitz, H. Beutel, T. Matthies, and J. Meyer. A Novel Cellpositioning Technique for Extracellular Recording and Impedance Measurements on Single Cells using Planar Electrode Substrates. *Proceedings of the 20th Annual Conference of the IEEE Engineering in Medicine and Biology Society*, 1998.
[66] K. Verhaegen, J. Simaels, W.V. Driessche, K. Baert, W. Sansen, B. Puers, L. Hermans, and R. Mertens. *Biomed. Microdev.*, 2(2):93–98, 1999.
[67] T. Vo-Dinh and B. Cullum. *Fresenius J. Analyt. Chem.*, 366(6–7):540–551, 2000.
[68] J. Voldman, R.A. Braff, M. Toner, M.L. Gray, and M.A. Schimidt. *Biophys. J.*, 80(1):531–41, 2001a.
[69] J.Voldman, R.A. Braff, M.Toner, M.L. Gray, and M.A. Schmidt. Quantitative design and analysis of single-particle dielectrophoretic traps. In V.D.B. et al. *Micro Total Analysis Systems 2000*, Kluwer Academic Publishers: 431–434, 2000.
[70] J. Voldman, M. Toner, M. L. Gray, and M.A. Schmidt. *Transducers*, 01:322–325, 2001b.
[71] J. Voldman, M. Toner, M.L. Gray, and M.A. Schmidt. *J. Electrostat.*, 57:69–90 2003.
[72] X.-B. Wang, Y. Huang, J.P.H. Burt, G.H. Markx, and R. Pethig. *J. Phys. D: Appl. Phys.*, 26:1278–1285, 1993.
[73] X.-B. Wang, Y. Huang, P.R.C. Gascoyne, and F.F. Becker. *IEEE Transact. Indust. Appl.*, 33(3):660–9, 1997.
[74] M. Washizu, T. Nanba, and S. Masuda. *IEEE Transact. Indust. Appl.*, 26:352–358, 1990.
[75] B.H. Weigl, R.L. Bardell, and C.R. Cabrera. *Advanc. Drug Del. Rev.* 55(3):349–377, 2003.
[76] G.M. Whitesides. *Nat. Biotechnol.* 21(10):1161–1165, 2003.
[77] K.D. Wise and K. Najafi. *Science*, 254:1335–1342, 1991.
[78] B. Wolf, M. Brischwein, W. Baumann, R. Ehret, and M. Kraus. *Biosens. Bioelectron.*, 13(5):501–509, 1998.
[79] J. Yang, Y. Huang, X.-B.Wang, F.F. Becker, and P.R.C. Gascoyne. *Analyt. Chem.*, 71(5):911–918, 1999.

10
Implantable Wireless Microsystems

Babak Ziaie
*School of Electrical and Computer Engineering, Purdue University,
W. Lafayette, IN 47907*

10.1. INTRODUCTION

The ability to use wireless techniques for measurement and control of various physiological parameters inside human body has been a long-term goal of physicians and biologists going back to the early days of wireless communication. From early on, it was recognized that this capability could provide effective diagnostic, therapeutic, and prosthetic tools in physiological research and pathological intervention. However, this goal eluded scientists prior to the discovery of transistor in 1947. Vacuum tubes were too bulky and power hungry to be of any use in implantable systems. During the late 50's, *MacKay* performed his early pioneering work on what he called "Endoradiosonde" [1]. This was a single-transistor blocking oscillator designed to be swallowed by a subject and was able to measure pressure and temperature in the digestive track. Following this early work came a number of other simple discrete systems each designed to measure a specific parameter (temperature, pressure, force, flow, etc.) [2]. By the late 60's, progress in the design and fabrication of integrated circuits provided an opportunity to expand the functionality of these early systems. Various hybrid single and multichannel telemetry systems were developed during the 70's and the 80's [3]. In addition, implantable therapeutic and prosthetic devices started to appear in the market. Cardiac pacemakers and cochlear prosthetics proved effective and reliable enough to be implanted in thousands of patients. Recent advances in microelectromechanical (MEMS) based transducer and packaging technology, new and compact power sources (high efficiency inductive powering and miniature batteries), and CMOS low-power wireless integrated circuits have provided another major impetus to the development of wireless implantable microsystems [4–9]. These advances have created new opportunities for increased reliability and functionality, which had been hard to achieve with pervious technologies. Furthermore, the burgeoning area of nanotechnology is poised to further enhance these capabilities beyond what have been achievable using MEMS techniques. This is particularly true in the biochemical sensing and chemical delivery areas and will undoubtedly have a major impact on the future generations of implantable wireless microsystems. In this article, we present some of these recent advances in the context of several devices currently being pursued in academia and industry. The fact that these systems are designed to be implanted creates regulatory concerns, which has contributed to their late arrival in the clinical market. In the following sections, after discussing several major components of such microsystems such as transducers, interface electronics, wireless communication, power sources, and packaging; we will present some selected examples to demonstrate the state of the art. Although we have separated these microsystems under diagnostic, therapeutic, and prosthetic categories; this division is not always representative and systems that can be considered diagnostic in one situation may represent a therapeutic device under another circumstance.

10.2. MICROSYSTEM COMPONENTS

For the purpose of current discussion implantable wireless microsystems can be defined as a group of medical microdevices that: 1) incorporate one or several MEMS-based transducers (i.e., sensors and actuators), 2) have an on-board power supply (i.e., battery) or are powered from outside using inductive coupling, 3) can communicate with outside (bi-directional or uni-directional) through an RF interface, 4) have on-board signal processing capability, 5) are constructed using biocompatible materials, and 6) use advanced MEMS-based packaging techniques. Although one microsystem might incorporate all of the above components, the demarcation line is rather fluid and can be more broadly interpreted. For example, passive MEMS-based microtransponders do not contain on-board signal processing capability but use advanced MEMS packaging and transducer technology and are usually considered to be wireless microsystems. We should also emphasize that the above components are inter-related and a good system designer must pay considerable attention from the onset to this fact. For example one might have to choose a certain power source or packaging scheme to accommodate the desired transducer, interface electronics, and wireless communication.

10.2.1. Transducers

Transducers are interfaces between biological tissue and readout electronics/signal processing and their performance is critical to the success of the overall microsystem [10–14]. Current trend in miniaturization of transducers and their integration with signal processing circuitry has considerably enhanced their performance. This is particularly true with respect to MEMS-based sensors and actuators where the advantages of miniaturization have been prominent. Development in the area of microactuators has been lagging behind the microsensors due to the inherent difficulty in designing microdevices that efficiently and reliably generate motion. Although some transducing schemes such as electrostatic force generation has advantageous scaling properties in the microdomain, problems associated with packaging and reliability has prevented their successful application. MEMS-based microsensors have been more successful and offer several advantages compared to the macrodomain counterparts. These include lower power consumption, increased sensitivity, higher reliability, and lower cost due to batch fabrication. However, they suffer from a poor signal to noise ratio hence requiring a close by interface circuit. Among the many microsensors designed and fabricated over the past two decades, physical sensors have been by and large more successful. This is due to their inherent robustness and isolation from any direct contact with biological tissue in sensors such as accelerometers and gyroscopes. Issues related to packaging and long-term stability have plagued the implantable chemical sensors. Long-term baseline and sensitivity stability are major problems associated with implantable sensors. Depending on the type of the sensor several different factors contribute to the drift. For example in implantable pressure sensors, packaging generated stresses due to thermal mismatch and long-term material creep are the main sources of baseline drift. In chemical sensors, biofouling and fibrous capsule formation is the main culprit. Some of these can be mitigated through clever mechanical design and appropriate choice of material, however, some are more difficult to prevent (e.g., biofouling and fibrous capsule formation). Recent developments in the area of anti-fouling material and controlled release have provided new opportunities to solve some of these long standing problems [15–17].

10.2.2. Interface Electronics

As was mentioned previously, most MEMS-based transducers suffer from poor signal to noise ratio and require on-board interface electronics. This of course is also more so essential for implantable microsystems. The choice of integrating the signal processing with the MEMS transducer on the same substrate or having a separate signal processing chip in close proximity depends on many factors such as process complexity, yield, fabrication costs, packaging, and general design philosophy. Except post-CMOS MEMS processing methods, which rely on undercutting micromechanical structures subsequent to the fabrication of the circuitry [18], other integrated approaches require extensive modifications to the standard CMOS processes and have not been able to attract much attention. Post-CMOS processing is an attractive approach although packaging issues still can pose roadblocks to successful implementation. Hybrid approach has been typically more popular with the implantable microsystem designers providing flexibility at a lower cost. Power consumption is a major design consideration in implantable wireless microsystems that rely on batteries for an energy source. Low-power and sub-threshold CMOS design can reduce the power consumption to nW levels [19–23]. Important analog and mixed-signal building blocks for implantable wireless microsystems include amplifiers, oscillators, multiplexers, A/D and D/A converters, and voltage references. In addition many such systems require some digital signal processing and logic function in the form of finite-state machines. In order to reduce the power consumption, it is preferable to perform the DSP functions outside the body although small finite-state machines can be implemented at low power consumptions.

10.2.3. Wireless Communication

Bi or uni-directional wireless communication is a central feature in all implantable wireless microsystems. In systems that are powered from outside using inductive link bi-directional communication can be achieved using several techniques [24]. The inward link can be easily implemented using amplitude modulation, i.e., the incoming RF signal that powers the microsystem is modulated by digitally varying the amplitude. It is evident that the modulation index can not be 100% since that would cut off the power supply to the device (unless a storage capacitor is used). The coding scheme is based on the pulse time duration, i.e., "1" and "0" have the same amplitude but different durations [5, 25]. This modulation technique requires a simple detection circuitry (envelope detector) and is immune to amplitude variations, which are inevitable in such systems. The outward link can be implemented using two different techniques. One relies on "load modulation", i.e., the outgoing digital stream of data is used to load the receiver antenna [26]. This can be picked up through the transmitter coil located outside the body. The second technique is more complex and requires an on-chip transmitter and a second coil to transmit the recorded data at a different frequency.

Battery operated implantable wireless microsystems rely on different communication schemes. Outward data transmission can be accomplished using any of the several modulation schemes (AM, FM, and other pulse modulation methods) which offer standard trade offs between transmitter and receiver circuit complexity, power consumption, and signal to noise ratio [27]. Inward transmission of data can also be accomplished in a similar fashion. Typical frequencies used in such systems are in the lower UHF range (100–500 MHz). Higher frequencies result in smaller transmitter antenna at the expense of increased tissue loss. Although tissue loss is a major concern in transmitting power to implantable microsystems; it is less of an issue in data transmission since a sensitive receiver outside the body can easily demodulate the signal. Recent explosive proliferation of wireless communication systems (cell phones, wireless PDAs, Wi-Fi systems, etc.) have provided a unique opportunity to piggy back major RFIC manufacturers and simplify the design of implantable microdevices [28–30]. This can not only increase the performance of the system but also creates a standard platform for many diverse applications. Although the commercially available wireless chips have large bandwidths and some superb functionality, their power consumption is higher than what is acceptable for many of the implantable microsystems. This however, is going to be changed in the future by the aggressive move towards lower power handheld consumer electronics.

10.2.4. Power Source

The choice of power source for implantable wireless microsystems depends on several factors such as implant lifetime, system power consumption, temporal mode of operation (continuous or intermittent), and size. Progress in battery technology is incremental and usually several generations behind other electronic components [31]. Although lithium batteries have been used in pacemakers for several years, they are usually large for microsystem applications. Other batteries used in hearing aids and calculators are smaller but have limited capacity and can only be used for low power systems requiring limited lifespan or intermittent operation. Inductive powering is an attractive alternative for systems with large power requirements (e.g., neuromuscular stimulators) or long lifetime (e.g., prosthetic systems with >5 years lifetime) [7]. In such systems a transmitter coil is used to power a microchip using magnetic coupling. The choice of the transmission frequency is a trade off between adequate miniaturization and tissue loss. For implantable microsystems the frequency range of 1–10 MHz is usually considered optimum for providing adequate miniaturization while still staying below the high tissue absorption region (>10 MHz) [32]. Although the link analysis and optimization methods have been around for many years [33], recent integration techniques that allow the fabrication of microcoils on top of CMOS receiver chip has allowed a new level of miniaturization [34]. For applications that require the patient carry the transmitter around, a high efficiency transmitter is required in order to increase the battery lifetime. This is particularly critical in implantable microsystem, where the magnetic coupling between the transmitter and the receiver is low (<1%). Class-E power amplifier/transmitters are popular among microsystem designers due to their high efficiency (>80%) and relatively easy design and construction [5, 35, 36]. They can also be easily amplitude modulated through supply switching.

Although ideally one would like to be able to tap into the chemical reservoir (i.e., glucose) available in the body to generate enough power for implantable microsystems (glucose based fuel cell), difficulty in packaging and low efficiencies associated with such fuel cells have prevented their practical application [37]. Thin-film batteries are also attractive, however, there still remain numerous material and integration difficulties that need to be resolved [38]. Another alternative is nuclear batteries. Although they have been around for several decades and were used in some early pacemakers, safety and regulatory concerns forced medical device companies to abandon their efforts in this area. There has been a recent surge of interest in microsystem nuclear batteries for military applications [39]. It is not hard to envision that due to the continuous decrease in chip power consumption and improve in batch scale MEMS packaging technology, one might be able to hermetically seal a small amount of radioactive source in order to power an implantable microsystem for a long period of time. Another possible power source is the mechanical movements associated with various organs. Several proposals dealing with parasitic power generation through tapping into this energy source have been suggested in the past few years [40]. Although one can generate adequate power from activities such as walking to power an external electronic device, difficulty in efficient mechanical coupling to internal organ movements make an implantable device hard to design and utilize.

10.2.5. Packaging and Encapsulation

Proper packaging and encapsulation of implantable wireless microsystems is a challenging design aspect of such microdevices. The package must accomplish two tasks simultaneously: 1) protect the electronics from the harsh body environment while providing access windows for transducers to interact with the desired measurand, and 2) protect the body from possible hazardous material in the microsystem. The second task is easier to fulfill since there is a cornucopia of various biocompatible materials available to the implant designer [41]. For example, silicon and glass, which are the material of choice in many MEMS applications, are both biocompatible. In addition, polydimethylsiloxane (PDMS) and several other polymers (e.g., polyimide, polycarbonate, parylene, etc.) commonly used in microsystem design are also accepted by the body. The first requirement is however more challenging. The degree of protection required for implantable microsystems depends on the required lifetime of the device. For short durations (several months) polymeric encapsulants might be adequate if one can conformally deposit them over the substrates (e.g., plasma deposited parylene) [42]. These techniques are considered non-hermetic and have limited lifetime. For long term operation, hermetic sealing techniques are required [43]. Although pacemaker and defibrillator industries have been very successful in sealing their systems in tight titanium enclosures; these techniques are not suitable for microsystem applications. For example a metallic enclosure prevents the transmission of power and data to the microsystem. In addition, these sealing methods are serial in nature (e.g., laser or electron beam welding) and not compatible with integrated batch fabrication methods used in microsystem design. Silicon-glass electrostatic and silicon-silicon fusion bonding are attractive methods for packaging implantable microsystems [44]. Both of these bonding methods are hermetic and can be performed at the wafer level. These are particularly attractive for inductively powered wireless microsystems since most batteries can not tolerate the high temperatures required in such substrate bondings. Other methods such as metal electroplating have also been used to seal integrated MEMS microsystems. However, their long term performance is usually inferior to the anodic and fusion bondings. In addition to providing a hermetic seal, the package must allow feeedthrough for transducers located outside the package [45]. In macrodevices such as pacemakers where the feedthrough lines are large and not too many, traditional methods such as glass-metal or ceramic-metal has been employed for many years. In microsystems such methods are not applicable and batch scale techniques must be adopted.

10.3. DIAGNOSTIC MICROSYSTEMS

Diagnostic wireless microsystems are used to gather physiological or histological information from within the body in order to identify pathology. In this category, we will discuss two recent examples. The first one is a microsystem designed to be implanted in the eye and measure the intraocular pressure in order to diagnose low-tension glaucoma. The second system although not strictly implanted is an endoscopic wireless camera-pill designed to be swallowed in order to capture images from the digestive track.

Figure 10.1 shows the schematic diagram of the intraocular pressure (IOP) measurement microsystem [46, 47]. This device is used to monitor the IOP in patients suffering from

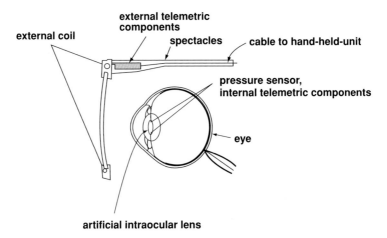

FIGURE 10.1. Schematic of the IOP measurement microsystem [46].

IMPLANTABLE WIRELESS MICROSYSTEMS

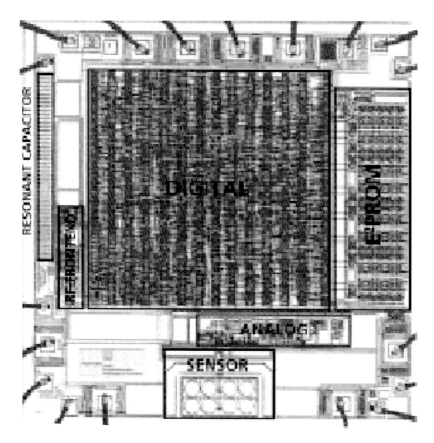

FIGURE 10.2. Micrograph of the IOP measurement microsystem receiver chip showing surface micromachined capacitive pressure sensors and other parts of the receiver circuitry [47].

low tension glaucoma, i.e., the pressure measured in the doctor's office is not elevated (normal IOP is ~10–20 mmHg) while the patient is showing optics nerve degeneration associated with glaucoma. There is a great interest in measuring the IOP in such patients during their normal course of daily activity (exercising, sleeping, etc). This can only be achieved using a wireless microsystem. The system shown in Figure 10.1 consists of an external transmitter mounted on a spectacle, which is used to power a microchip implanted in the eye. A surface micromachined capacitive pressure sensor integrated with CMOS interface circuit is connected to the receiving antenna. The receiver chip implemented in an n-well 1.2 μm CMOS technology has overall dimensions of 2.5×2.5 mm^2 and consumes 210 μW, Figure 10.2. The receiver polyimide-based antenna is however much larger (1 cm in diameter and connected to the receiver using flip chip bonding) requiring the device to be implanted along with an artificial lens. The incoming signal frequency is 6.78 MHz while the IOP is transmitted at 13.56 MHz using load-modulation scheme. This example illustrates the levels of integration that can be achieved using low power CMOS technology, surface micromachining, and flip chip bonding.

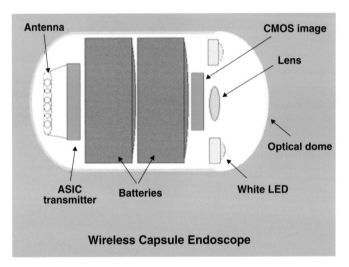

FIGURE 10.3. A photograph (top) and internal block diagram (bottom) of Given Imaging wireless endoscopic pill. (Courtesy Given Imaging)

The second example in the category of diagnostic microsystems is an endoscopic wireless pill shown in Figure 10.3 [48–50]. This pill is used to image small intestine, which is a particularly hard area to reach using current fiber optic technology. Although these days colonoscopy and gastroscopy are routinely performed, they can not reach the small intestine and many disorders (e.g., frequent bleeding) in this organ have eluded direct examination. A wireless endoscopic pill can not only image the small intestine, but also will reduce the pain and discomfort associated with regular gastrointestinal endoscopies. The endoscopic pill is a good example of capabilities offered by advanced consumer microelectronics. Although

IMPLANTABLE WIRELESS MICROSYSTEMS

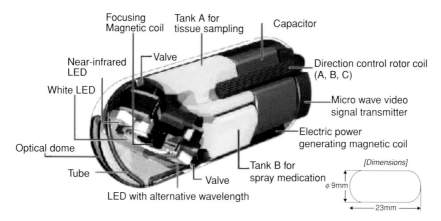

FIGURE 10.4. A schematic of the RF Systems wireless video pill. (Courtesy RF Systems Lab)

the idea of a video pill is not new, before the development of low-power microelectronics, white LED, CMOS image sensor, and wide band wireless communication, fabrication of such a device was not feasible. The video pill currently marketed by Given Imaging is 11 mm in diameter and 30 mm in length (size of a large vitamin tablet) and incorporates: 1) a short focal length lens, 2) a CMOS image sensor (90,000 pixel), 3) four white LEDs, 4) a low power ASIC transmitter, and 5) two batteries (enough to allows the pill to go though the entire digestive track). The pill can capture and transmit 2 images/second to an outside receiver capable of storing up to 5 hours of data. Another company (RF Systems Lab.) is also developing a similar microsystem using a higher resolution CCD (410,000 pixel) camera (30 images/second) and inductive powering, Figure 10.4 [51].

10.4. THERAPEUTIC MICROSYSTEMS

Therapeutic microsystems are designed to alleviate certain symptoms and help in the treatment of a disease. In this category, we will describe two such microsystems. The first one is a drug delivery microchip designed to administer small quantities of potent drugs upon receiving a command signal from the outside. The second device is a passive micromachined glucose transponder, which can be used to remotely monitor glucose fluctuations allowing a tighter blood glucose control through frequent measurements and on-demand insulin delivery (pump therapy or multiple injections)

Figure 10.5 shows the central component of the drug delivery microchip [52, 53]. It consists of several micro-reservoirs (25 nl in volume) etched in a silicon substrate. Each micro-reservoir contains the targeted drug and is covered by a thin gold membrane (3000 Å), which can be dissolved through the application of a small voltage (1 V vs. SCE), Figure 10.5. The company marketing this technology (MicroCHIPS Inc.) is in the process of designing a wireless transceiver that can be used to address individual wells and release the drug upon the reception of the appropriate signal [54]. Another company (ChipRx Inc.) is also aiming to develop a similar microsystem (Smart Pill), Figure 10.6 [55]. Their release approach however is different and is based on conductive polymer actuators acting similar to a

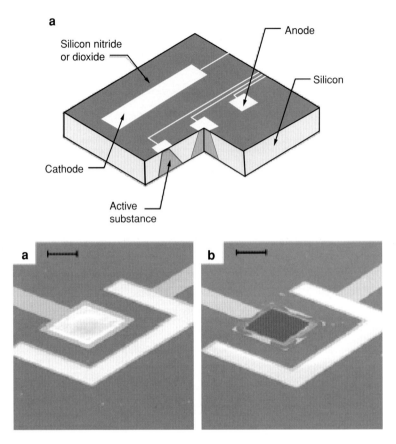

FIGURE 10.5. MicroCHIP drug delivery chip (top), a reservoir before and after dissolution of the gold membrane (bottom) [52].

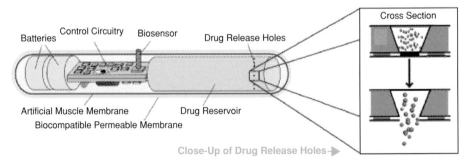

FIGURE 10.6. Schematic of the ChipRx smart drug delivery capsule. (Courtesy ChipRx)

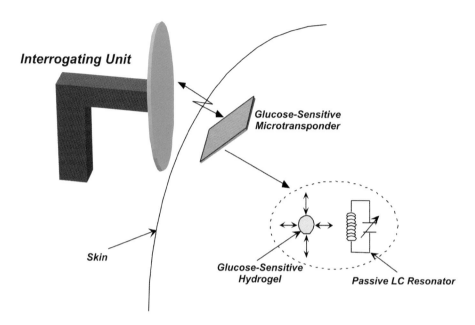

FIGURE 10.7. Basic concept behind the glucose-sensitive microtransponder.

sphincter, opening and closing a tiny reservoir. Due to the potency of many drugs, safety and regulatory issues are more stringent in implantable drug delivery microsystems and will undoubtedly delay their appearance in the clinical settings.

Figure 10.7 shows the basic concept behind the glucose-sensitive microtransponder [56]. A miniature MEMS-based microdevice is implanted in the subcutaneous tissue and an interrogating unit remotely measures the glucose levels without any hardwire connection. The microtrasponder is a passive LC resonator, which is coupled to a glucose-sensitive hydrogel. The glucose-dependent swelling and de-swelling of the hydrogel is coupled to the resonator causing a change the capacitor value. This change translates into variations of the resonant frequency, which can be detected by the interrogating unit. Figure 10.8 shows

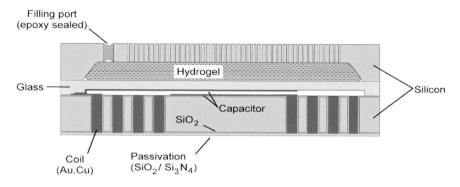

FIGURE 10.8. Cross section of glucose microtransponder.

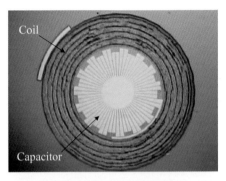
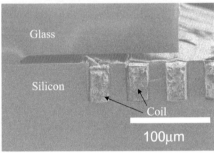

FIGURE 10.9. Optical micrograph and SEM cross section of glucose sensitive microtransponders.

the schematic drawing of the micro-transponder with a capacitive sensing mechanism. The glucose sensitive hydrogel is mechanically coupled to a glass membrane and is separated from body fluids (in this case interstitial fluid) by a porous stiff plate. The porous plate allows the unhindered flow of water and glucose while blocking the hydrogel from escaping the cavity. A change in the glucose concentration of the external environment will cause a swelling or de-swelling of the hydrogel, which will deflect the glass membrane and change the capacitance. The coil is totally embedded inside the silicon and can achieve a high quality factor and hence increased sensitivity by utilizing the whole wafer thickness (reducing the series resistance). The coil-embedded silicon and the glass substrate are hermetically sealed using glass-silicon anodic bonding. Figure 10.9 shows the optical micrograph of several devices and an SEM cross section showing the glass membrane and embedded coil.

10.5. REHABILITATIVE MICROSYSTEMS

Rehabilitative microsystems are used to substitute a lost function such as vision, hearing, or motor activity. In this category, we will describe two microsystems. The first one is a single channel neuromuscular microstimulator used to stimulate paralyzed muscle groups in paraplegic and quadriplegic patients. The second microsystem is a visual prosthetic device designed to stimulate ganglion cells in retina in order to restore vision to people afflicted with macular degeneration or retinitis pigmentosa.

Figure 10.10 shows a schematic of the single channel microstimulator [25]. This device is $10 \times 2 \times 2$ mm^3 in dimensions and receives power and data through an inductively coupled link. It can be used to stimulate paralyzed muscle groups using thin-film microfabricated

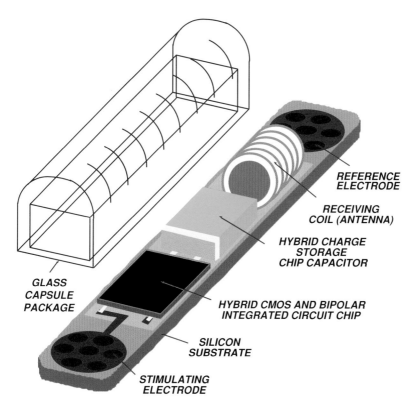

FIGURE 10.10. Schematic of a single channel implantable neuromuscular microstimulator.

electrodes located at the ends of a silicon substrate. A hybrid capacitor is used to store the charge in between the stimulation pulses and to deliver 10 mA of current to the muscle every 25 msec. A glass capsule hermetically seals a BiCMOS receiver circuitry along with various other passive components (receiver coil and charge storage capacitor) located on top of the silicon substrate. Figure 10.11 shows a photograph of the microstimulator in the bore of a gauge 10 hypodermic needle. As can be seen, the device requires a complicated hybrid assembly process in order to attach a wire-wound coil and a charge storage capacitor to the receiver chip. In a subsequent design targeted for direct peripheral nerve stimulation (requiring smaller stimulation current), the coil was integrated on top of the BiCMOS electronics and on-chip charge storage capacitors were used thus considerably simplifying the packaging process. Figure 10.12 shows a micrograph of the chip with the electroplated copper inductor [57].

Figure 10.13 shows the schematic of the visual prosthetic microsystem [46]. A spectacle mounted camera is used to capture the visual information followed by digital conversion and transmission of data to a receiver chip implanted in the eye. The receiver uses this information to stimulate the ganglion cells in the retina through a microelectrode array in sub or epi-retinal location. This microsystem is designed for patients suffering from macular degeneration or retinitis pigmentosa. In both diseases the light sensitive retinal cells (cones and rods) are destroyed while the more superficial retinal cells, i.e., ganglion cells, are still viable and can be stimulated. Considering that macular degeneration is an

FIGURE 10.11. Photograph of the microstimulator in the bore of a gage 10 hypodermic needle.

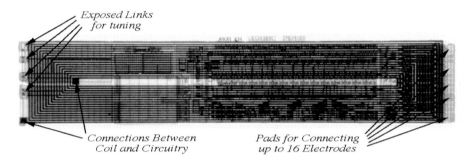

FIGURE 10.12. Microstimulator chip with integrated receiver coil and on-chip storage capacitor [57].

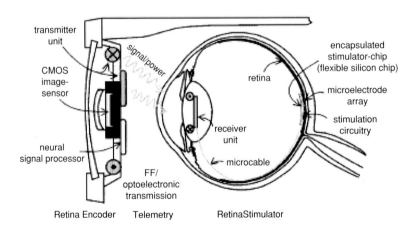

FIGURE 10.13. Schematic of a visual prosthetic microsystem [46].

IMPLANTABLE WIRELESS MICROSYSTEMS

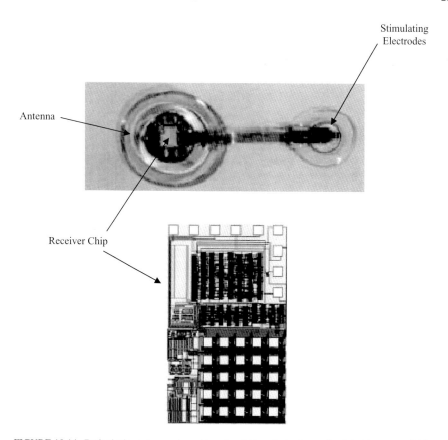

FIGURE 10.14. Retinal stimulator receiver chip, stimulating electrodes, and polyimide antenna [46].

age related pathology and will be afflicting more and more people as the average age of the population increases, such a microsystem will be of immense value in the coming decades. There are several groups pursuing such a device with different approaches to electrode placement (epi or sub retinal), chip design, and packaging. A German consortium which has also designed the IOP measurement microsystem is using a similar approach in antenna placement (receiver antenna in the lens), chip design, and packaging technology to implement a retinal prosthesis [46]. Figure 10.14 shows photographs of the retinal stimulator receiver chip, stimulating electrodes, and polyimide antenna. The effort in the United States is moving along a similar approach [58, 59].

10.6. CONCLUSIONS AND FUTURE DIRECTIONS

In this article, we reviewed several implantable wireless microsystems currently being developed in the academia and industry. Recent advances in MEMS-based transducers, low-power CMOS integrated circuit, wireless communication transceivers, and advanced batch scale packaging have provided a unique opportunity to develop implantable wireless microsystems with advanced functionalities not achievable previously. These microsystems will be indispensable to the 21st century physician by providing assistance in diagnosis and treatment. Future research and development will probably be focused on three areas: 1) nano-transducers, 2) self-assembly, and 3) advanced biomaterials. Although MEMS-based sensors and actuators have been successful in certain areas (particularly physical sensors), their performance could be further improved by utilizing nano-scale fabrication technology. This is particularly true in the area of chemical sensors where future diagnostic depends on detecting very small amounts of chemicals (usually biomarkers) well in advance of any physical sign. Nanosensors capable of high sensitivity chemical detection will be part of the future implantable microsystems. In the actuator/delivery area, drug delivery via nanoparticles is a burgeoning area which will undoubtedly be incorporated into future therapeutic microsystems. Future packaging technology will probably incorporate self-assembly techniques currently being pursued by many micro/nano research groups. This will be particularly important in microsystems incorporating multitude of nanosensors. Finally, advanced nano-based biomaterials will be used in implantable microsystems (wireless or not) in order to enhance biocompatibility and prevent biofouling. These will include biocompatible surface engineering and interactive interface design (e.g., surfaces that release anti-inflammatory drugs in order to reduce post implant fibrous capsule formation).

REFERENCES

[1] R.S. MacKay and B. Jacobson. *Nature*, 179:1239–1240, 1957.
[2] R.S. MacKay. *Eng. Med. Biol. Mag.*, 2:11–17, 1983.
[3] J.W. Knutti, H.V. Allen, and J.D. Meindl. *Eng. Med. Biol. Mag.*, 2:47–50, 1983.
[4] K.D. Wise (ed.). *Proceedings of the IEEE*, 86:1998.
[5] P.R. Troyk and M.A.K. Schwan. *IEEE Transact. Biomed. Eng.*, 39:589–599, 1992.
[6] Z. Hamici, R. Itti, and J. Champier. *Meas. Sci. Technol.*, 7:192–201, 1996.
[7] W.J. Heetderks. *IEEE Trans. Biomed. Eng.*, 35:323–327, 1988.
[8] P.R. Gray and R.G. Meyer. *Proceedings of the Custom Integrated Circuits Conference*, pp. 83–89, 1995.
[9] A.A. Abidi. *IEEE Microw. Mag.*, 4:47–60, 2003.
[10] W. Gopel, J. Hesse, and J.N. Zemel. *Sensors: A Comprehensive Survey*, Vols. 1–8, VCH Publishers, New York, 1989.
[11] J.M. Hohler and H.P. Sautz (ed.). *Microsystem Technology: A Powerful Tool for Biomolecular Studies*. Birkhauser, Boston, 1999.
[12] R.F. Taylor and J.S. Schultz. *Handbook of Chemical and Biological Sensors*. IOP Press, Bristol, 1996.
[13] E.K. Rogers (ed.). *Handbook of Biosensors and Electronic Nose*. CRC Press, Boca Raton, 1997.
[14] G.A. Hak (ed.). *The MEMS Handbook*. CRC Press, Boca Raton, 2001.
[15] M. Zhang, T. Desai, and M. Ferrari. *Biomaterials*, 19:953–960, 1998.
[16] D.W. Branch, B.C. Wheeler, G.J. Brewer, and D.E. Leckband. *Biomaterials*, 22:1035–1047, 2001.
[17] N.A. Alcantar, E.S. Aydil, and J.N. Israelachvili. *J. Biomed. Mater. Res.*, 51:343–351, 2000.
[18] H. Baltes, O. Paul, and O. Brand. *Proc. IEEE*, 86:1660–1678, 1998.
[19] L.J. Stotts. *IEEE Cir. Dev. Mag.*, 5:12–18, 1989.
[20] T. Stouraitis and V. Paliouras. *IEEE Cir. Dev. Mag.*, 17:22–29, 2001.
[21] Y. Tsividis, N. Krishnapura,Y. Palakas, and L. Toth. *IEEE Cir. Dev. Mag.*, 19:63–72, 2003.
[22] L. Benini, G. De Micheli, and E. Macii. *IEEE Cir. Sys. Mag.*, 1:6–25, 2001.
[23] S.S. Rajput and S.S. Jamuar. *IEEE Cir. Sys. Mag.*, 2:24–42, 2002.
[24] P. Troyk. *Ann. Rev. Biomed. Eng.*, 1:177–209, 1999.
[25] B. Ziaie, M. Nardin, A.R. Coghlan, and K. Najafi. *IEEE Trans. Biomed. Eng.*, 44:909–920, 1997.
[26] K. Finkenzeller. *RFID Handbook*, John Wiley, New York, 2003.
[27] J.G. Proakis and M. Salehi. *Communication System Engineering*. Pearson Education, 2001.

[28] T.H. Lee. *Design of CMOS Radiofrequency Integrated Circuits*. Cambridge University Press, Cambridge, 1998.
[29] B. Razavi. *IEEE Cir. Dev. Mag.*, 12:12–25, 1996.
[30] L.E. Larson. *IEEE J. Solid-State Cir.*, 33:387–399, 1998.
[31] D. Linden and T. Reddy. *Handbook of Batteries*. McGraw-Hill, New York, 2001.
[32] K.R. Foster and H.P. Schwan. In C. Polk and E. Postow (eds.), *Handbook of Biological Effects of Electromagnetic Fields*. CRC Press, Boca Raton, 1996.
[33] W.H. Ko, S.P. Liang, and C.D.F. Fung. *Med. Biol. Eng. Comput.*, 15:634–640, 1977.
[34] K.B. Ashby, I.A. Koullias,W.C. Finley, J.J. Bastek, and S. Moinian. *IEEE J. Solid-State Cir.*, 31:4–9, 1996.
[35] N.O. Sokal and A.D. Sokal. *IEEE J. Solid-State Cir.*, 10:168–176, 1975.
[36] B. Ziaie, S.C. Rose, M.D. Nardin, and K. Najafi. *IEEE Trans. Biomed. Eng.*, 48:397–400, 2001.
[37] V. Mehta and J.S. Cooper. *J. Power Sou.*, 114:32–53, 2003.
[38] D. Singh, R. Houriet, R. Giovannini, H. Hofmann, V. Craciun and R.K. Singh. *J. Power Sou.*, 97–98:826–831, 2001.
[39] H. Guo and A. Lal. *Proceeding of the Transducers Conference*, pp. 36–39, 2003.
[40] T. Starner. *IBM J. Sys.*, 35:618–629, 1996.
[41] B.D. Ratner, F.J. Schoen, A.S. Hoffman, and J.E. Lemons. *Biomaterials Science: An Introduction to Materials in Medicine*. Elsevier Books, New York, 1997.
[42] G.E. Loeb, M.J. Bak, M. Salcman, and E.M. Schmidt. *IEEE Trans. Biomed. Eng.*, 24:121–128, 1977.
[43] M.F. Nichols. *Critical Rev. Biomed. Eng.*, 22:39–67, 1994.
[44] M.A. Schmidt. *Proc. IEEE*, 86:1575–1585, 1998.
[45] B. Ziaie, J.A. Von Arx, M.R. Dokmeci, and K. Najafi. *IEEE J. Microelectromech. Sys.*, 5:166–179, 1996.
[46] W. Mokwa and U. Schenakenberg. *IEEE Trans. Instrument. Measure.*, 50:1551–1555, 2001.
[47] K. Stangel, S. Kolnsberg, D. Hammerschmidt, B.J. Hosticka, H.K. Trieu, and W. Mokwa. *IEEE J. Solids-State Cir.*, 36:1094–1100, 2001.
[48] G. Iddan, G. Meron, A. Glukhovsky, and P. Swain. *Nature*, 405:417, 2000.
[49] D.G. Adler and C.J. Gostout. *Hospital Physician*, 14–22, 2003.
[50] http://www.givenimaging.com.
[51] http://www.rfnorika.com/.
[52] J.T. Santini, M.J. Cima, and R. Langer. *Nature*, 397:335–338, 1999.
[53] J.T. Santini, A.C. Richards, R. Scheidt, M.J. Cima, and R. Langer. *Angewandte Chemie.*, 39:2396–2407, 2000.
[54] http://www.mchips.com.
[55] http://www.chiprx.com.
[56] M. Lei, A. Baldi, T. Pan, Y. Gu, R.A. Siegel, and B. Ziaie. *Proc. IEEE MEMS*, 391–394, 2004.
[57] J.A. Von Arx and K. Najafi. *IEEE Solid-State Circuits Conference.* pp. 15–17, 1999.
[58] W. Liu, E. McGucken, K. Vichienchom, S.M. Clements, S.C. Demarco, M. Humayun, E. de Juan, J. Weiland, and R. Greenberg. *IEEE Systems, Man, and Cybernetics, Conference*, pp. 364–369, 1999.
[59] M.S. Humayun, J. Weiland, B. Justus, C. Merritt, J. Whalen, D. Piyathaisere, S.J. Chen, E. Margalit, G. Fujii, R.J. Greenberg, E. de Juan, D. Scribner, and W. Liu. *Proceedings of the 23rd Annual IEEE EMBS Conference.* pp. 3422–3425, 2001.

11

Microfluidic Tectonics*

J. Aura Gimm and David J. Beebe
*Department of Biomedical Engineering, University of Wisconsin-Madison,
WI 53706*

11.1. INTRODUCTION

Microfluidics has the potential to significantly change the way modern biology is performed. Microfluidic devices offer the ability to work with smaller reagent volumes, shorter reaction times, and the possibility of parallel operation. They also hold the promise of integrating an entire laboratory onto a single chip [23]. In addition to the traditional advantages conferred by miniaturization, the greatest potential lies in the physics of the scale. By understanding and leveraging micro scale phenomena, microfluidics can be used to perform techniques and experiments not possible on the macroscale allowing new functionality and experimental paradigms to emerge. Two examples of devices commonly considered microfluidic are gene chips and capillary electrophoresis. While gene chips take advantage of some of the benefits of miniaturization, they are not technically microfluidic devices. Chip-based capillary electrophoresis devices are now commercially available and reviews are available elsewhere [19, 28]. Certain fluid phenomena are dominant at the microscale and affect how devices can be made and used. Current techniques for making the devices will be outlined and examples will be given with an emphasis on a recently developed organic technology platform called microfluidic tectonics. Components of microdevices capable of actuating, sensing, and measuring within microfluidic systems will be discussed. Finally, complete systems that have been developed to perform functions in biology will be described.

---

* The research was generously funded by grants from DARPA-MTO (#F30602-00-1-0570)(Program manager: Dr. Michael Krihak).

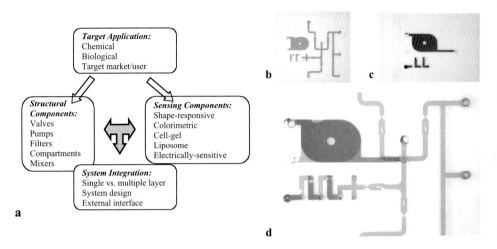

FIGURE 11.1. Microtectonic toolbox (a) and top view of layers fabricated for an integrated analysis system (d) where two separate layers (b, c) are connected through punched holes [59]. Channel width is 1 mm.

11.2. TRADITIONAL MANUFACTURING METHODS

The traditional techniques used for fabricating microfluidic devices include micromachining, embossing, and injection molding. Each technique has advantages and disadvantages and the most suitable method of device fabrication often depends on the specific application of the device [9].

11.2.1. Micromachining

Silicon micromachining is widely used in microelectromechanical systems (MEMS) and was one of the first techniques to be applied microfluidics. Complex systems can be manufactured out of silicon [49]. Recent advances in nanotechnology can also create nanometer structures for microfluidic applications [14]. Although micromachining techniques are widely used, silicon is often not the ideal material for microfluidic applications due to optical opacity, cost, difficulty in component integration, and surface characteristics that are not well suited for biological applications. The needs of many microfluidic applications do not require the precision that micromachining can offer. In addition, micromachining techniques are costly, labor intensive, and require highly specialized skills, equipment, and facilities. Silicon and glass based microfluidic devices are, however, well suited to some chemistry applications that require strong solvents, high temperatures, or chemically stable surfaces. Chip-based capillary electrophoresis is still largely the domain of glass machining because of the surface properties provided by glass.

11.2.2. Micromolding

Injection molding is a very promising technique for low cost fabrication of microfluidic device [18]. Thermoplastic polymer materials are heated past their glass transition temperature to make them soft and pliable. The molten plastic is injected into a cavity that contains the master. Since the cavity is maintained at a lower temperature than the plastic, rapid cooling of the plastic occurs, and the molded part is ready in only a few minutes. The only time consuming step is creating the master that shapes the plastics. This master, often referred to as the molding tool, can be fabricated in several ways including metal micromachining, electroplating, and silicon micromachining. The methods of fabricating the molding tool are similar to those used for making the master for hot embossing and thus, the same issues of cost apply. However, the injection molding process is considerably faster than hot embossing and is the preferred method, from a cost perspective, for high volume manufacturing. Limitations of injection molding for microfluidics include resolution and materials choices.

11.3. POLYMERIC μFLUIDIC MANUFACTURING METHODS

11.3.1. Soft Lithography

In order to promote widespread use of microfluidic devices in biology, a faster, less expensive, and less specialized method for device fabrication was needed. Elastomeric micromolding was first developed at Bell Labs in 1974 when researchers developed a technique of molding a soft material from a lithographic master [5]. The concepts of soft lithography have been used to pattern surfaces via stamping and fabricate microchannels using molding and embossing. Several advances were made in Japan in the 1980s that demonstrated micromolded microchannels for use in biological experiments [51]. More recently, Whitesides [20, 52, 77] and others [36, 68] have revolutionized the way soft lithography is used in microfluidics.

Soft lithography typically refers to the molding of a two-part polymer (elastomer and curing agent), called polydimethylsiloxane (PDMS), using photoresist masters. A PDMS device has design features that are only limited by the master from which it is molded. Therefore, techniques used to create multidimensional masters using micromachining or photolithography can also be used to create complex masters to mold PDMS microstructures. A variety of complex devices have been fabricated, including ones with multidimensional layers [1, 46]. Soft lithography is faster, less expensive, and more suitable for most biological applications than glass or silicon micromachining. The application of soft lithography to biology is thoroughly reviewed elsewhere [77].

The term "soft lithography" can also be used to describe hot embossing techniques [27, 50]. Hot embossing usually refers to the transfer of a pattern from a micromachined quartz or metal master to a pliable plastic sheet. Typically, the polymer substrate and master-mold are heated separately under vacuum to an equal and uniform temperature higher than the glass transition temperature (Tg) of the polymer material. The master-mold is then pressed against the polymer substrate by a precisely controlled force. After a certain time the substrate and the mold are cooled to a temperature below the Tg while still applying the embossing force. The subsequent step of the process is deembossing, where the master-mold is separated from the substrate. During recent years a variety of micro- and nanostructures have been fabricated using hot embossing process [10, 34, 60, 71, 72]. The most commonly used polymeric materials for hot embossing are polycarbonate (PC), polystyrene (PS), polymethylmethacrylate (PMMA), polyvinylbutyral (PVB), and polyethylene (PE). Hot embossing offers low cost devices but does not offer a timely method for changing designs. In order to create new features or channel sizes, a new micromachined master is required which is costly and time consuming. Hot embossing is appropriate for device designs that do not have to undergo changes and offers more material options than the elastomeric-based soft lithography techniques described above.

11.3.2. Other Methods

Another method of forming microfluidic devices is laser ablation of polymer surface [29, 37, 38, 66, 70] with subsequent bonding to form channels. The process can easily be adapted to create multi-layer channel networks. Limitations include throughput due to the "writing" nature of the cutting process.

11.3.3. Liquid Phase Photopolymerization—Microfluidic Tectonics (μFT)

Recently, a new method for in situ construction of microfluidic devices using photodefinable polymers, called microfluidic tectonics, was introduced [11]. The concept uses liquid phase photopolymerizable materials, lithography, and laminar flow to create microfluidic devices. The liquid prepolymer is confined to a shallow cavity and exposed to UV light through a mask (Figure 11.2). The prepolymer polymerizes in less than a minute. Channel walls are formed by the exposed polymer, which is a hard, clear, chemically resistant solid. Any unpolymerized monomer is flushed out of the channel [42]. Once the walls have been formed, other types of photopolymerizable materials can be flowed into the channel and polymerized through masks to form components such as valves [78] and filters [58]. The process is fast, typically requiring only a few minutes to create a simple device. Also, there is no need for cleanroom facilities, specialized skills, or expensive equipment. This method

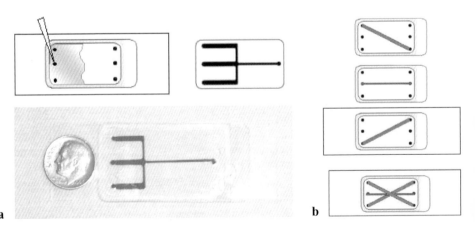

FIGURE 11.2. μFT device fabrication. A polycarbonate film with an adhesive gasket and predrilled holes was placed on a microscope slide forming a cartridge. The cartridge is filled with prepolymer mixture and a photomask is placed on top. The cartridge is exposed to UV light and the polymerized material forms the channel network (a) [42]. Using similar technique but with multiple cartridges components of the device could be build in isolation. The multilayer technique increases overall surface area for device fabrication (b).

may prove to be useful for researchers wanting to enter the field of microfluidics without investing in expensive equipment or cleanroom facilities. The method also eliminates the bonding step (often the yield limiting step in manufacturing) associated with other methods. Although this method provides a reasonably low cost alternative, the device dimensions are limited by the resolution of the mask and polymerization effects of the polymer. Several materials have been used for microfluidic tectonics, including an isobornyl acrylate (IBA) based polymer [11], as well as other UV-curable polymers [12, 31].

One definition of 'tectonics' is the science of assembling and shaping in construction. 'Microfluidic tectonics' (μFT) refers to the fabrication and assembly of microfluidic components in a universal platform. In μFT, one starts with a "blank slate" (shallow cavity) and proceeds to shape micro channels and components within the cavity via liquid phase photo polymerization.

In μFT, the channel walls and the microfluidic components are created from three-dimensional (3D) polymeric structures. Liquid phase photo-polymerization allows for fabrication of these structures directly inside a shallow cavity (or blank slate), which is formed by bonding a polycarbonate film to a glass substrate via an adhesive gasket. We refer to the polycarbonate/gasket/glass system as the "universal cartridge". The universal cartridge is filled with a pre-polymer mixture consisting of monomer, cross-linker and a photo-initiator, the type and composition of the pre-polymer mixture dictates the physical and chemical properties of the resulting polymeric structure (e.g. cross-linker concentration influences the rigidity and mechanical strength of the polymer). A transparency mask is placed on top of the cartridge and light of appropriate wavelength (usually UV) is irradiated to initiate polymerization of the monomer in the exposed regions, to form polymerized structures inside the cartridge. The polymerization time ranges between 10 s to about 5 minutes, depending on the nature of the pre-polymer mixture and channel depth. The unpolymerized mixture is removed from the cartridge via suction, leaving an open channel network or a desired component of the device. Photo-polymerization allows the components to be fabricated in any location in the microsystem (*in situ* fabricated). Moreover, by stacking polycarbonate layers, fabrication of a 3D channel network is possible. Utilizing the third dimension allows more efficient space utilization as well as increased functionality (e.g. 3D chaotic mixer designs, sheath flow). In liquid phase photo-polymerization, a blank slate of any shape and a variety of materials can be utilized; the main requirements include transparency to polymerizing wavelengths of light and compatibility with pre-polymer mixtures. The fabrication of a channel network or component inside a universal cartridge is now limited only by the time required to draw the layout on the computer, thus allowing real-time μFT. The rest of this chapter will discuss the capabilities and tools created using μFT.

11.3.4. Systems Design

A multi-disciplinary approach has been adopted for the design and fabrication of microsystems. The inspiration for the design of components and choice of materials comes from biological systems, rigid polymers to provide framework and responsive materials to provide functional qualities of the microsystem. While the device framework and components are made from organic materials, the control of fabrication can be achieved by using physical phenomena (e.g. pattern with laminar flow and diffusion). Engineering techniques are employed to optimize and improve the efficiency of the fabrication process. A brief overview of materials and components, created using μFT methods, for handling fluids and carrying out processes is given in following sections.

11.3.4.1. Structural Components A typical lab-on-a-chip microsystem consists of analytical processes connected by fluidic pathways or channel networks. The channel walls also forms the skeleton of the microsystem, and are therefore fabricated from mechanically strong hydrophobic polymers. Since these materials provide the basic structure of the device, they are referred to as 'construction materials'. Poly(iso-bornyl acrylate) and poly(bis-GMA) are the main members of this group of materials. These materials can be fabricated using liquid phase photopolymerization or laminar flow method, and display minimal change (5–10%) in volume after polymerization. Moreover, the polymerized structures show high tolerance to common organic solvents (like methanol, iso-propanol and acetone). These materials are also utilized to form 'pillars' or 'posts' that can provide mechanical support for other microfluidic components. The channel network can also be designed to function as a passive component. For example, a 3D serpentine channel network was used to improve the extent of mixing between laminar flow streams.

11.3.5. Valves

The ability to manipulate fluid flow using valves is essential in many microfluidic applications. There are two types of valves: passive valves that require no energy and active valves that use energy for operation. The type of valve used in a device depends on the amount and type of control needed for the application. Active valves often use external macroscale devices that control the actuation and provide energy. Some recent designs include an electromagnetically actuated microvalve [16] and an air-driven pressure vale [74]. Other active valve designs use energy from direct chemical to mechanical conversions or from the driving fluid, eliminating the need for external power. Rehm has demonstrated a hydrogel slug valve in which the driving force of the fluid moves a passive hydrogel slug to open or close an orifice [69]. Others have used stimuli responsive hydrogel materials that undergo volume changes through direct chemical to mechanical energy conversion. A variety of responsive hydrogel post valves [46] have been demonstrated. A responsive biomimetic hydrogel valve resembling the check valves found in veins has also been fabricated [79] as well as a hydrogel-based flow sorter device [11] that directs flow autonomously based on the pH of the stream (Figure 11.3). The hydrogel valves described here are practical due to the physics of the microscale. Since diffusion determines the

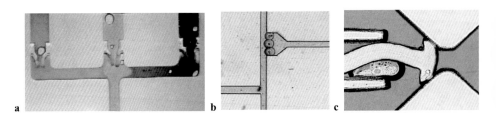

FIGURE 11.3. Example of valves. Simple check valves (a), three hydrogel posts in swelling condition where they function as valves (b) [11], and "hammerhead" valve with spring force to physically closing the channel (c) [43].

response of the hydrogel, scaling effects make the hydrogel respond faster (on the order of seconds) when constructed with smaller dimensions and larger surface area to volume ratios.

Passive valves can be used to limit flow to one direction, to remove air, or to provide a temporary flow stop. Passive one-way valves (similar to the responsive bi-strip valve described above) have been constructed from both silicon and elastomers [35]. An alternative method of constructing passive valves involves the use of porous hydrophobic materials or surface treatments to create selective vents or flow stops, respectively. Vents control fluid movement by allowing air to pass but not the liquid being moved [2]. Hydrophobic surface patterning can also be used to create valve by making a section of channel hydrophobic [39].

An *in situ* polymerized hydrogel plug was utilized to create a 'mobile' valve inside a microchannel. The materials for the channel and the plug were chosen so that there is minimum adhesion between them. Upon applying high pressure, the hydrogel plug was moved in the microchannel to close specific channels. Although a high external pressure is required to move the plug, this approach gives the user freedom to open or shut channels based on the application. The deformability and mobility of hydrogel was combined to fabricate a 'flat worm' check-valve using μFT [43]. In this design, a hydrogel strip was anchored at one end, while the strip was free to move in and out of a constricted region or 'valve neck'. This movement was brought about by the direction and internal pressure of the fluid stream. Thus by using responsive materials, various types of valves for can be created inside the microchannel.

11.3.6. Pumps

Pumping schemes incorporate many different physical principles [67]. The different types of pumps have drastically different features including flow rate, stability, efficiency, power consumption, and pressure head. A few examples of pumping schemes that use external control include a shape memory alloy micropump [13], a valve-less diffuser pump [3], a fixed valve pump [7] that uses piezoelectric actuation, and a self-filling pump based on printed circuit board technology [76]. Pumps can also be injection molded [15] to form inexpensive disposable pumping chambers that are externally actuated Magnetically driven pumps include a magnetically embedded silicon elastomer [41, 33], a magnetohydrodynamic micropump [45], and pumps driven by ferrofluidic movement [26, 62]. A micromotor that can valve, stir, or pump fluids was also developed that was controlled by external magnetic forces [6].

The physical processes that dominate at the microscale allow the creation of pumps that are not feasible on the macroscale. Some designs require no moving parts like a bubble pump [24] that relies on the formation of a vapor bubble in a channel, an osmotic-based pump [73], and an evaporation based pump that relies on a sorption agent to wick fluid through the channel [22]. The surface tension present in small drops of liquid can also be used to pump fluid. The passive pumping technique provides a means of moving fluid by the changes in internal pressure of liquid drops [76]. A smaller drop has a higher internal pressure than a larger drop. When a small drop is fluidically connected to a larger drop (i.e., through a microchannel), the fluid in the small drop will move towards the larger drop. In this manner, fluid can be passively pumped through microchannels simply by controlling the size of the drops on top of the microchannels.

FIGURE 11.4. Ferromagnetic pump. The major components of the pump are external magnet, an actuator and microchannel (a). A closer view of the actuator that is composed of an iron bar and two separate layers of polymers (b). The pump in action where the sizes of water drops change significantly in less than 5 seconds (c, d) [4].

Eddington et al. have demonstrated a self-regulating pump utilizing the pH responsive nature of the poly (hydroxyethyl methacrylate-acrylic acid) (HEMA-co-AA) hydrogel post where the feedback system is relayed to the responsive gel post on an orifice upstream [21]. Another recent pump is the oscillating ferromagnetic micropump that utilizes the centrifugal force (Figure 11.4) [4]. This pump relies on a magnetically driven actuator. The actuator is a direct drive as it converts the energy provided by a rotating magnetic field into linear propulsion of liquid without gears or additional parts. The volumetric flow can be easily controlled changing the spinning velocity of the external motor. Smaller micropumps and greater volumetric flow can be obtained by optimizing the geometry and position of the inlet and outlet channels.

11.3.7. Filters

In a microsystem, a filter is useful in sample preparation (e.g. remove blood cells from whole blood) and purification (e.g. chromatographic separation). Filters made of porous materials also provide a large area where surface-catalyzed reaction or detection of a sample may be carried out. Another function of a filter can be to provide docking stations to 'hold' a cell (e.g. ova, embryo) or other objects (e.g. bead, vesicle) in a known section of the microchannel. Using μFT methods a porous filter was prepared inside the microchannel by 'emulsion photo-polymerization' of a mixture consisting of monomer, porogen (e.g. water, salts), a cross-linker and a photo-initiator. By agitating the mixture, an emulsion consisting of monomer droplets was formed. Upon polymerization and further processing (e.g. drying to remove water), a contiguous polymer network surrounded by interconnected paths (pores) was formed. The size and distribution of pores, and the mechanical properties of the filter are dependent on a number of factors; including composition of pre-polymer mixture, polymerization technique and the surface energy of the channel walls. The large parameter space allows the filter property to be varied to fit the application.

The utilization of the fabricated filter to perform biological separation was explored in preparation of whole blood samples for diagnosis studies. Separation of blood cells from whole blood is required if the diagnostic device is to assay the body fluid directly from

MICROFLUIDIC TECTONICS 231

FIGURE 11.5. Photopolymerizable filter made with HEMA. The blood cells are separated from the serum (a) [58]. SEM image of internal filter structure (b).

the patient (or end user). The current methodology for separation is to use centrifuging techniques. However, this process is difficult for very small sample volumes (nL to a few mL) as the inertial forces diminish with reducing size. Moreover, centrifugation requires external power for operation and may be impractical in a portable diagnostic device. Separation by the porous filter was found to be as efficient as centrifuge techniques while retaining the advantage to address small volumes (Figure 11.5). A number of polymer-based monoliths that were initially targeted for capillary electrophoresis [63, 64] is being integrated into microfluidic platforms [58, 78]. Some have demonstrated novel encapsulation of proteins by functionalizing the monoliths [65].

11.3.8. Compartmentalization: "Virtual Walls"

Compartmentalization is a general term for creating compartments or separated units in a microsystem. In a broad sense, the construction of channel networks also creates compartments. Our focus is on temporary compartments, wherein the walls can be removed by an external stimulus. Temporary isolation will be important to prevent contamination from another process or reactant. The compartments can be formed either by fabricating a responsive polymeric structure or by changing the surface energy of an existing channel wall. Dissolvable physical walls have been fabricated by liquid phase photopolymerization using chemo-responsive hydrogels that contain disulfide crosslinkers (Figure 11.6) [80]. The disulfide bonds are cleaved in the presence of a reducing agent (e.g. tris (2-carboxyethyl) phosphine hydrochloride) causing the polymeric structure to dissolve away.

Since the surface energy of the channel walls can influence the flow profile inside the microchannels, by patterning hydrophobic and hydrophilic regions, temporary compartments

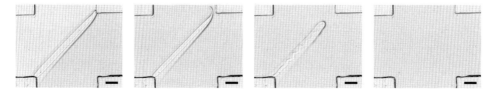

FIGURE 11.6. Demonstration of dissolvable hydrogels as selective sacrificial structures in microfluidic channel. The walls with disulfide crosslinkers dissolves away in presence of a reducing agent thereby controlling the direction of the flow. Scale bars 500 μm.

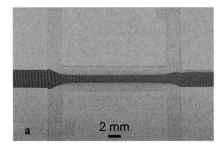
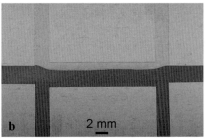

FIGURE 11.7. "Virtual" compartmentalization. Virtual wall compartmentalization can be achieved by multistream laminar flow surface patterning (a) or by UV photopatterning of photocleavable SAMs on glass surface (b) [81].

can be created. At low pressures, aqueous fluids are confined to hydrophilic regions, with the interfaces between the patterns acting as 'virtual' walls. However, the walls break down when the pressure is increased past the threshold allowing the fluid to flow throughout the channel. Such virtual walls have been realized by patterning hydrophobic regions with self-assembled monolayers (SAMs) using laminar flow method [81] or by photopatterning UV-sensitive SAMs [82] (Figure 11.7). The compartments created were removed by increasing the pressure of the fluid. The threshold pressure to break the wall is dependent on the difference of surface energies between the hydrophobic and hydrophilic adjacent regions. Multiple compartments with different threshold pressures can be created by patterning the channel walls with different surface energies.

11.3.9. Mixers

Mixing at the microscale is an ongoing challenge [40]. A unique characteristic of fluids at the microscale is the presence of laminar flow. However with laminar flow, there is no turbulence, and mixing does not readily occur. At the microscale, the channel dimensions lead to low Reynolds numbers where mixing occurs only by diffusion. Most mixers fall into two categories, active and passive. Passive mixers generally utilize set channel geometry to enhance diffusion and have the benefit of no moving parts (Figure 11.8). Active mixers

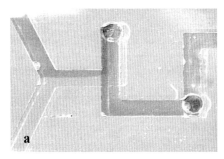
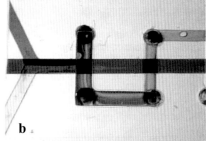

FIGURE 11.8. Passive chaotic mixer. Two-layer passive mixer (a) and three-layer passive mixer around another channel (the straight channel going left right) (b). Such a 'wrap' around a channel could be used to regulate temperature [55]. Channel width 500 μm.

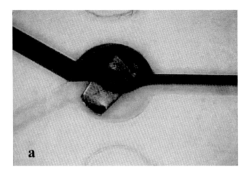 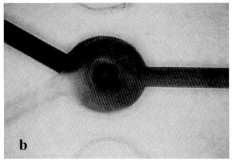

FIGURE 11.9. Active magnetic mixer shown before mixing begins (a), during mixing (b) at a flow rate of 2 mL/min [54]. Channel width 1 mm.

generally require an external power source and moving parts to accomplish mixing. Active mixers have the advantage that the user has the ability to control if and when mixing occurs. Traditionally, active mixers have been based on MEMS technology that requires expensive fabrication techniques and cleanroom facilities. Inline magnetically actuated stirrers have been previously described in the literature using an electromagnetic micromotor [6] and micromachined mixers [48]. While these mixers show magnetic actuation using external fields, the size of the stir bar limits the effective volume that can be mixed. Recently another mixer based on magnetic actuation has also been presented [32] based on a wire placed inside tubing.

Using liquid phase photopolymerization an active, magnetically controlled micromixer can be made that is inexpensive and easy to fabricate without the need for cleanroom facilities [53]. A magnetic mixing device is made by positioning a blade inside a cavity, filling the cavity with prepolymer, and exposing the device to UV to form the channel network, followed by polymerization of a post inside the hole of the blade. The blade is actuated by a common stir-plate, giving the user a convenient method of controlling the mixing in the device (Figure 11.9). The ease of fabrication lets the user customize the mixer so that the mixer operates efficiently within the constraints of the channel network. This type of mixer has also been shown to lyse cells due to the high shear [53].

11.3.9.1. Sensing Components The development of microchannels resulted in the need for sensing and measuring capabilities at the microscale. The need for sensing in microfluidics falls into two general categories.

First, one needs to measure the output of the device or system. Reducing volumes for chemical or biological assays to the microscale is of little use if there is no way to determine results quantitatively as in the macroscale. Reducing the sample size means reducing the amount of material to detect and increases the need for greater sensitivity. Creating sensors or sensing capabilities that are more responsive and smaller in size is an ongoing challenge at the microscale.

Second, one needs to measure the physics and chemistry of flow in microfluidic devices in order to understand and improve device and system designs. Quantifying both electrokinetic and pressure driven flow characteristics inside micro channels is critical to providing a basic science foundation upon which the field of microfluidics can grow [17].

Within the framework of μFT several approaches to sensing have been explored. In following sections we will focus on methods and techniques that have been developed within the framework.

11.3.10. Hydrogel as Sensors

Responsive materials, such as hydrogels, have the ability to change their properties based on environmental conditions. These materials have been explored for fabrication of microfluidic components that can function autonomously, *i.e.* requiring no external control.

Hydrogels have been around for about fifty years and recently, these materials are being extensively studied for use in drug delivery and as tissue scaffolds [30]. Hydrogels are a class of cross-linked polymers that have the ability to 'absorb' water. Responsive hydrogels can undergo phase transitions, wherein large changes in the volume can occur due to an external stimulus. The stimulus can be the presence of specific ions (e.g. pH), chemical or biological agents [56], or a change in temperature or an applied electric field [61]. The stimulus (of a responsive hydrogel) changes the polymer backbone, which then affects the movement of water and ions in and out of the polymer matrix. Two well-studied responsive hydrogels are those that are sensitive to temperature and pH changes. While in the temperature sensitive hydrogel (e.g. poly(NIPAm)), the movement of water is initiated by change in hydropathy of the backbone, in the pH sensitive hydrogel (e.g. poly(HEMA-co-AA)), the movement of water is initiated by ionization of the backbone. The time scales for the volume change will depend on the distance traveled or the initial size of the hydrogel. The change in volume can provide a mechanical force; thus transducing a chemical stimulus into a mechanical action. The factors affecting the force are the dimension of the hydrogel structure, chemical composition of the polymer matrix and the environmental conditions. Responsive hydrogels have been explored for use as sensors, or as actuators, or both. While hydrogels undergo phase transition with changes in environmental conditions, there also exist materials that can self-assemble in various geometries depending.

11.3.11. Sensors That Change Shape

The phase transition in a responsive hydrogel is brought about by changes in specific groups on the polymer backbone. In pH responsive hydrogel, this change is the ionization of a chemical group (e.g. carboxylic acid, amine) (Figure 11.10). Another way to engineer responsiveness is to incorporate cross-linker that can be cleaved by chemical or enzymatic reactions; resulting in volume change or disintegration of the hydrogel. Yu et al. have demonstrated a chemo-responsive hydrogel in the μFT platform. The cross-linker (N, N'-cystamine-bisacrylamide) contains disulphide bonds, which was broken in the presence of a reducing agent (e.g. dithiothritol); leading to disintegration or 'dissolving' of the hydrogel [80]. Yet another way to realize detection (as a structural change) is to create hydrogels where the matrix is held by specific interaction. For example, Miyata and coworkers have developed a bio-responsive hydrogel where the cross-links are formed by antigen-antibody interaction. In the presence of a free antigen, the cross-links are 'dissolved', resulting in volumetric expansion of the structure, and thus recognition of the specific antigen. The selectivity of detection thus depends on the type of cross-linker and the sensitivity will depend on the density of the cross-linker in the polymer matrix. The disappearance or volume

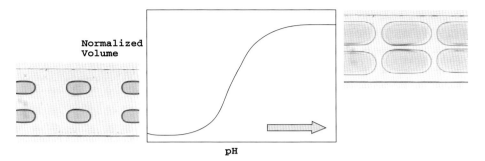

FIGURE 11.10. pH-sensitive hydrogel sensors. Typical volume change seen with pH responsive hydrogels like HEMA-AA (2-hydroxyethyl methacrylate—acrylic acid). Channel width 1 mm.

change in the hydrogel sensor can be easily visualized without other instrumentation. An unaided human eye can detect the disappearance of a 100 um circle or a few 10's of μm change in diameter of a post within a microchannel. Alternatively, the action (dissolution or volume change) can be used to trigger a color producing reaction.

11.3.12. Sensors That Change Color

Color change is easily perceived by the human eye and this simple detection mechanism can be exploited by entrapping ion-sensitive dyes in a hydrogel matrix. Both the dye and the hydrogel have a specific response function to the local environment, changes in which are reflected in the color and size of the readout structure. A combinatorial readout display consisting of a layout of different dyes was fabricated using liquid phase photopolymerization, and tested to display pH changes inside the microchannel (Figure 11.11) [57]. The basic colorimetric format was extended to include biomolecule detection and chemical reactions by entrapping proteins in the polymer (unpublished). One advantage of using the

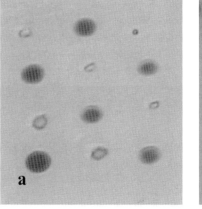
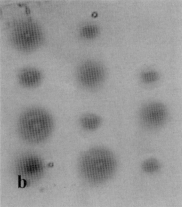

FIGURE 11.11. Colorimetric pH-sensitive readouts. Combinatorial pH sensor with alternate patterns of Congo Red and Phenophthalein dyes immobilized in hydrogel mix, in acidic solution (a) or in basic solution (b) [57].

dye-immobilized-gel construct is that the high volume support of the hydrogel provides sufficient color signal intensity to allow perception by the naked eye, unlike surface immobilized dyes that typically require optical/electronic detection due to low intensities. The wide availability of dyes sensitive to both chemical and biological agents will allow extension of this idea to many applications such as rapid screening of combinatorial libraries. Moreover, polymer matrices responsive to other stimuli like temperature can be chosen to provide readouts that are sensitive to multiple parameters.

11.3.13. Cell-gel Sensors

Cells respond to various types of stimuli—physical forces, temperature, and chemical and biological molecules. Cells are being explored as potential biosensors for the recognition of pathogens. However, using cells comes with a cost, maintenance of appropriate environments and supply of nutrients. To produce sensors that have the similar capability as the cells, "artificial cells" have been developed in the μFT platform by overlaying a monolayer of amphiphilic molecules around a responsive hydrogel [44]. The lipid molecules protect the hydrogel from external environment in a cell-like manner. When the monolayer is disrupted due to mechanical stress or chemical molecules; the hydrogel is exposed to the environment (Figure 11.12). If the environment favors phase transition or a color change, a response can be detected. The cell-gel system can be modified to detect specific molecules by embedding surface receptors in the lipid layer so that upon recognition, the monolayer is disrupted. As the lipid layer and the hydrogel are responsive to different stimuli, cell-gel sensors are activated only in the presence of both stimuli. This mechanism may help ensure that there are minimal false positives. Moreover, this construct provides with a large parameter space, as various combinations of the stimuli-sensitive materials can be engineered into the component. A similar mechanism can be found in many biological processes, where at least two

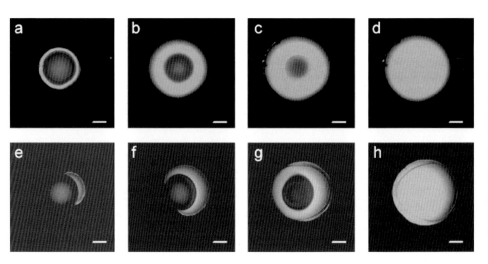

FIGURE 11.12. Effect of lipid modification on μgel sensor. The presence of elevated pH solution is signaled by an increase in fluorescence emission. With an unmodified μgel, the expansion starts at the exterior of the gel (a), and moves inward (b, c, d). In contrast, when lipid-modified μgel, permeation began in a localized site (e) and spread asymmetrically (f, g, h). Scale bar 200 μm.

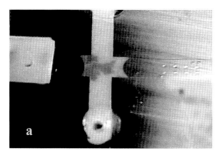 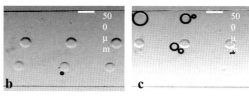

FIGURE 11.13. Liposome sensor relay. Functionalized liposomes are captured by the emulsion filter. Normal and responsive gel posts are seen down stream (a). As liposomes were lysed upstream the responsive gel posts with disulfide crosslinkers dissolve away (b, c after 14 minutes) [25]. Channel width approximately 2 mm.

signals are required to elicit a response. For example, T-cells (in the immune system) require recognition of two signals simultaneously to activate it against tumor or virus-infected cells.

11.3.14. Liposome Sensor

Relying completely on the responsive hydrogel can limit the types of signals and molecules that may be detected. For each type of signal, the raw materials for the hydrogel must be created individually, which can be both difficult and time-consuming. A more efficient method would be to use a 'signal transfer' mechanism commonly seen in signaling cascades of biological systems. This mechanism allows for the sensitivity of one interaction to initiate a common cascade; and in most cases, the signal is amplified during the transfer. For example, in cells, while there are various receptors in the cell membrane that are sensitive to specific molecules, the binding event is relayed to the nucleus via a common kinase pathway. A similar strategy has been developed using the μFT platform, wherein functional liposomes (lipid vesicles) were used in conjugation with a responsive hydrogel (Figure 11.13). The liposome contained the stimulus for the hydrogel and was held by a porous filter, upstream from the hydrogel. In the presence of an external signal (chemical or biological), specific reactions were initiated at the surface of the liposome, leading to its lysis and subsequent spilling of the contents, thus relaying the signal to the hydrogel [25].

11.3.15. E-gel

The volume change of a hydrogel under the influence of electricity has been reported previously, however their application for use in microfluidic systems has only been recently investigated. Although one of the main advantages of hydrogel actuators is their ability to change volume without electronic controls, it would be shortsighted to entirely dismiss electronics integrated with hydrogels due to the ubiquitous nature of electronics. By coupling simple electronic circuits with hydrogel actuators, we can combine the main advantages of both platforms such as ease of fabrication with precise control over system performance. Bassetti et al. demonstrated the use of square voltage waveforms with varying pulse widths to precisely control the volume of a poly(HEMA-AA) hydrogel actuator (Bassetti, Submitted) as shown in Figure 11.14. The voltages used for the study were low (5-12 V) and could be easily integrated into a microfluidic system.

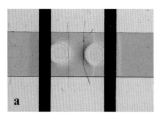 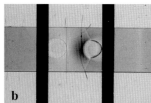 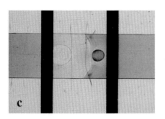

FIGURE 11.14. Electrically responsive hydrogel under low DC voltage (anode is on right side). The change in color is due to change in pH. As time progresses (c is after 960 seconds after), a low pH front diffuses away from the anode allowing change of the pH indicator [8].

Present limitations include asymmetric swelling and bubble formation at electrodes. However, improved electrode materials and designs should mitigate these limitations. The volume change is controlled by varying the duty cycle of the pulse width and the volume change occurs within seconds of changing the duty cycle.

The ability to finely tune the volume of the hydrogel with an electric field opens the door to electrically controllable valves and micropumps for flow control in microsystems; further broadening the potential uses of hydrogels in microfluidics. A device could be made to vary the fluidic resistance of a microchannel through modulation of the hydrogel volume with an electric field. If the hydrogel were positioned on a flexible membrane above a second channel, as described in a previous section, the flow could be regulated through pulse width modulation. The time response of electrically stimulated hydrogels is superior (seconds) to chemically stimulated hydrogels (minutes) (for similar diffusion distances). The reason for the improved time response is complex and is described elsewhere.

11.3.15.1. Systems Integration The previous sections examined design and fabrication of individual microfluidic components that were able to function in an autonomous manner due to the utilization of responsive materials. Now, to realize a microsystem capable of performing a complete assay, the components must be integrated to allow the various analytical process steps within the assay to be performed sequentially in an autonomous and continuous manner. To accommodate for the variation in the sequence of the process steps between assays, μFT provides the end-user with the flexibility to design and fabricate the microsystem on an ad-hoc basis. Moreover, the connectivity between the analytical processes can be improved by incorporating a decision mechanism, wherein an end or by-product of a preceding process activates subsequent component or process. For example, a physical wall separating two reagents can be 'dissolved' by the end product of a preceding reaction step, thus initiating the next step of the assay. Furthermore, since the components are *in situ* fabricated, the integration process is a part of the design and fabrication processes. Therefore, the tasks for development of a microsystem is reduced to designing the layout of the components, choosing appropriate materials, and fabrication of the components via liquid phase photopolymerization or laminar flow method; all of which can be performed by the end-user. However, since most of the components are created from monomer solutions and require solvents to remove unpolymerized materials, compatibility between polymerized structures and monomer solutions of the next component to be fabricated (or solvent) must be addressed. By judiciously choosing the sequence in which the components are fabricated, such compatibility issues can be averted. Additionally, temporary valves / walls (e.g. virtual wall)

MICROFLUIDIC TECTONICS   239

can be included in the design layout to separate polymerized components from monomer or solvent.

As an initial step towards integration of microfluidic components, a biochemical signal transduction detection system (refer to earlier section on detection via signal transfer) was fabricated. A porous filter was used to trap liposomes while a dissolvable hydrogel was used as the readout. To minimize fabrication issues the components were created in the following order—channel network, hydrogel readout and filter. Presently, we are developing an integrated microsystem that can perform sample preparation (dilution and separation of serum) and detection of a bioagent via ELISA (Enzyme Linked Immunoassay). The various components include reservoir, chaotic mixer, check-valves, filter and a detection unit. A detail description of the integrated ELISA device is published elsewhere (Moorthy, Submitted).

The ease and versatility of liquid phase micro fabrication facilitates the creation of microstructured devices using low cost materials and equipment. The ability to add multiple layers without bonding allows for added geometry and increases the functional density. The multilayer technique provides a method of interconnecting layers or combining separate layers to form a truly integrated multilayered microfluidic device. Because this method is based on the fundamentals of µFT, all components (valves, mixers, filters) compatible with µFT can be integrated into multilayer channel networks. An example of multilayered devices is shown in Figure 11.15.

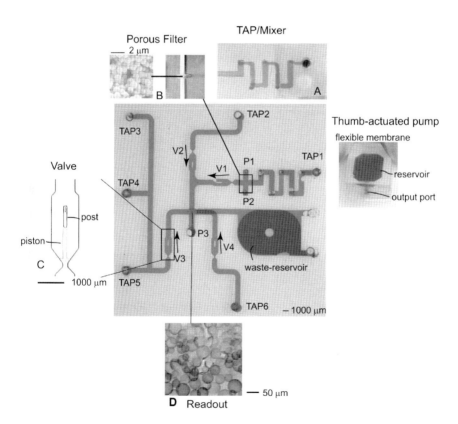

FIGURE 11.15. Layout of an integrated device composed of mixer (a), filter (b), valve (c), and readout (d) [59].

11.4. CONCLUDING REMARKS

There are many approaches to designing and fabricating microsystems and the choice of approach will ultimately depend upon the specific requirements of the end application. In this chapter, we have described a technology platform and design approach called microfluidic tectonics. The approach utilizes liquid phase photopolymerization to create all the components required to perform many microfluidic operations. The methods are very rapid and easy to implement with a minimum investment in equipment. The all organic approach eliminates the need for external power in some applications. Because a set of common fabrication methods is used to make all the components, the integration of multiple components is straightforward. Finally, the use of stimuli responsive materials allows for autonomous operation.

REFERENCES

[1] J.R. Anderson, D.T. Chiu, R.J. Jackman, O. Cherniavskaya, J.C. McDonald, H.K. Wu, S.H. Whitesides, and G.M. Whitesides. *Anal. Chem.*, 72(14):3158–3164, 2000.
[2] R.C. Anderson, G.J. Bogdan, Z. Bamiv, T.D. Dawes, J. Winkler, and K. Roy. *Proceedings of International Solid State Sensors and Actuators Conference (Transducers '97)*, 16–19 June 1997, IEEE, Chicago, IL, USA, 477–480, 1997.
[3] H. Andersson, W. van derWijngaart, P. Nilsson, P. Enoksson, and G. Stemme. *Sens. Actu. B-Chem.*, 72(3):259–265, 2001.
[4] J. Atencia and D.J. Beebe. *Micro Total Analysis Systems*. Kluwer Academics, Squaw Valley, USA, 883–886, 2003.
[5] G.D. Aumiller, E.A. Chandross, W.J. Tomlinson, and H.P. Weber. *J. Appl. Phys.*, 45(10):4557–4562, 1974.
[6] M. Barbic, J.J. Mock, A.P. Gray, and S. Schultz. *Appl. Phys. Lett.*, 79(9):1399–1401, 2001.
[7] R.L. Bardell, N.R. Sharma, F.K. Forster, M.A. Afromowitz, and R.J. Penney. *Proceedings of the 1997 ASME International Mechanical Engineering Congress and Exposition*, Nov 16–21, 1997, ASME, Fairfield, NJ, USA, Dallas, TX, USA, pp. 47–53, 1997.
[8] M.J. Bassetti, A.N. Chatterjee, N.R. Aluru, and D.J. Beebe. *J. Microelectomech. Sys.*, in press.
[9] H. Becker and C. Gartner. *Electrophoresis*, 21(1):12–26, 2000.
[10] H. Becker, U. Heim, and O. Roetting. *Proceedings of SPIE—The International Society for Optical Engineering*, pp. 74–79, 1999.
[11] D.J. Beebe, J.S. Moore, Q. Yu, R.H. Liu, M.L. Kraft, B.H. Jo, and C. Devadoss. *Proc. Natl. Acad. Sci. USA*, 97(25):13488–13493, 2000.
[12] K.D. Belfield, K.J. Schafer, Y.U. Liu, J. Liu, X.B. Ren, and E.W. Van Stryland. *J. Phys. Org. Chem.*, 13(12):837–849, 2000.
[13] W.L. Benard, H. Kahn, A.H. Heuer, and M.A. Huff. *J. Microelectromech. Sys.*, 7(2):245–251, 1998.
[14] G. Bernstein, H. Goodson, and G. Snier. Fabrication technoloiges for nanoelectromechanical systems. *The MEMS Handbook*. M. Gad-el-Hak, New York, CRC, Vol. 36, pp. 1–24, 2002.
[15] S. Bohm, W. Olthuis, and P. Bergveld. *Sens. Actu. A-Phys.*, 77(3):223–228, 1999.
[16] M. Capanu, J.G. Boyd, and P.J. Hesketh. *J. Microelectromech. Sys.*, 9(2):181–189, 2000.
[17] W. Chang, D. Tzebotich, L.P. Lee, and D. Liepmann. *1st Annual International IEEE-EMBS Special Topic Conference on Microtechnologies in Medicine and Biology. Proceedings*. 12–14 Oct. 2000, IEEE, Lyon, France, 311–315, 2000.
[18] J. Choi, S. Kim, R. Trichur, H. Cho et al. *Micro Total Analysis Systems*, Kluwer Academics, Boston, USA, pp. 411–412, 2001.
[19] V. Dolnik, S.R. Liu, and S. Jovanovich. *Electrophoresis*, 21(1):41–54, 2000.
[20] D.C. Duffy, J.C. McDonald, O.J.A. Schueller, and G.M. Whitesides. *Anal. Chem.*, 70(23):4974–4984, 1998.
[21] D.T. Eddington, R.H. Liu, J.S. Moore, and D.J. Beebe. *Lab on a Chip*, 1(2):96–99, 2001.

[22] C. Effenhauser, H. Harttig, and P. Kramer. *Micro Total Analysis Systems*, Kluwer Academics, Boston, USA, 397–398, 2001.
[23] D. Figeys and D. Pinto. *Anal. Chem.*, 72(9):330a–335a, 2000.
[24] X. Geng, H. Yuan, H.N. Oguz, and A. Prosperetti. *J. Micromech. Microeng.*, 11(3):270–276, 2001.
[25] J.A. Gimm, A.E. Ruoho, and D.J. Beebe. *Micro Total Analysis Systems*, Kluwer Academics, Nara, Japan, 922–924, 2002.
[26] A. Hatch, A.E. Kamholz, G. Holman, P. Yager, and K.F. Bohringer. *J. Microelectromech. Sys.*, 10(2):215–221, 2001.
[27] M. Heckele, W. Bacher, and K.D. Muller. *Microsys. Technol.*, 4(3):122–124, 1998.
[28] C. Heller. *Electrophoresis*, 22(4):629–643, 2001.
[29] A.C. Henry, E.A. Waddell, R. Shreiner, and L.E. Locascio. *Electrophoresis*, 23(5):791–798, 2002.
[30] A.S. Hoffman. *Adv. Drug Deliv. Rev.*, 54(1):3–12, 2002.
[31] R.J. Jackman, S.T. Brittain, A. Adams, H.K. Wu, M.G. Prentiss, S. Whitesides, and G.M. Whitesides. *Langmuir*, 15(3):826–836, 1999.
[32] W.C. Jackson, T.A. Bennett, B.S. Edwards, E. Prossnitz, G.P. Lopez, and L.A. Sklar. *Biotech.* 33(1):220–226, 2002.
[33] W.C. Jackson, H.D. Tran, M.J. O'Brien, E. Rabinovich, and G.P. Lopez. *J. Vacul. Sci. Technol. B*, 19(2):596–599, 2001.
[34] R.W. Jaszewski, H. Schift, J. Gobrecht, and P. Smith. *Microelectron. Eng.*, 42:575–578, 1998.
[35] N.L. Jeon, S.K.W. Dertinger, D.T. Chiu, I.S. Choi, A.D. Stroock, and G.M. Whitesides. *Langmuir*, 16(22):8311–8316, 2000.
[36] B.H. Jo, L.M. Van Lerberghe, K.M. Motsegood, and D.J. Beebe. *J. Microelectromech. Sys.*, 9(1):76–81, 2000.
[37] T.J. Johnson, D. Ross, M. Gaitan, and L.E. Locascio. *Anal. Chem.*, 73(15):3656–3661, 2001a.
[38] T.J. Johnson, E.A. Waddell, G.W. Kramer, and L.E. Locascio. *Appl. Sur. Sci.*, 181(1–2):149–159, 2001b.
[39] V.W. Jones, J.R. Kenseth, M.D. Porter, C.L. Mosher, and E. Henderson. *Anal. Chem.*, 70(7):1233–1241, 1998.
[40] A. Kakuta, F.G. Bessoth, and A. Manz. *Chem. Rec.*, 1(5):395–405, 2001.
[41] M. Khoo and C. Liu. *Sens. Actu. A-Phys.*, 89(3):259–266, 2001.
[42] C. Khoury, G.A. Mensing, and D.J. Beebe. *Lab on a Chip*, 2(1):50–55, 2002.
[43] D. Kim and D.J. Beebe. *Micro Total Analysis Systems*, Kluwer Academics, Squaw Valley, USA, 527–530, 2003.
[44] M.L. Kraft and J.S. Moore. *J. Am. Chem. Soc.*, 123(51):12921–12922, 2001.
[45] A.V. Lemoff and A.P. Lee. *Sens. Actu. B (Chemical)*, B63(3):178–185, 2000.
[46] R.H. Liu, M.A. Stremler, K.V. Sharp, M.G. Olsen, J.G. Santiago, R.J. Adrian, H. Aref, and D.J. Beebe. *J. Microelectromech. Sys.*, 9(2):190–197, 2000.
[47] R.H. Liu, Q. Yu, and D.J. Beebe. *J. Microelectromech. Sys.*, 11(1):45–53, 2002.
[48] L.H. Lu, K.S. Ryu, and C. Liu. *J. Microelectromech. Sys.*, 11(5):462–469, 2002.
[49] M. Madou. MEMS fabrication. *The MEMS Handbook*. M. Gad-el-Hak, New York, CRC, Vol. 16, pp. 1–183, 2002.
[50] L. Martynova, L.E. Locascio, M. Gaitan, G.W. Kramer, R.G. Christensen, and W.A. MacCrehan. *Anal. Chem.*, 69(23):4783–4789, 1997.
[51] S. Masuda, M. Washizu, and T. Nanba. *IEEE Trans. Ind. Appl.*, 25(4):732–737, 1989.
[52] J.C. McDonald, D.C. Duffy, J.R. Anderson, D.T. Chiu, H.K. Wu, O.J.A. Schueller, and G.M. Whitesides. *Electrophoresis*, 21(1):27–40, 2000.
[53] G. Mensing, T. Pearce, and D.J. Beebe. *2nd Annual International IEEE-EMB Special Topic Conference on Microtechnologies in Medicine & Biology*, pp. 531–534, 2002.
[54] G. Mensing, T. Pearce, M. Graham, and D.J. Beebe. *Philosoph. Trans.: Math., Phys., and Eng. Sci.*, 362(1818):1059–1068, 2004.
[55] G. Mensing, T. Pearce, and D.J. Beebe. *J. Assoc. Lab Automat.*, 10:24–28, 2005.
[56] T. Miyata, T. Uragami, and K. Nakamae. *Adv. Drug Deliv. Rev.*, 54(1):79–98, 2002.
[57] J. Moorthy and D.J. Beebe. *Lab on a Chip*, 2(2):76–80, 2002.
[58] J. Moorthy and D.J. Beebe. *Lab on a Chip*, 3(2):62–66, 2003.
[59] J. Moorthy, G.A. Mensing, D. Kim, S. Mohanty, D.T. Eddington, W.H. Tepp, E.A. Johnson, and D.J. Beebe. *Electrophoresis*, 25:1705–1713, 2004.

[60] M. Ornelas-Rodriguez, S. Calixto, Y.L. Sheng, and C. Turck. *Appl. Optics.*, 41(22):4590–4595, 2002.
[61] Y. Osada and S.B. Ross-Murphy. *Sci. Am.*, 268:82–87, 1993.
[62] R. Perez-Castillejos, J. Esteve, M. Acero, and J. Plaza. *Micro Total Analysis Systems*, Kluwer Academics, Boston, USA, 492–494, 2001.
[63] E.C. Peters, M. Petro, F. Svec, and J.M.J. Frechet. *Anal. Chem.*, 69(17):3646–3649, 1997.
[64] E.C. Peters, M. Petro, F. Svec, and J.M.J. Frechet. *Anal. Chem.*, 70(11):2288–2295, 1998.
[65] D.S. Peterson, T. Rohr, F. Svec, and J.M.J. Frechet. *Anal. Chem.*, 74(16):4081–4088, 2002.
[66] R. Pethig, J.P.H. Burt, A. Parton, N. Rizvi, M.S. Talary, and J.A. Tame. *J. Micromech. Microeng.*, 8(2):57–63, 1998.
[67] N.A. Polson and M.A. Hayes. *Anal. Chem.*, 73(11):312a–319a, 2001.
[68] S.R. Quake and A. Scherer. *Science*, 290(5496):1536–1540, 2000.
[69] J. Rehm, T. Shepodd, and E. Hesselbrink. *Micro Total Analysis Systems*, Kluwer Academics, Boston, 227–227, 2001
[70] M.A. Roberts, J.S. Rossier, P. Bercier, and H. Girault. *Anal. Chem.*, 69(11):2035–2042, 1997.
[71] H. Schift, C. David, M. Gabriel, J. Gobrecht, L.J. Heyderman, W. Kaiser, S. Koppel, and L. Scandella. *Microelectron. Eng.*, 53(1–4):171–174, 2000.
[72] I. Simdikova, A. Kueper, I. Sbarski, E. Harvey, and J.P. Hayes. *Proceedings of SPIE—The International Society for Optical Engineering*, pp. 82–92, 2002.
[73] Y.-C. Su, L. Lin, A.P. Pisano. *Proceedings of the IEEE Micro Electro Mechanical Systems (MEMS)*, Vol. 393, 2001.
[74] M.A. Unger, H.P. Chou, T. Thorsen, A. Scherer, and S.R. Quake. *Science*, 288(5463):113–116, 2000.
[75] G.M. Walker and D.J. Beebe. *Lab on a Chip*, 2(3):131–134, 2002.
[76] A. Wego and L. Pagel. *Sens. Actu. A-Phys.*, 88(3):220–226, 2001.
[77] G.M. Whitesides, E. Ostuni, S. Takayama, X.Y. Jiang, and D.E. Ingber. *Ann. Rev. Biomed. Eng.*, 3:335–373, 2001.
[78] C. Yu, M.H. Davey, F. Svec, and J.M.J. Frechet. *Anal. Chem.*, 73(21):5088–5096, 2001a.
[79] Q. Yu, J.M. Bauer, J.S. Moore, and D.J. Beebe. *Appl. Phys. Lett.*, 78(17):2589–2591, 2001b.
[80] Q. Yu, J.S. Moore, and D.J. Beebe. *Micro Total Analysis Systems*, Kluwer Academics, Nara, Japan, 712–714, 2002.
[81] B. Zhao, J.S. Moore, and D.J. Beebe. *Science*, 291(5506):1023–1026, 2001.
[82] B. Zhao, J.S. Moore, and D.J. Beebe. *Anal. Chem.*, 74(16):4259–4268, 2002.

12

AC Electrokinetic Stirring and Focusing of Nanoparticles

Marin Sigurdson, Dong-Eui Chang, Idan Tuval,
Igor Mezic, and Carl Meinhart
Department of Mechanical Engineering, University of California—Santa Barbara

12.1. INTRODUCTION

Immunoassay-based sensors rely on specific antigen-antibody binding for identification of proteins. These sensors have applications in both clinical laboratories for medical diagnostics, and in research laboratories for highly-multiplexed testing. In these cases, throughput is a key consideration. One factor limiting test duration is diffusion of analyte to the reporter. An incubation step of minutes to hours is required for diffusion-limited reactions to reach detectable levels. These tests are usually performed at centralized labs where high throughput is achieved through robotics and highly parallel assays. However, if the assay could be moved from a centralized lab to the point of care, the test could be much faster, as well as smaller, while maintaining high sensitivity.

In response to this need, microfluidic assays for diagnostics have developed dramatically in recent years. This facilitates the use of the lab-on-a-chip concepts for point-of-care diagnosis, and high throughput screening for molecular diagnostics. The small length scales associated with microfluidic devices permit small sample sizes and shorter assay incubation times. In addition, on-chip sample preparation reduces fluid handling steps. Though greatly aided by their small length scales, these assays can still be diffusion limited. Ac electrokinetic stirring can potentially reduce incubation times, and can be adaptable to a wide variety of assay configurations.

12.2. AC ELECTROKINETIC PHENOMENA

Ac electrokinetics refers to induced particle or fluid motion resulting from externally applied ac electric fields. Dc electrokinetics has been widely successful for lab-on-a-chip applications such as capillary zone electrophoresis (Aclara and Caliper [1, 5], capillary gel electrophoresis for DNA fractionation [19] and electroosmotic pumping [3, 4]. However, ac electrokinetics has received relatively little attention. Ac electrokinetics have the advantages over its dc counterpart by (1) largely avoiding electrolysis, and (2) operating at relatively lower voltages (1 $\sim$ 20 V). Ac electrokinetics can be classified into three broad areas: dielectrophoresis (DEP), electrothermal flow, and AC electro-osmosis [18].

Dielectrophoresis is a force arising from differences in polarizability between the particle and the fluid medium in the presence of a non-uniform electric field. DEP has been used to separate blood cells and to capture DNA molecules [7, 12, 21, 23, 24], provides an overview). However, since the force scales with the cube of particle radius, it has limited effectiveness for manipulating nanoscale molecules (such as 10 nm-scale antigen).

AC Electroosmosis arises when the tangential component of the electric field interacts with a field-induced double layer along a surface. It becomes less important for sufficiently large electric field frequencies. For example, in an aqueous saline solution with an electrical conductivity of $\sigma = 2 \times 10^{-3}$ S/m, it is predicted that AC electroosmosis is not important above 100 kHz [17].

Transport enhancement for small proteins may be most successful through electrothermally driven flow (ETF). A non-uniform electric field produces non-uniform Joule heating of the fluid, which gives rise to spatial variations in electrical conductivity and permittivity. These variations create electrical charge density variations, even for electrically neutral fluids. The electrical charge density coupled with the applied electric field gives rise to Coulomb body forces in the fluid. The Coulomb body forces induce local fluid stirring. These characteristic swirling flow patterns can be used to transport suspended molecules towards a heterogeneous binding region, or for non-local focusing of particles away from the electrode surface. This can increase the binding rate of immuno-assays, and therefore can improve the response time and overall sensitivity of microfluidic-based sensors.

12.3. DEP: A SYSTEM THEORY APPROACH

If a dielectric particle is suspended in an ac electric field, acting within a dielectric medium, it will polarize. The magnitude and direction of the induced dipole will depend on the frequency and the magnitude of the applied electric field and the dielectric properties of the particle and the medium. A nonhomogeneous electric field acting on the induced dipole in turn produces a force on the dipole, called the dielectrophoretic (DEP) force. Thus, dielectrophoresis is the force exerted on a particle in the presence of a non-uniform electric field [16] (see Fig. 12.1).

To explain this in more detail we describe a systems theory of dielectrophoresis, as developed in Chang et al. 2003. The induced dipole moment, $m(q, t)$, in a particle due to an external electric field, $E(q, t)$, depends linearly on the electric field [6, 10]. This linear

AC ELECTROKINETIC STIRRING AND FOCUSING OF NANOPARTICLES

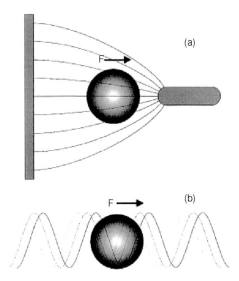

FIGURE 12.1. Particles suspended in a nonhomogeneous ac electric field experience a force due to the interaction of the induced dipole moment and the applied electric field. In a) force due to magnitude gradient is represented. In b) force due to phase gradient of the electric field is shown (figure from [9]).

relation can be written as

$$\hat{\mathbf{m}}(q, s) = G(s)\hat{\mathbf{E}}(q, s), \qquad (12.1)$$

where $\hat{\mathbf{m}}(q, s)$, $\hat{\mathbf{E}}(q, s)$ are the Laplace transforms of $\mathbf{m}(q, t)$, $\mathbf{E}(q, t)$, respectively, and $G(s)$ is the transfer function. When a spherical particle with the permittivity ε_p, the conductivity σ_p and radius r, lies in a medium with the permittivity ε_m and the conductivity σ_m, the transfer function $G(s)$ is given by

$$G(s) = 4\pi r^3 \varepsilon_m \frac{\left(\varepsilon_p + \frac{\sigma_p}{s}\right) - \left(\varepsilon_m + \frac{\sigma_m}{s}\right)}{\left(\varepsilon_p + \frac{\sigma_p}{s}\right) + 2\left(\varepsilon_m + \frac{\sigma_m}{s}\right)}, \qquad (12.2)$$

where $G(s)/(4\pi r^3 \varepsilon_m)$ is the so-called Clausius-Mossotti function [10, 78]. Notice that the transfer function depends on the electric properties both of the particle and of the medium. The dielectrophoretic force, $\mathbf{F}_{dep}$, on the particle due to the interaction between the induced dipole and the electric field, is given by

$$\mathbf{F}_{dep}(q, t) = (\mathbf{m}(q, t) \cdot \nabla)\mathbf{E}(q, t)., \qquad (12.3)$$

The time-averaged force, $\langle \mathbf{F}_{dep} \rangle$, is defined by

$$\langle \mathbf{F}_{dep} \rangle(q) = \lim_{T \to \infty} \frac{1}{T} \int_0^T \mathbf{F}_{dep}(q, t) dt \qquad (12.4)$$

assuming this limit exists. These equations give the relationship between the electric field and the resultant dielectrophoretic force on particles.

We illustrate this formalism with the computation of dielectrophoretic forces corresponding to various (curl-free) electric fields; similar computations can be used to compute DEP forces of various geometries and time-dependencies. We will consider the following four cases:

Case 1. The electric field is:

$$\mathbf{E}(q, t) = \mathbf{E}_1(q) \cos(\omega t) \tag{12.5}$$

Giving:

$$\mathbf{m}(q, t) = |G(j\omega)| \cos(\omega t + \angle G(j\omega))\mathbf{E}_1(q),$$

$$\mathbf{F}_{\text{dep}}(q, t) = \frac{1}{2}|G(j\omega)| \cos(\omega t + \angle G(j\omega)) \cos(\omega t) \nabla |\mathbf{E}_1(q)|^2,$$

$$\langle \mathbf{F}_{\text{dep}} \rangle (q) = \frac{1}{4}\text{Re}[G(j\omega)]\nabla |\mathbf{E}_1(q)|^2. \tag{12.6}$$

Notice that $\langle \mathbf{F}_{dep} \rangle$ moves particles toward the maxima of the magnitude of the electric field if $\text{Re}[G(j\omega)] > 0$, see Fig. 12.1. The maxima of the magnitude of electric fields usually occur at the edge of electrodes. This is known as positive DEP or p-DEP. Negative DEP occurs when the DEP force is away from intense electric fields, and is denoted by n-DEP.

Case 2. The electric field is periodic with period $T > 0$ as:

$$\mathbf{E}(q, t) = \mathbf{E}(q, t + T) \tag{12.7}$$

Here, we can express the electric field as a Fourier series:

$$\mathbf{E}(q, t) = \mathbf{E}_0(q) + \sum_{n=1}^{\infty} (\mathbf{E}_n^c(q) \cos(n\omega t) + \mathbf{E}_n^s(q) \sin(n\omega t)). \tag{12.8}$$

Then,

$$\langle \mathbf{F}_{\text{dep}} \rangle (q) = \frac{1}{2}G(0)\nabla |\mathbf{E}_0|^2 + \sum_{n=1}^{\infty} \frac{1}{2}\text{Re}[G(jn\omega)]\nabla(|\mathbf{E}_n^c|^2 + |\mathbf{E}_n^s|^2)$$

$$+ \sum_{n=2}^{\infty} \frac{1}{2}\text{Im}[G(jn\omega)]\nabla \times (\mathbf{E}_n^c \times \mathbf{E}_n^s). \tag{12.9}$$

Writing the periodic field in the following form

$$\mathbf{E}(q, t) = \begin{bmatrix} E_{0,x}(q) \\ E_{0,y}(q) \\ E_{0,z}(q) \end{bmatrix} + \sum_{n=1}^{\infty} \begin{bmatrix} E_{n,x}(q) \cos(n\omega t + \phi_{n,x}(q)) \\ E_{n,y}(q) \cos(n\omega t + \phi_{n,y}(q)) \\ E_{n,z}(q) \cos(n\omega t + \phi_{n,z}(q)) \end{bmatrix}, \tag{12.10}$$

we obtain the following form of the averaged dielectrophoretic force:

$$\langle \mathbf{F}_{\text{dep}} \rangle (q) = \frac{1}{2}G(0)\nabla |\mathbf{E}_0|^2 + \sum_{n=1}^{\infty} \frac{1}{4}\text{Re}[G(jn\omega)]\nabla(E_{n,x}^2 + E_{n,y}^2 + E_{n,z}^2)$$

$$+ \sum_{n=1}^{\infty} \frac{1}{2}\text{Im}[G(jn\omega)](E_{n,x}^2 \nabla \phi_{n,x} + E_{n,y}^2 \nabla \phi_{n,y} + E_{n,z}^2 \nabla \phi_{n,z}). \tag{12.11}$$

In dielectrophoresis literature, the time-dependence of the force is typically sinusoidal. However, it is sometimes convenient to use non-sinusoidal periodic signals such as square waves, saw-tooth waves, to achieve a desired effect. The formulas above allow us to compute the corresponding time-averaged force. Notice that the electric field is not only periodic but also traveling. In addition, the dielectrophoretic force depends on the imaginary part of the transfer function and the gradient of the phases. This results in traveling wave DEP, or tw-DEP, illustrated in Fig. 12.1b.

Case 3. An *almost-periodic* electric field of the form:

$$\mathbf{E}(q,t) = \mathbf{E}_0(q) + \sum_{n=1}^{\infty} (\mathbf{E}_n^c(q)\cos(\omega_n t) + \mathbf{E}_n^s(q)\sin(\omega_n t)) \quad (12.12)$$

where all the nonzero ω_n are distinct. The averaged dielectric force is given by

$$\langle \mathbf{F}_{\text{dep}} \rangle (q) = \frac{1}{2} G(0) \nabla |\mathbf{E}_0|^2 + \sum_{n=1}^{\infty} \frac{1}{4} \text{Re}[G(j\omega_n)] \nabla (|\mathbf{E}_n^c|^2 |\mathbf{E}_n^s|^2)$$

$$+ \sum_{n=1}^{\infty} \frac{1}{2} \text{Im}[G(j\omega_n)] \nabla \times (\mathbf{E}_n^c \times \mathbf{E}_n^c). \quad (12.13)$$

Case 4. A general time-varying electric field $\mathbf{E}(q,t)$:
The corresponding dielectric force can be written in a compact form as follows:

$$\mathbf{F}_{\text{dep}}(q,t) = \int_0^t g(t-\tau)(\mathbf{E}(q,\tau) \cdot \nabla)\mathbf{E}(q,t) d\tau \quad (12.14)$$

where $g(t)$ is the impulse response of the dipole system, $G(s)$.

12.4. NON-LOCAL DEP TRAPPING

The above theory is valid when the fluid flow is negligible. However, if ac electrokinetically-induce fluid flow (such as electrothermal or ac electroosmotic flow) is present, it can induce both desirable and undesirable effects. In the case of n-DEP, particles can be trapped close to the electrodes, instead of being induced away from the electrodes. In the case of p-DEP, it may not be desirable for particles to collect at the electrodes.

By utilizing carefully the effects of electrokinetically-induced fluid motion, one can focus particles at a non-local region away from the electrode surfaces using p-DEP, leading to orders of magnitude increase in local concentration of particles. Here, we discuss the theory behind this focusing phenomenon, based on the work in [20].

As described in the previous sections, an electric field can induce fluid motion through an electrothermal force. Experimental evidence, as well as full numerical simulations, show convective rolls centered at the electrode edges [11, 18, 22]. The fluid velocity ranges from 1-100 μm/s, with an exponential decay as we move away from the electrodes. The boundary conditions are: no-slip at the bottom of the device, and both the horizontal component of the velocity and the normal derivative of the vertical velocity are zero at the symmetry planes.

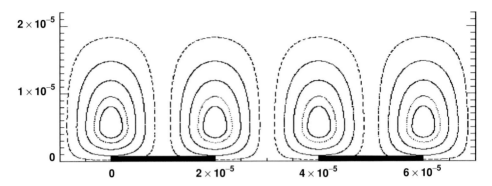

FIGURE 12.2. Streamlines of the cellular flow used in the model.

A simple model that captures these ideas was described in [20]. For an interdigitated array of electrodes a cellular flow is produced, and is depicted in Fig. 12.2.

One possible stream function is

$$\psi_{steady} = \mathbf{u_0} \cdot \mathbf{y}^2 e^{-y/\beta} \cos(\pi x), \tag{12.15}$$

where the flow velocity naturally satisfies the incompressibility condition. The parameter β determines the position of the center of the rolls.

Inertia in micron-size devices can be neglected, and the velocity of the particles can be obtained directly from the DEP force, buoyancy, drag force and Brownian motion, and can be described by the following stochastic ordinary differential equation.

$$dq = (u(q) + \frac{\langle \mathbf{F}_{dep} \rangle(q)}{6\pi \eta r} + (\rho_p - \rho_f) \cdot \frac{2r^2}{9\eta} \cdot g)dt + dW_t \tag{12.16}$$

where $u(q)$ is the fluid velocity at q and W_t is the Brownian motion of variance $2D = kT/3\pi \eta r$, where k is the Boltzmann constant, η is the viscosity of the fluid and T is the temperature. The relative importance of the first three (deterministic force) terms, given the particle and fluid physical properties, depends basically on three parameters: the applied voltage V, the radius of the particle r, and the size of the electrode device d. The relative influence of fluid flow and Brownian motion gets progressively larger for smaller particles, and the buoyancy term becomes important only far from the electrodes where both the flow and DEP force are small.

For particles larger than a few microns, Brownian motion becomes less important and under certain circumstanced may be neglected [20]. The discussion in Tuval et al. [20] is in the context of dynamical systems methods. Two effects of different nature must be noted. Far from the electrodes, where the fluid velocity is smaller, the flow acts only as a small perturbation of the no-flow state. Therefore, the fixed points that exist due to the balance between negative DEP force and positive buoyancy, persist under the perturbation. The basic change is the accumulation of most of the particles in a small trapping area above the electrodes. This main effect has been pointed in several experiments [8, 13]. Under positive DEP, particles tend to accumulate at the edges of the electrodes. But fluid flow, that

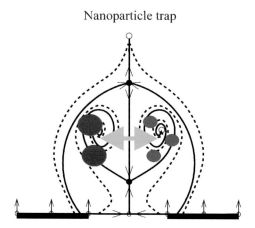

FIGURE 12.3. Sketch of the dynamical behavior of particles in the trapping zone.

is also stronger in that region, can influence the dynamics and move the particles across the electrode surface, finally collecting them above the center of the electrodes [15, 18].

A second effect that takes place close to the electrode surfaces is the creation of a closed zone from which particles can not escape There are two qualitatively different behaviors; some particles are trapped in a closed area above the electrodes or in the gap between electrodes depending on the sense of rotation of the flow, whereas others escape from the flow influence under negative DEP. A sketch of its dynamical structure of the trapping zone is depicted in Fig. 12.3.

Particles in the trapping zone are attracted towards two foci. Particles outside the trapping zone escape from the flow influence and finally reach and equilibrium position due to positive buoyancy, as in the absence of flow. One problem of interest that can be addressed using control theory methods is stabilization of the trapping zone.

12.5. ELECTROTHERMAL STIRRING

The finite element simulation software Femlab (*Comsol; Stokholm, Sweden*) is used for analysis of electrothermally-induced flow and subsequent enhanced binding in the cavity. First, the two-dimensional quasi-static potential field for two electrodes along the cavity wall is calculated, according to Laplace's Equation, $\nabla^2 V = 0$. **The resulting base electric field, given by** $\vec{E} = -\nabla V$ gives rise to a non-uniform temperature field through Joule heating. Ignoring unsteady effects and convection (low Peclet number), and balancing thermal diffusion with Joule heating yields

$$k\nabla^2 T + \sigma E^2 = 0, \qquad (12.17)$$

where T is temperature, E is the magnitude of the electric field, and k and σ are the thermal and electrical conductivities. Thermal boundary conditions are insulating on the channel surfaces. The metal electrodes are isothermal. The treatment of the electrodes

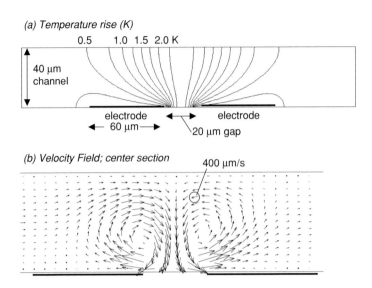

FIGURE 12.4. Simulation of electrothermally-driven flow in a 40 μm channel using Femlab software. (a) Non-uniform temperature distribution created by Joule heating, and (b) Electrothermally-driven fluid motion. The pressure driven channel flow is moving from left to right at an average velocity of 100 μm/s. The velocity of the electrothermally-driven flow is of order 400 μm/s and is characterized by a pair of counter rotating vortices.

as isothermal is appropriate for electrodes of sufficient thickness relative to length. The resulting temperature field is shown in Fig. 12.4a.

Gradients in temperature produce gradients in permittivity and conductivity in the fluid. For water $(1/\sigma)(\partial\sigma/\partial T) = +2\%$ and $(1/\varepsilon)(\partial\varepsilon/\partial T) = -0.4\%$ per degree Kelvin. These variations in electric properties produce gradients in charge density and perturb the electric field. Assuming the perturbed electric field is much smaller than the applied electric field, and that advection of electric charge is small compared to conduction, the time-averaged electrothermal force per unit volume for a non-dispersive fluid can be written as [18]

$$\vec{F}_{ET} = -0.5\left[\left(\frac{\nabla\sigma}{\sigma} - \frac{\nabla\varepsilon}{\varepsilon}\right)\vec{E}_{rms}\frac{\varepsilon\vec{E}_{rms}}{1+(\omega\tau)^2} + 0.5|\vec{E}_{rms}|^2\nabla\varepsilon\right], \quad (12.18)$$

where $\tau = \varepsilon/\sigma$ is the charge relaxation time of the fluid medium and the incremental temperature-dependent changes are

$$\nabla\varepsilon = \left(\frac{\partial\varepsilon}{\partial T}\right)\nabla T, \qquad \nabla\sigma = \left(\frac{\partial\sigma}{\partial T}\right)\nabla T. \quad (12.19)$$

The first term on the right hand side of Eq. (12.18) is the Coulomb force, and is dominant at low frequencies. The second term is the dielectric force, and is dominant at high frequencies. The crossover frequency scales inversely with the charge relaxation time of the fluid; an aqueous solution with conductivity 10^{-2} S/m has a crossover frequency around 14 MHz.

AC ELECTROKINETIC STIRRING AND FOCUSING OF NANOPARTICLES 251

The electrothermal force shown in Eq. (12.18) is a body force on the fluid. The motion of the fluid can determined by solving the Stokes' equation for zero Reynolds number fluid flow, such that

$$0 = -\nabla p + \mu \nabla^2 \vec{u} + \vec{F}_{ET}, \quad (12.20)$$

where $\vec{u}$ is the fluid velocity, p is the pressure in the fluid, and μ is the dynamic viscosity of the fluid. Figure 12.4b shows the resulting velocity field. The velocity of the ETF is of order 400 μm/s, and characterized by a pair of counter rotating vortices, which may circulate the fluid effectively. The velocity field is similar to the streamlines shown in Fig. 12.2.

12.6. ENHANCEMENT OF HETEROGENEOUS REACTIONS

The effect of electrothermally-driven motion upon heterogeneous binding rates is examined in the following section. The convective scalar equation is solved to predict the suspended concentration $C(x,y)$ of antigen within the microchannel:

$$\frac{\partial C}{\partial t} + \vec{u} \cdot \nabla C = D \nabla^2 C, \quad (12.21)$$

where $\vec{u}$ is the fluid velocity and $D = 2 \times 10^{-11}\ m^2\ s^{-1}$ is the diffusivity of an antigen. An antigen concentration of $C_0 = 0.1$ nM is introduced into the left hand side of the channel, for time $t > 0$. Since the base flow is parabolic, analyte will be transported downstream most rapidly at the channel center (see Fig. 12.5a). After 1 second, the highest analyte concentration extends into the center of the channel, but no analyte concentration has yet reached the sensor binding site. When an alternating electrical potential of 7Vrms is applied

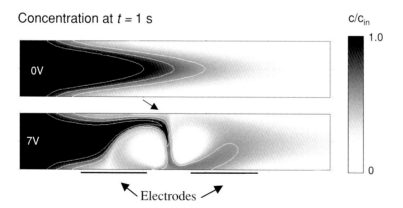

FIGURE 12.5. Concentration plots of electrothermally modified channel flow with applied voltages of 0V and 7V. With optimal size and placement of electrodes, the electrothermal eddies can be engineered to span width of the channel, as is the case here, for a 40 micron channel. High concentration gradients and therefore an increase in diffusive flux in the vertical direction near the top channel wall indicate a favorable alternative location for the sensor here.

to the electrodes, the electrothermally-induced motion transports the analyte close to the upper surface of the channel (Fig. 12.5b). This suggests that for these flow conditions and electrode configurations, an excellent sensor location is opposite the electrode gap.

Assuming a 1st order heterogeneous reaction, the rate of binding is $k_{on}C_w(R_T - B)$, where $k_{on} = 1e8\ M^{-1}\ s^{-1}$ is the on-rate constant. The quantity, $R_T - B$, is the available antibody concentration, and $C_w(x)$ is the suspended concentration of antigen along the wall [14]. The off-rate is $k_{off}B$, where $k_{off} = .02\ s^{-1}$ is the off-rate constant, and B is the concentration of bound antigen. The time rate of change of antigen bound to the immobilized antibodies is equal to the rate of association minus the rate of dissociation

$$\frac{\partial B}{\partial t} k_{on}C_w(R_T - B) - K_{off}B. \tag{12.22}$$

The rate of antigen binding to immobilized antigen, $\partial B/\partial$, must be balanced by the diffusive flux of antigen at the binding surface, $y = 0$, such that

$$\frac{\partial B}{\partial t} = D\frac{\partial C}{\partial Y}\bigg|_{y=0}. \tag{12.23}$$

Equations (12.21), (12.22) & (12.23) are solved with an immobilized antibody concentration $R_T = 1.7$ nM cm (i.e. one molecule per 100 nm^2). The binding rates for three conditions, 0, 7 and 14 Vrms, are shown in Fig. 12.6. The 0 Vrms case corresponds to the passive case, which is the result of pure diffusion. This is the standard mode of most immobilized assays. The 7 and 14 Vrms curves correspond to the result of electrothermally-driven flow enhancing the transport of antigen to the immobilized antibodies. The curves in

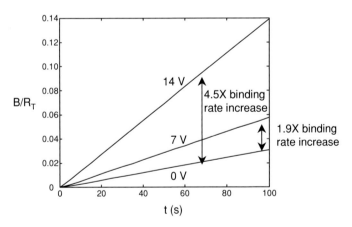

FIGURE 12.6. Numerical simulation of normalized bound concentration for a microchannel assay. The binding rate is increased by a factor of 2, when 7 Vrms is applied to the electrodes. The binding rate is increased by a factor of 4.5, when 14 Vrms is applied to the electrodes. These results suggest that electrothermally induced flow can significantly improve immunoassay performance by increasing binding rates. Parameters: Diffusivity, $D = 2 \times 10^{-11}\ m^2s^{-1}$ (corresponding to 20 nm spherical particle); inlet velocity is parabolic with average 100 μm s^{-1}; inlet concentration $c_0 = 0.1$ nM; $\sigma_w = .00575\ Sm^{-1}$; $k_{on} = 1e8\ M^{-1}\ s^{-1}$; $k_{off} = .02\ s^{-1}$; $R_t = 1.67e\text{-}11\ Mm$.

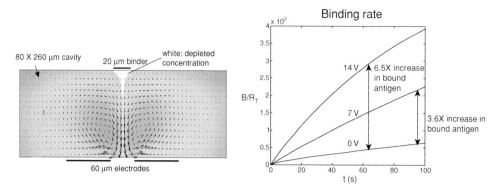

FIGURE 12.7. Microcavity (i.e. no flow-through) simulations (a) Velocity and concentration fields. The binder is centered above the electrodes; depleted concentration (white) is drawn down into the cavity. (b) Binding curves for non-enhanced (0 V) and enhanced (7V, 14V) transport. The differences in the two curves show an increase in binding rate which yields a **factor of 3.6** higher binding for 7 V and a **factor of 6.5** higher binding after 60 seconds for 14 V applied root-mean square potential. Paramters: Diffusivity, 10^{-11} m^{2}s^{-1}; zero net flow; all other parameters identical to those in Fig. 12.3. Initial condition: $C(x, y) = C_0$ at $t = 0$.

Fig. 12.6 show that a factor of 4.5 improvement in binding rate is obtained using ac electrokinetics in combination with a flowing microchannel.

Electrothermally-driven stirring can be used to improve assays for ELISA tests, microarray assays, and microtitre plates, where there is zero net flow. In these assays, the sample is often pre-mixed with a fluorescent or a chemiluminescent reporter, which increases the effective size of the analyte thereby decreasing its diffusivity. We simulate this effect by reducing the diffusivity by a factor of two, such that $D = 10^{-11}$ m^2 s^{-1}.

The numerical simulation results are shown in Figure 12.7. The recirculating velocity field, which is characteristic of electrothermally-driven flow, is shown in Fig. 12.7a. This corresponds approximately to the streamlines shown in Fig. 12.2. For an applied voltage of 7 and 14 Vrms, the binding rate is increased by a factor of 3.6 and 6.5, respectively (see Fig. 12.7b). In the current simulation, the ac frequency is $f = 100$ kHz. In this range, the electrothermal velocity is not sensitive to changes in electrical frequency. At much lower frequencies (~100 Hz), ac electroosmosis and electrode polarization typically dominate. At much higher frequencies (~MHz), the dielectric component of the electrothermal force (last term in Eq. 12.18) dominates [17].

12.7. CONCLUSIONS

An analytical theory is presented that suggests that the combination of positive DEP and electrothermal fluid motion can produce non-localize trapping zones away from electrodes. The results suggest that this phenomenon could be used to focus bio-molecules in microfluidic sensors, when molecular diffusivity is low.

Numerical simulations are used to show how electrothermally-generated forces can be used to stir fluids in microchannels. Fluid velocities of approximately ~400 μm/s are generated by applying potentials of $V = 14$ Vrms at $f = 100$ kHz. Precision stirring can be used to enhance the transport molecules towards functionalized surfaces. The results indicate that the binding rates of heterogeneous diffusion-limited reactions can be improved by a factor of 2–6, by applying 7–14 V_{rms} electrical potentials.

ACKNOWLEDGMENTS

This work has been supported by DARPA/ARMY DAAD19-00-1-0400, DARPA/Air Force F30602-00-2-0609, NSF CTS-9874839 and NSF ACI-0086061, and through the Institute for Collaborative Biotechnologies through grant DAAD19-03-D-0004 from the U.S. Army Research Office.

REFERENCES

[1] L. Bousse, C. Cohen, T. Nikiforov, A. Chow, A.R. Kopf-Sill, R. Dubrow, and J.W. Parce. Electrokinetically controlled microfluidic analysis systems [Review]. *Ann. Rev. Biophys. Biomol. Struct.*, 29:155–181, 2000.
[2] D.-E. Chang, S. Loire, and I. Mezic. Closed-Form Solutions in the Electrical Field Analysis for Dielectrophoretic and Travelling Wave Inter-Digitated Electrode Arrays. *Proceedings of the Conference on Decision and Control*. Maui, HI, 2003.
[3] C.H. Chen and J.G. Santiago. A planar electroosmotic micropump. *J. MEMS.*, 11(6):672–683, 2002.
[4] L.X. Chen, J.P. Ma, F. Tan, and Y.F. Guan. Generating high-pressure sub-microliter flow rate in packed microchannel by electroosmotic force: potential application in microfluidic systems. *Sens. Actu. B-Chem.*, 88(3):260–265, 2003.
[5] C. Ring-Ling et al. Simultaneous hydrodynamic and electrokinetic flowcontrol. *Micro Total Analysis Systems 2002*, 1:386–388, Nov. 2002.
[6] V.V. Daniel. *Dielectric relaxation*. Academic Press, New York, 1967.
[7] P.R.C. Gascoyne and J. Vykoukal. Particle separation by dielectrophoresis [Review]. *Electrophoresis*, 23(13):1973–1983, 2002.
[8] N.G. Green, A. Ramos, and H. Morgan. AC electrokinetics: a survey of sub-micrometre particle dynamics, *J. Phys. D*, 33:632–641, 2000.
[9] M.P. Hughes. *Electrophoresis*, 23(16):2569, 2002.
[10] T.B. Jones. *Electromechanics of Particles*, Cambridge University Press, 1995.
[11] C.D. Meinhart, D. Wang, and K. Turner. Measurement of AC electrokinetic flows. *J. Biomed. Microdev.*, 5(2):139–145, 2003.
[12] R. Miles, P. Belgrader, K. Bettencourt, J. Hamilton, and S. Nasarabadi. Dielectrophoretic manipulation of particles for use in microfluidic devices, *MEMS-Vol. 1, Microelectromechanical Systems (MEMS). Proceedings of the ASME International Mechanical Engineering Congress and Exposition*, Nashville, TN, Nov. 14–19, 1999.
[13] T. Muller, A. Gerardino, T. Schnelle, S.G. Shirley, F. Bordoni, G. DeGasperis, R. Leoni, and G. Fuhr. Trapping of micrometre and sub-micrometre particles by high-frequency electric fields and hydrodynamic forces. *J. Phys. D.: Appl. Phys.*, 29, 340–349, 1996.
[14] D.G. Myszka. Survey of the 1998 optical biosensor literature. *J. Mol. Recognit.*, 12:390–408, 1998.
[15] R. Pethig, Y. Huang, X.B. Wang, and J.P.H. Burt. Positive and negative dielectrophoretic collection of colloidal particles using interdigitated castellated microelectrodes. *J. Phys. D.: App. Phys.*, 25:881–888, 1992.
[16] H.A. PohL. *Dielectrophoresis*. Cambridge University Press, 1978.
[17] A. Ramos, A. Castellanos, A. Gonzales, H. Morgan, and N. Green. Manipulation of Bio-Particles in Microelectrode Structures by means of Non-Uniform AC Electric Fields. *Proceedings of ASME International Mechanical Engineering Conngress & Exposition*. New Orleans, LA Nov. 17–22, 2002.
[18] A. Ramos, H. Morgan, N.G. Green, and A. Castellanos. AC electrokinetics: a review of forces in microelectrode structures. *J. Phys. D: Appl. Phys.*, 31:2338–2353, 1998.

[19] W. Thormann, I. Lurie, B. McCord, U. Mareti, B. Cenni, and N. Malik. Advances of capillary electrophoresis in clinical and forensic analysis (1999–2000). *Electrophoresis*, 22:4216–4243, 2001.
[20] I. Tuval, I. Mezic, and O. Piro. Control of particles in micro-electrode devices. UCSB preprint 2004.
[21] X-B. Wang, J. Vykoukal, F. Becker, and P. Gascoyne. Separation of polystyrene microbeads using dielectrophoretic/gravitational field-flow-fractionation. *Biophys. J.*, 74:2689–2701, 1998.
[22] D. Wang, M. Sigurdson, and C. Meinhart. Experimental analysis of particle and fluid motion in AC electrokinetics. Accepted in *Exp. in Fluids*. 2004.
[23] M. Washizu, O. Kurosawa, I. Arai, S. Suzuki, and N. Shimamoto. Applications of electrostatic stretch and positioning of DNA. *IEEE Trans. Ind. Appl.*, 32(3):447–445, 1995.
[24] J. Yang, Y. Huang, X. Wang, X.-B. Wang, F. Becker, and P. Gascoyne. Dielectric properties of human leukocyte subpopulations determined by electrorotation as a cell separation criterion. *Biophys. J.*, 76:3307–3314, 1999.

III

Micro-fluidics and Characterization

13

Particle Dynamics in a Dielectrophoretic Microdevice

S.T. Wereley and I. Whitacre

*Purdue University, School of Mechanical Engineering, West Lafayette,
IN 47907-2088, USA*

13.1. INTRODUCTION AND SET UP

13.1.1. DEP Device

A dielectrophoretic device has been designed to trap, separate, and concentrate biological components carried in solution. The operating principle of the device is the dielectrophoretic interaction between the spheres and the fluid. The device was designed and manufactured by at Purdue University [6]. The device consists of a microchannel with a depth of 11.6 µm, width of 350 µm, and length of 3.3 mm. The channel was anisotropically etched in silicon to produce a trapezoidal cross-section. The channel was covered by a piece of anodically bonded glass. A schematic view and digital photo of the device are shown in Figure 13.1. Bright regions represent platinum electrodes and the dark regions represent the electrode gaps. The electrodes are covered by a 0.3 µm thick layer of PECVD silicon dioxide, which insulates the electrodes from the liquid medium, suppressing electrolysis. The electrodes are arranged in interdigitated pairs so that the first and third electrodes from Figure 13.1 are always at the same potential. The second and fourth electrodes are also at the same potential, but can be at a different potential than the first and third electrodes. An alternating electric potential is applied to the interdigitated electrodes to create an electromagnetic field with steep spatial gradients. Particle motion through the resulting electric field gradients causes polarization of the suspended components, resulting in a body force that repels particle motion into increasing field gradients. In the experiments, sample solutions were injected into the chamber using a syringe pump (World Precision Instruments Inc., SP200i) and a 250µl gas-tight luer-lock syringe (ILS250TLL, World Precision

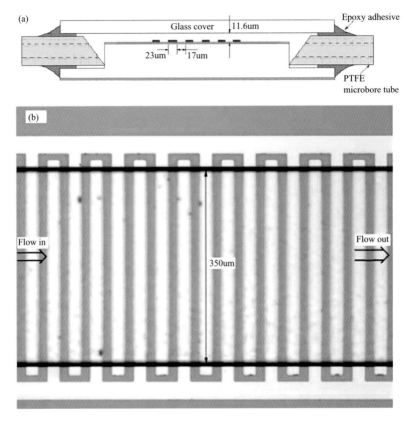

FIGURE 13.1. (a) Schematic view of experimental apparatus and (b) photo of apparatus.

Instruments Inc.). The flow rate could be adjusted and precautions were taken to avoid air bubbles. An HP 33120A arbitrary waveform generator was used as the AC signal source to produce sinusoidal signal with frequency specified at 1MHz.

13.1.2. Dielectrophoresis Background

Dielectrophoretic forces arise from a difference in induced dipole moments between a particle and a surrounding fluid medium caused by an alternating electric field. Differences in permittivity between two materials result in a net force which depends on the gradient of the electric field [11]. This force can be used to manipulate particles suspended in a fluid [5, 7, 11]. The DEP force can either attract or repel particles from regions of large field gradient. A particle with permittivity smaller than that of the suspension medium will be repelled from entering regions of increasing field density, as is the case for the device observed in this study. The equation for the time-averaged DEP force [5] is as follows:

$$F_{DEP}(\omega) = 2\pi \cdot \varepsilon_m \cdot r^3 \operatorname{Re}(f_{cm}) \nabla |E_{RMS}|^2 \qquad (13.1)$$

This force depends on the volume of a particle, the relative permittivities of a particle and surrounding medium, and the absolute gradient of average electric field squared. The

Clausius-Mossotti factor f_{cm} contains the permittivities and frequency dependence of the DEP force. The factor is defined by

$$f_{cm} = \frac{\varepsilon_p^* - \varepsilon_m^*}{\varepsilon_p^* + 2\varepsilon_m^*} \qquad (13.2)$$

where ε^* denotes a complex permittivity based on the conductivity and frequency of excitation. The subscripts m and p denote properties related to a medium and particle respectively. The complex permittivity of the medium is given as

$$\varepsilon_m^* = \varepsilon_m - i\frac{\sigma_m}{\omega} \qquad (13.3)$$

where σ and ω represent conductivity and angular frequency, respectively. The form of Eq. 13.3 is analogous for the complex permittivity of a particle. The conductivity of the particles used in this study, polystyrene microspheres, is approximately zero, thus ε_p^* is equal to ε_p which is found to be equal to 2.6.

Since the dielectrophoretic force scales with the cube of particle size, it is effective for manipulating particles of order one micron or larger. DEP has been used to separate blood cells and to capture DNA molecules [9, 12]. DEP has limited effectiveness for manipulating proteins that are of order 10–100 nm [3]. However, for these small particles, DEP force may be both augmented and dominated by the particle's electrical double layer, particularly for low conductivity solutions [4]. DEP has been used to manipulate macromolecules and cells in microchannels. For example, Miles et al. [9] used DEP to capture DNA molecules in microchannel flow. Gascoyne & Vykoukal [4] presents a review of DEP with emphasis on manipulation of bioparticles.

13.1.3. Micro Particle Image Velocimetry

Images of the particles were acquired using a standard μPIV system. In these experiments a mercury lamp is used to illuminate the 0.7 μm polystyrene latex (PSL) microspheres (Duke Scientific) that are suspended in de-ionized water in concentrations of about 0.1% by volume. The particles are coated with a red fluorescing dye ($\lambda_{abs} = 542$ nm, $\lambda_{emit} = 612$ nm). The images were acquired using a Photometrics CoolSNAP HQ interline transfer monochrome camera (Roper Scientific). This camera is capable of 65% quantum efficiency around the 610 nm wavelength. The largest available image size that can be accommodated by the CCD array is 1392 by 1040 pixels, but the camera has the capability of pixel binning, which can drastically increase the acquisition frame rate by reducing the number of pixels that need to be digitized. A three-by-three pixel binning scheme was used in this experiment, producing images measuring 464 by 346 pixels, which were captured at a speed of 20 frames per second. The average focused particle diameter in the images was approximately 3 pixels. Sample images are shown in Figure 13.2.

Shallow Channel Considerations When performing μPIV measurements on shallow microchannels, the depth of focus of the microscope can be comparable in size to the depth of the flow. A PIV cross-correlation peak, the location of which is the basis for conventional PIV velocity measurements, is a combination of the velocity distribution in the interrogation region and some function of average particle shape. PIV velocity measurements containing velocity gradients can substantially deviate from the ideal case of depthwise uniform flow.

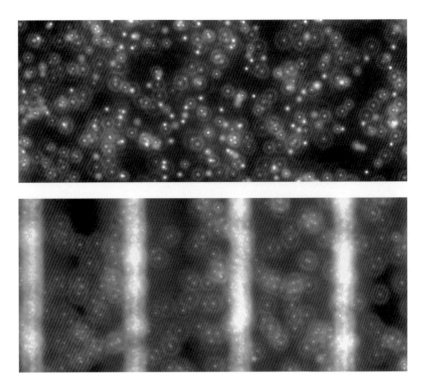

FIGURE 13.2. A photo of PSL particles with 0.0 volts (top) and 4.0 (bottom).

Gradients within the light sheet plane have been addressed by image correction techniques [13], but gradients in the depthwise direction remain problematic. They can cause inaccurate velocity measurements due to the presence of multiple velocities within an interrogation region that are independent of mesh refinement. One problem with depthwise velocity gradients is cross-correlation peak deformation which reduces the signal to noise ratio of a PIV measurement [2]. Cross-correlation peak deformation can also reduce the effectiveness of subpixel peak fitting schemes which are based on a particular cross-correlation shape, such as a common five point Gaussian fit. There are both hardware and software approaches toward resolving these problems. In situations where a large in plane region must be imaged in a relatively thin device, the physics of the imaging system dictate that the entire depth of the channel will be focused. Hence a software approach must be used. Two different approaches are explored: deconvolving the cross correlation with the autocorrelation and image processing to replace the original particle images with unit impulse particle images.

13.2. MODELING/THEORY

13.2.1. Deconvolution Method

Simulated PIV experiments show a qualitative similarity between the cross-correlation function from a test flow containing velocity gradients in the depthwise direction and the corresponding velocity histogram for that depthwise gradient, suggesting that the deconvolution procedure may work. The major hypothesis of the deconvolution method is that the PIV cross-correlation function can be approximated by the convolution of a particle image autocorrelation with the velocity distribution in the interrogation region. This hypothesis is based on observations of cross-correlations from experimental situations. For the case of a uniform flow, the velocity distribution is an impulse, and the resulting cross-correlation can be approximated by a position-shifted autocorrelation. This suggests that the cross-correlation can be approximated by a convolution of the impulse velocity distribution and an image autocorrelation.

Olsen and Adrian [10] approximated the cross-correlation as a convolution of mean particle intensity, a fluctuating noise component, and a displacement component. Deconvolution of a cross-correlation with an autocorrelation is used by Cummings (1999) to increase the signal-to-noise ratio for a locally uniform flow. The new idea is to extract velocity distributions by deconvolving a PIV cross-correlation function with its autocorrelation. The result is a two-dimensional approximation of the underlying velocity distribution. One drawback to deconvolution procedures is sensitivity to noise resulting from a division operation in frequency space. Thus, it is important to have high information density in both the cross-correlation and autocorrelation. The information density can be increased by correlation averaging both the cross-correlation and autocorrelation [8].

13.2.2. Synthetic Image Method

Since deconvolution is inherently sensitive to noise, it would be beneficial to eliminate the deconvolution step from the velocity profile extraction process. This could be done if the autocorrelation were an impulse or delta function. Since the deconvolution of a function with an impulse is the original function, this would render the deconvolution operation trivial and unnecessary.

The autocorrelation is related to the particle intensity distributions from a set of PIV recordings, i.e. the shape of particles in an image pair. So, the most practical way to affect the autocorrelation is to alter the imaged shape of these particles. If the particle intensity distributions in a set of PIV recordings are reduced to single pixel impulses, the autocorrelation appears as a single pixel impulse. This desirable autocorrelation trivializes the deconvolution method such that the cross-correlation alone is equal to the deconvolution of the cross-correlation and the autocorrelation. This is the basis for the synthetic image method.

Experimentally PIV recordings containing uniformly illuminated single pixel particle images can be approximately obtained by illuminating very small seed particles with a high intensity laser sheet, such that most particles are imaged by a single pixel by a digital camera. However, this approach can only be an approximation because even very small particles located near the edge of a pixel would be imaged over two neighboring pixels. Furthermore, in μPIV the particles are already very small, so making them any smaller could render them invisible to the camera; imaged particle intensity decreases proportionally as the cube of particle radius. Also, the image of a very small particle is dominated by diffraction optics, thus reducing the physical particle size will have very little impact on the imaged particle size due to a finite diffraction limited spot size. The typical point response function associated with microscope objectives used in μPIV is 5 pixels, so the desired particle intensity distribution cannot easily be obtained in raw experimental images.

13.2.3. Comparison of Techniques

Three cases were examined to compare and validate the two methods of velocity distribution extraction from cross-correlation PIV. The cases involve three different velocity profiles that are frequently encountered in μPIV. The first is a uniform flow (depth of field small compared to flow gradients), the second is a linear shear (near wall region of a channel flow), and the third is a parabolic channel flow (depth averaged pressure-driven flow in a shallow micro device).

Uniform Flow The velocity profile of uniform flow is an ideal situation for making accurate PIV measurements. This case also has the simplest velocity histogram, a delta function at the uniform velocity value. The uniform one dimensional velocity profile simulated in this case study is given by $V_x = 6.294$. Both the deconvolution and the synthetic image methods generate the expected histograms and are shown in Figure 13.3. Since the histograms are calculated only at integer pixel values, they both show a peak at 6 pixels and a lower, but non-zero value at 7 pixels. By taking the average of the histogram values at 6 and 7 pixels, the true value of the uniform displacement can be found. For example in the case of the synthetic image method, $V_{x,mean} = (6 \times 1 + 7 \times 0.46)/1.46 = 6.32$, which is very near the input value of 6.294.

Linear Shear Another common flow profile is a linear shear. In microfluidics, this type of flow can be seen in the near wall region of a parabolic flow profile. Since the

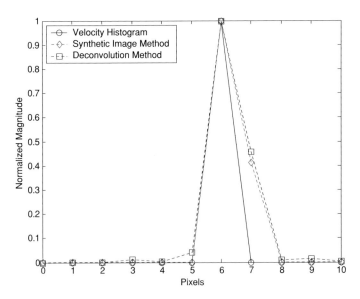

FIGURE 13.3. Uniform flow simulation results, $V_x = 6.294$.

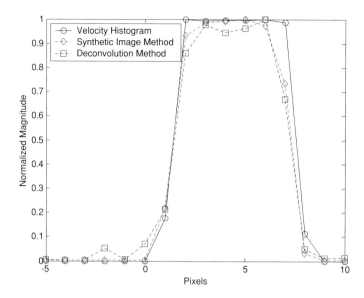

FIGURE 13.4. Linear shear simulation results, $V_x = 6.294 \cdot Z + 1.248$.

displacement probability density function for a linear shear has the simple shape of a top hat, this type of flow presents a good case to further benchmark the methods of deconvolution and synthetic images. This simulation case study used the velocity profile given by $V_x = 6.294 \times Z + 1.248$ where Z is the position along the axis of the imaging system and can assume values between 0 and 1. Consequently we expect a top hat with the left edge at 1.248 and the right edge at 7.552. The results are plotted in Figure 13.4. The synthetic image method is a slightly better predictor of the velocity histogram than the deconvolution method by virtue of the steeper gradients at the edges of the distribution. This behavior is expected because images analyzed by deconvolution contain particle diameter and particle intensity variations, as well as slight readout noise.

Parabolic or Poiseuille Flow The final common velocity profile to be considered is parabolic or Poiseuille flow. This type of flow is found in pressure-driven microchannel devices. Furthermore, it is the profile expected from the LOC experimental device being considered here with zero voltage applied to the DEP electrodes. This simulation used the velocity profile is given by $V_x = 50.352 \times (Z - Z^2)$, with Z varying between 0 and 1. Hence, we expect a velocity distribution varying between a minimum of 0 pixels at $Z = 0$ and $Z = 1$ and a maximum of 12.558 at $Z = 1/2$. The results are plotted in Figure 13.5. These results clearly show the increased accuracy of the synthetic image method over the deconvolution method. The synthetic image data very closely agrees with the velocity histogram. The deconvolution method suffers spurious oscillations but still gives a reasonable approximation of the velocity histogram. These spurious oscillations were initially attributed to a lack of statistical convergence, so additional image sets were added, eventually totaling ten thousand sets. The same spurious oscillations were found, so they must be inherent to this particular case. It is not typical for such oscillations to occur, but as demonstrated, the deconvolution method has definite limitations.

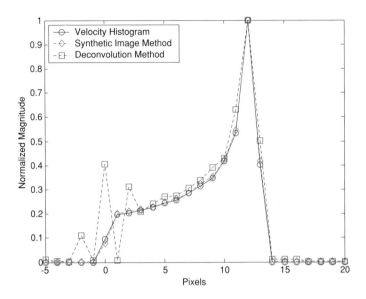

FIGURE 13.5. Parabolic flow simulation results, $V_x = 50.352 \cdot (Z - Z^2)$.

13.3. EXPERIMENTAL RESULTS

The experiments presented here are designed to quantify the dielectrophoretic performance of the device. The experiments used six sets of 800 images each to analyze the effect of dielectrophoresis on particle motion in the test device. These images are high quality with low readout noise, as can be seen in the example fluorescent image of Figure 13.6. The top image demonstrates the many different particle intensity distributions which are typically present in a μPIV image in which the particles are distributed randomly within the focal plane. The bottom figure shows how, as the result of the DEP force, the particles migrate to the top of the channel and all have nearly identical images. The top image also shows that when a significant DEP force exists, the particles are trapped at the electrode locations by the increase in DEP force there. In general, the observed particle image shape is the convolution of the geometric particle image with the point response function of the imaging system. The point response function of a microscope is an Airy function when the point being imaged is located at the focal plane. When the point is displaced from the focal plane, the Airy function becomes a Lommel function [1]. For a standard microscope the diffraction limited spot size is given by

$$d_e = \frac{1.22 \cdot \lambda}{NA} \qquad (13.4)$$

A numerical aperture (or NA) of 1.00 and an incident light wavelength λ of 540nm results in a diffraction limited spot size of 0.66μm, while the particles used are 0.69μm. Consequently the particle intensity distributions as recorded by the camera are partly due to the geometric image of the particle and partly due to diffraction effects. Hence, the distance of any particle from the focal plane can be determined by the size and shape of the diffraction rings.

PARTICLE DYNAMICS IN A DIELECTROPHORETIC MICRODEVICE 267

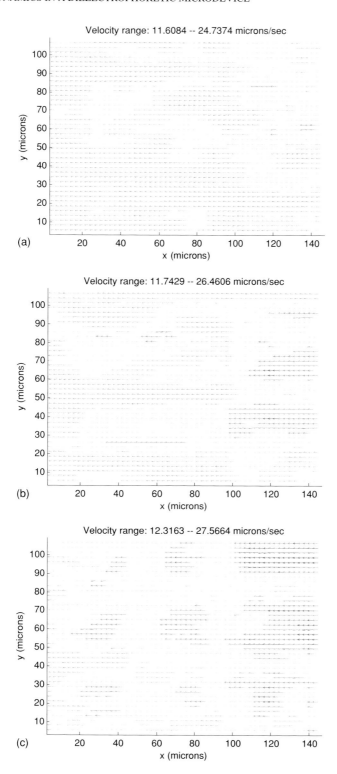

FIGURE 13.6. PIV vector plots for electrode voltages of a) 0.5 volts, b) 1.0 volts, c) 2.0 volts, d) 2.5 volts, e) 3.5 volts and f) 4.0 volts.

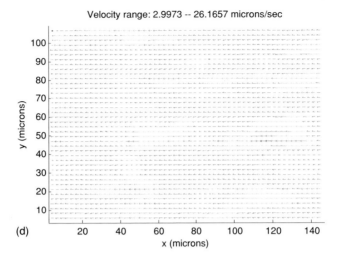
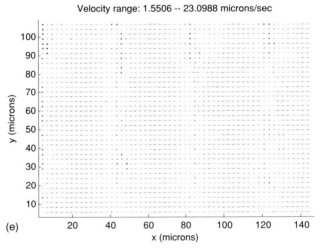
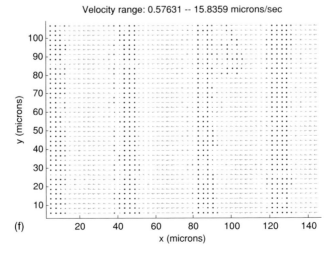

FIGURE 13.6. (*Continued*)

Conventional μPIV analysis Two different PIV analyses were performed. Initially a conventional μPIV analysis was performed to obtain an estimate of the average particle velocity field. Because the goal of the conventional μPIV analysis is not to extract velocity distribution, only the median velocity is reported. Figure 13.6 shows vector plots for the experimental cases ranging from 0.5 volts to 4.0 volts. A uniform color scale was applied among all the figures representing velocity magnitude, so that velocity changes between cases can be more easily interpreted. An interrogation region of 32 by 32 pixels and a grid spacing of 8 by 8 pixels were used. While this grid spacing over samples the data, it provides a good means of interpolating between statistically independent measurements. The advanced interrogation options included central difference interrogation (CDI), continuous window shifting (CWS), ensemble correlation averaging, and a multiple iteration scheme set to two passes. In all cases, the electrode voltage frequency was set to 580 kHz. This was determined by sweeping between frequencies of 100 kHz to 10MHz and qualitatively determining the most effective trapping frequency.

Velocity measurement statistics are reported in Table 13.1. The statistics show a maximum velocity of 27.5 microns/sec; this results in a Reynolds number of $3.3 * 10^{-4}$. The statistics also show that the average transverse velocity is negligible in comparison to axial velocities. A particle velocity increase can be observed as electrode voltage is increased from 0.5 volts to 2.0 volts. A velocity decrease is observed as electrode voltage is further increased from 2.0 volts to 4.0 volts. In the 2.5 volt case, low velocity regions between electrodes are first noticeable. In the 3.5 volt and 4.0 volt cases, large numbers of trapped particles result in near zero velocities.

The PIV results are summarized in Figure 13.7 which is a plot of the average axial velocities for the six electrode voltage cases. The three lowest voltages share a trend of decreasing particle velocity in the downstream direction. This phenomenon is evident in the curve fit parameters for linear slope given in Table 13.2. One explanation for this behavior may be that with each electrode a particle encounters, it lags the fluid velocity a little more. The cumulative effect results in a gradual slowing of the particle.

TABLE 13.1. PIV measurement statistics, all measurements are in (microns/second).

Voltage	Velocity Component	Average	RMS	Maximum	Minimum	Deviation
0.5V	axial	−18.4	18.6	−11.6	−24.7	2.83
	transverse	0.1622	0.272	0.988	−0.465	0.218
1.0V	axial	−020.3	20.5	−11.7	−26.4	2.87
	transverse	0.166	0.276	0.869	−0.755	0.220
2.0V	axial	−22.6	22.7	−12.3	−27.5	2.24
	transverse	0.201	0.277	1.03	−0.384	0.190
2.5V	axial	−16.0	16.4	−3.00	−26.1	3.48
	transverse	0.193	0.328	1.53	−0.635	0.265
3.5V	axial	−11.6	12.0	−1.55	−23.1	3.29
	transverse	0.113	0.241	0.906	−0.631	0.213
4.0V	axial	−7.19	8.35	0.576	−15.8	4.25
	transverse	0.0393	0.170	0.660	−0.648	0.165

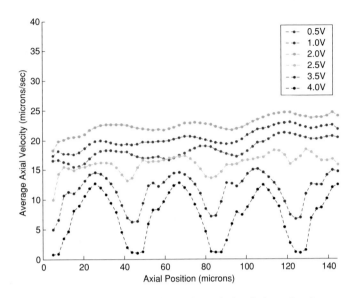

FIGURE 13.7. Average axial velocity from PIV results for all electrode voltage cases.

Another interesting result apparent from Figure 13.7 is that initially the average particle velocity increases as the voltage increases, 0.5 volts to 2.0 volts. This phenomenon is explained by particles being displaced from the channel bottom into faster areas of the fluid flow. This biases the velocity distribution toward higher velocities, altering the shape of the cross-correlation peak to favor higher velocities even though the fluid flow is constant. For higher voltages the effect of particles being hindered by axial field gradients is compounded by particles being forced beyond the high speed central portion of the flow profile by the DEP force. It can be qualitatively confirmed that particles migrate to the top of the channel by observing particle shapes in the images from the higher voltage cases, i.e. comparing the many particle shapes found in Figure 13.2 (top) which is acquired at 0.5 volts with the single particle shape found in Figure 13.2 (bottom) which is acquired at 4.0 volts. It is confirmed that the particles are indeed at the top of the channel by moving the focal plane throughout the measurement volume.

TABLE 13.2. Curve fit parameters ($V = A^*x + B + C^* \sin(2^*pi/D + E)$) for averaged axial velocities measured in microns per second.

Voltage	A	B	C	D	E
0.5V	0.0326	16.0	0.869	42.3	2.83
1.0V	0.0334	17.9	0.654	43.2	3.72
2.0V	0.0269	20.6	0.795	42.1	3.16
2.5V	0.0228	14.3	1.08	44.9	4.91
3.5V	0.00800	11.0	3.50	39.6	3.95
4.0V	0.00400	6.97	5.47	40.1	3.83

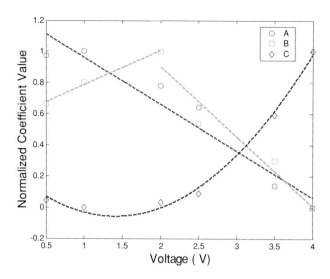

FIGURE 13.8. Curve fit parameters from (V = A*x + B + C*sin(2*pi/D + E) versus electrode voltage; dashed lines represent linear, piecewise linear, and parabolic curve fits for parameters A, B, and C, respectively.

The experimental results in Figure 13.7 were fit to the function

$$V_{axial} = A \cdot x + B + C \cdot \sin\left(\frac{2\pi}{D} + E\right) \qquad (13.5)$$

by minimizing the root mean squared error between the function and the experimental data. The resulting constants capture important dynamics of the particle velocity. The linear portion captures particles speeding up or slowing down over the total length of the device while the sinusoidal portion captures the dynamics of particles responding to each electrode. These results are tabulated in Table 13.2 and shown graphically in Figure 13.8. The parameters for the cases of 0.5 to volts 2.5 volts are very similar, with the exception of a gradual increase in "B". In the two highest voltage cases the linear slope, "A" is reduced to nearly zero, and the magnitude of the sinusoidal component, "C" is greatly increased. From these results it is apparent that the particle velocities are reduced as particles travel downstream for low electrode voltages, and that the particles are dominated by a periodic trajectory for higher electrode voltages.

Synthetic Image Analysis The PIV images acquired from the dielectrophoretic device were analyzed a second time using the synthetic image method to extract velocity distributions at various cross-sections of the device. The analysis of the device showed a displacement distribution of roughly zero to ten pixels. This distribution was used to approximate the maximum range of velocities, and correspondingly an interrogation region with a minus five pixel integer window shift was used to center the distribution in the cross-correlation plane; centering the measurement minimizes bias errors.

From the PIV analysis it can be seen that particles move perpendicular to the electrodes or axially, so an interrogation region of 48 pixels axially by 346 pixels transversely was used. The effective measurement volume is the full channel depth by the imaged channel width by 53 pixels in the axial direction. A measurement was made every ten pixels in the axial direction, considerably over-sampled. The results are given in Figure 13.9, with the fluid flowing from right to left. This analysis reports the particle velocity distributions, which offer a better characterization of the flow than the conventional PIV analysis. The velocity distributions offer an additional dimension of information that is unavailable from a conventional PIV analysis.

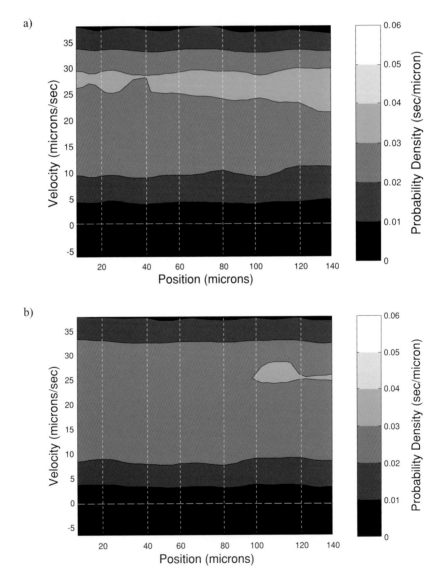

FIGURE 13.9. Probability density function for axial particle speed as a function of axial position within the DEP particle trap for electrode voltages of a) 0.5 volts, b) 1.0 volts, c) 2.0 volts, d) 2.5 volts, e) 3.5 volts and f) 4.0 volts.

PARTICLE DYNAMICS IN A DIELECTROPHORETIC MICRODEVICE 273

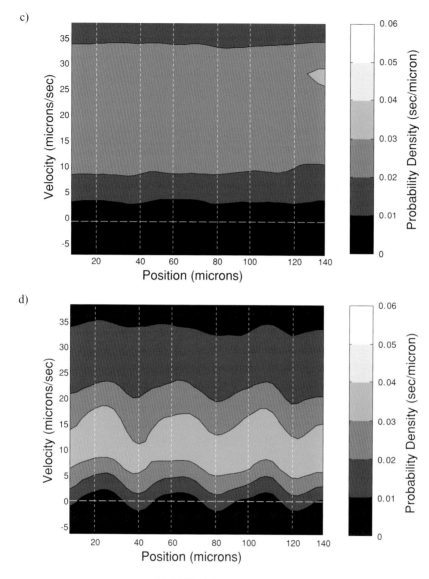

FIGURE 13.9. (*Continued*)

For the cases from 0.5 to 2.0 volts, the particle velocity distribution is as expected for flow in a slot, with the exception of a slight inclination of the contour lines that is consistent with Figure 13.7. The transition of the velocity distributions between the cases of 2.0 volts and 2.5 volts is quite dramatic. It can be seen that the bulk particle velocity is substantially reduced, while a fair number of particles still acquire high velocities. As the voltage is further increased the percentage of high speed particles is greatly reduced.

Particles travel from right to left so it can be seen from Figure 13.9 (d) that particles reach their maximum velocity faster than they slow to their minimum velocity. This is expected

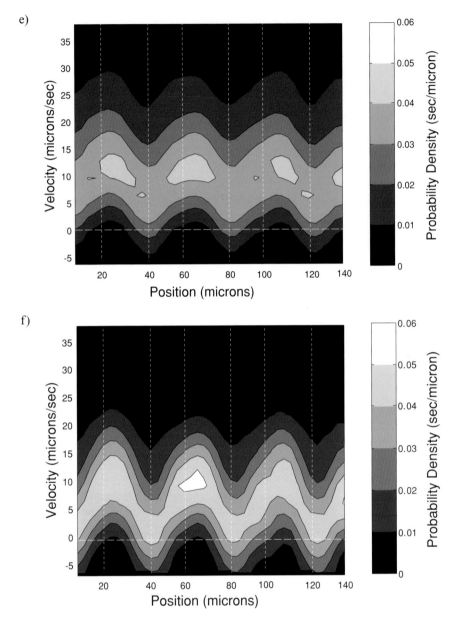

FIGURE 13.9. (*Continued*)

because the dielectrophoretic effect decelerates particles moving toward an electrode gap and accelerates particles moving away from a gap. The deceleration is hindered by a drag force on the particles while the acceleration is aided by the drag force.

This analysis shows that particles move fastest over the centers of the electrodes and slowest on the leading edge of the electrodes. The velocity distributions for all voltage cases are plotted at the fastest and slowest positions in Figure 13.9 (e) and (f), respectively. These plots show how the velocity distributions change as voltage is increased. For the "fast" position, as voltage is increased the range of particle velocities is reduced. The range of velocities is decreased further at the "slow" position for high voltages. Another interesting observation at the "slow" position is that some particles are predicted to move in the negative axial direction. This is caused by Brownian motion of near zero velocity particles.

13.4. CONCLUSIONS

The dynamics of particles traveling through the device described in this paper are very complicated, exhibiting migration normal to the electrodes as well as trapping behavior in the plane of the electrode. Several novel μPIV interrogation techniques are applied to shed light on the particle dynamics. The results discussed in this paper provide insight into how particles respond to DEP forces. Work is in progress to generalize the results of this paper to assess whether current DEP models are sufficient for predicting how particles will respond to electrical forces.

ACKNOWLEDGEMENTS

Thanks to Professor Rashid Bashir, Haibo Li, and Rafael Gomez for providing the chip used for these experiments and also for many useful discussions about DEP. This work was partially supported by the National Science Foundation Nanoscale Science and Engineering program.

REFERENCES

[1] M. Born and E. Wolf. *Principles of Optics*. Oxford Press, Pergamon, 1997.
[2] E.B. Cummings. An image processing and optimal nonlinear filtering technique for PIV of microflows. *Exper. Fluids*, 29(Suppl.):S42–S50, 2001.
[3] J. Deval, P. Tabeling, and C.-M. Ho. A dielectrophoretic chaotic mixer. *Proc. IEEE MEMS Workshop*, 36–39, 2002.
[4] P.R.C. Gascoyne and J. Vykoukal. Particle separation by dielectrophoresis [Review]. *Electrophoresis*, 23(13):1973–1983, July 2002.
[5] N.G. Green and H. Morgan. Dielectrophoretic investigations of sub-micrometer latex spheres. *J. Phys. D: Appl. Phys.*, 30:2626–2633, 1997.
[6] H.B. Li, Zheng, D. Akin, and R Bashir. Characterization and modeling of a micro-fluidic dielectrophoresis-filter for biological species. submitted to *J. Microelectromech. Sys.*, 2004.
[7] C.D. Meinhart, D. Wang, and K. Turner. Measurement of AC electrokinetic flows. *Biomed. Microdev.*, 5(2): 139–145, 2002.
[8] C.D. Meinhart, S.T. Wereley, and J.G. Santiago. A PIV algorithm for estimating time-averaged velocity fields. *J. Fluids Eng.*, 122:809–814, 2000.
[9] R. Miles, P. Belgrader, K. Bettencourt, J. Hamilton, and S. Nasarabadi. Dielectrophoretic manipulation of particles for use in microfluidic devices. *MEMS-Vol. 1, Microelectromechanical Systems (MEMS), Proceedings of the ASME International Mechanical Engineering Congress and Exposition*, Nashville, TN, Nov. 14–19, 1999.

[10] M.G. Olsen and R.J. Adrian. Out-of-focus effects on particle image visibility and correlation in microscopic particle image velocimetry. *Exper. Fluids*, (Suppl.):S166–S174, 2000.
[11] A. Ramos, Morgan, H., Green, N.G., and A. Castellanos. AC electrokinetics: a review of forces in microelectrode structures. *J. Phys. D: Appl. Phys.*, 31:2338–2353, 1998.
[12] X.-B., Wang, J. Vykoukal, F. Becker, and P. Gascoyne. Separation of polystyrene microbeads using dielectrophoretic/gravitational field-flow-fractionation. *Biophys. J.*, 74:2689–2701, 1998.
[13] S.T. Wereley, L. Gui, and C.D. Meinhart. Advanced algorithms for microscale velocimetry. *AIAA J.*, 40(6): 1047–1055, 2002.

14

Microscale Flow and Transport Simulation for Electrokinetic and Lab-on-Chip Applications

David Erickson* and Dongqing Li**
*Sibley School of Mechanical and Aerospace Engineering, Cornell University Ithaca, NY, 14853
**Department of Mechanical Engineering, Vanderbilt University Nashville, TN, 37235

14.1. INTRODUCTION

The proliferation of manufacturing techniques for building micro- and nano-scale fluidic devices has led to a virtual explosion in the development of microscale chemical and biological analysis systems, commonly referred to as integrated microfluidic devices or Labs-on-a-Chip [14]. Application areas into which these systems have penetrated include: DNA analysis [47], separation based detection [10, 36], drug development [59], proteomics [22], fuel processing [31] and a host of others, many of which are extensively covered in this book series. The development of these devices is a highly competitive field and as such researchers typically do not have the luxury of large amounts of time and money to build and test successive prototypes in order to optimize species delivery, reaction speed or thermal performance. Rapid prototyping techniques, such as those developed by Whitesides' group [11, 44], and the shift towards plastics and polymers as a fabrication material of choice [8] have significantly helped to cut cost and development time once a chip design has been selected.

Computational simulation used in parallel with these rapid prototyping techniques can serve to dramatically reduce the time from concept to chip even further. Simulation allows researchers to rapidly determine how design changes will affect chip performance and thereby reduce the number of prototyping iterations. More importantly "numerical prototyping" applied at the concept stage can generally provide reasonable estimates of potential chip performance (*e.g.* rate of surface hybridization, speed of thermal cycling for PCR or separation performance in capillary electrophoresis) enabling the researcher to take a fruitful path from the beginning or conversely abandon a project that does not show promise.

In this chapter the fundamentals of microscale flow and transport simulation for Lab-on-Chip and other electrokinetic applications will be presented. The first section will provide an overview of the equations governing the analyses and numerical challenges, focusing on the aspects which separate the above from their macroscale counterparts. The physical length scales relevant to most current Lab-on-Chip devices range from 10s of nanometers to a few centimeters and thus we limit ourselves here to continuum based modeling. Following that three illustrative case studies will be presented that demonstrate the appropriate computational interrogation of these equations (*i.e.* assumptions and solution techniques) in order to capture all the desired information with the minimum degree of complexity.

14.2. MICROSCALE FLOW AND TRANSPORT SIMULATION

14.2.1. Microscale Flow Analysis

On the microscale fluid flow can be accomplished by several means, the most common of which are traditional pressure driven flow and electroosmotic flow [33, 41]. Another mechanism of recent interest is magnetohydrodynamic flow (in which the interaction of a magnetic field and electric field induces a Lorenz force and thereby fluid transport) [37] has many advantages however is beyond the scope of what will be covered here. In either of the two former cases fluid motion is governed by the momentum Eq. (14.1a) and continuity equations Eq. (14.1b) shown below,

$$\rho\left(\frac{\partial v}{\partial t} + v \cdot \nabla v\right) = -\nabla p + \eta \nabla^2 v - \rho_e \nabla \Phi, \tag{14.1a}$$

$$\nabla \cdot v = 0 \tag{14.1b}$$

where v, t, p, η, and ρ are velocity, time, pressure, viscosity and density respectively. Inherent in Eqs. (14.1) are assumptions of incompressibility and constant viscosity and thus they are effectively limited to the description of constant temperature liquid flows. While this is somewhat restrictive, particularly considering the importance of Ohmic heating in polymeric microfluidic systems [16], the approximation is usually reasonably good in high thermal conductivity chips (*e.g.* glass or silicon) or for pressure driven flows. The final term in Eq. (14.1a) represents the electrokinetic body force and is equivalent to the product of the net charge density in the double layer, ρ_e, multiplied by the gradient of the total electric field strength, Φ. In general, the relatively small channel dimensions and flow velocities, typical of microfluidic and biochip applications, limit microscale liquid flows to small Reynolds numbers ($Re = \rho L v_o/\eta$ where L is a length scale and v_o the average velocity). A typical flow velocity of 1mm/s in a channel with a 100 μm hydraulic diameter yields a Reynolds number on the order of 0.1. As a result the transient and convective terms (those on the left-hand side of Eq. (14.1a)) tend to be negligibly small and thus are typically ignored in order to simplify the formulation. In the majority of cases this assumption is quite good however it does have some consequences. The most important of which is that the lack of a transient term necessarily implies that the flow field reaches a steady state instantaneously. To quantify this, let's consider the timescale used in the above definition of the Reynolds number, L/v_o, which for the above example yields 0.1s. As such if the other quantities of interest (*e.g.* species transport and surface reactions) which are being studied occur on timescales greater than 0.1s, which most do, the assumption that the flow field reaches an instantaneous steady state is valid. In sections 5 and 6 of this chapter we will provide examples of higher Reynolds number and transient situations where the above assumptions are not met.

Along solid walls, the general proper boundary conditions on Eq. (14.1a) is a no slip condition as described by Eq. (14.2a),

$$\boldsymbol{v} = 0. \quad \text{at solid walls} \tag{14.2a}$$

Inflow and outflow boundary conditions are generally more involved, depending on the situation. When pure electroosmotic flow is to be considered (that is to say there is no externally applied pressure, which does not preclude the possibility of internally induced pressure gradients), a simple zero gradient boundary condition normal to the inlet, Eq (14.2b)

$$\boldsymbol{n} \cdot \nabla \boldsymbol{v} = 0 \quad \text{at inflow boundaries in pure EO flow and outflow boundaries} \tag{14.2b}$$

where $\boldsymbol{n}$ is the unit normal to the surface, and a no flow boundary condition parallel with the inlet can be applied. When an external pressure driven flow component is present the inlet boundary condition is typically represented by the summation of a parabolic velocity profile (accounting for the pressure driven component) and a plug flow profile (accounting for the electroosmotic component) applied at the inlet boundaries. The form of the pressure driven flow component of the boundary condition varies significantly based on the inlet geometry and thus the reader is referred to standard fluid mechanics references such as Panton [48], White [61] or Batchelor [2] for exact forms. Another possibility is to apply a pressure based body force to the fluid continuum, as will be done in Section 14.6. This is particularly useful when the inlet/outlet boundary conditions are not apparent (*e.g.* periodic cases) but it does require prior knowledge of the pressure distribution and is thus not practical in most cases. The proper outflow boundary condition for all these cases is a zero gradient in the normal direction and no flow tangential to the boundary, similar to that for the inlet in the pure EO flow case, Eq. (14.2b). The above boundary conditions are very specific to confined liquid flows and are not sufficient for all systems (*e.g.* those involving free surfaces and capillarity). For a more general formulation the reader is referred to Reddy and Gartling [51].

As is apparent from Eq. (14.1a) evaluation of the momentum equation requires a description of the net charge density, ρ_e, and the total electrical field, Φ. The total electric field is commonly split into two components: the electrical double layer field (EDL), ψ, and the applied electric field, ϕ, as per Eq. (14.3) below,

$$\Phi = \phi + \psi. \tag{14.3}$$

The decoupling of these two terms is contingent on a number of assumptions that are well described in the often-quoted work by Saville [54]. The relatively high ionic strength buffers used in most lab-on-chip applications typically yield very thin double layers and thus this decoupling is almost always valid within a reasonable degree of error. It is important to note that there are cases of both theoretical and practical interest to microfluidics where these two components cannot be fully decoupled as will be demonstrated in Section 14.6. For the remainder of this section however we will assume that the decoupling conditions are met and return to the more general case in that section.

14.2.2. Electrical Double Layer (EDL)

An electrical double layer is a very thin region of non-zero net charge density near the interface (in this case a solid-liquid interface) and is generally the result of surface adsorption of a charged species and the resulting rearrangement of the local free ions in solution so as to maintain overall electroneutrality [39, 40]. It is the interaction of the applied/induced electric field, discussed below, with the charge in the double layer that results in the electrokinetic effects discussed here. The double layer potential field and the net charge density are related via the Poisson equation,

$$\nabla \cdot (\varepsilon_w \varepsilon_0 \nabla \psi) + \rho_e = 0, \quad (14.4)$$

where ε_o and ε_w are the dielectric permittivity of a vacuum ($\varepsilon_o = 8.854 \times 10^{-12}$ C/Vm) and the local relative dielectric permittivity (or dielectric constant) of the liquid respectively. The ionic species concentration field within the double layer is given by the Nernst-Planck conservation equation,

$$\nabla \cdot \left(-D_i \nabla n_i - \frac{D_i z_i e}{K_b T} n_i \nabla \psi + n_i v\right) = 0, \quad (14.5)$$

where D_i and z_i are the diffusion coefficient and valence of the ith species and e ($e = 1.602 \times 10^{-19}$ C), k_b ($k_b = 1.380 \times 10^{-23}$ J/K) and T are the elemental charge, Boltzmann constant and temperature respectively. The two equations are coupled by the definition of the net charge density given by,

$$\rho_e = \sum_i z_i e n_i \quad (14.6)$$

In principal the system of Eq. (14.4) through Eq. (14.6) are coupled to the flow field by the convective (3rd) term in Eq. (14.5) and as such they must, in principal, all be solved simultaneously. This results in a highly unstable system that is difficult to solve numerically. Therefore it is commonly assumed that the convective term is small and can be ignored (thereby decoupling the double layer equations from the flow field). The second major difficulty is that in principal all charged species must be accounted for in order to accurately determine the net charge density. This tends to be exceedingly difficult particularly when one wishes to examine the multispecies buffers which are commonly used in most actual Lab-on-Chip devices. Thus it is typical to implement a two species model based on the most highly concentrated ions (dominant) in the buffer solution and ignore the rest.

By far the most commonly applied boundary condition along solid walls to the Poisson equation is a fixed potential represented by the zeta potential, ζ,

$$\psi = \zeta \quad \text{along solid walls} \tag{14.7a}$$

The zeta potential is a property of the solid/liquid interface and in most cases remains constant at all points in the computational domain, however there are several theoretically and practically important cases where this is not the case (Sections 14.4 and 14.6). Another possible boundary condition on Eq. (14.4) is a gradient condition related to the surface charge density, however this is not commonly used in Lab-on-Chip type applications and is often difficult to handle computationally, as the solution is no longer fixed at a point. At inflow and outflow boundaries it is common to apply a zero gradient condition,

$$\boldsymbol{n} \cdot \nabla \psi = 0 \quad \text{at inflow and outflow boundaries} \tag{14.7b}$$

The proper boundary condition on the ionic species conservation equations are a zero flux condition, Eq. (14.8), which is typically applied at both solid walls and inflow and outflow boundaries,

$$-D_i(\boldsymbol{n} \cdot \nabla n_i) - \frac{D_i z_i e}{k_b T} n_i (\boldsymbol{n} \cdot \nabla \psi) + n_i \boldsymbol{v} = 0 \quad \text{at all boundaries} \tag{14.8}$$

This general condition is significantly less complicated than it first appears. At the solid walls $\boldsymbol{v} = 0$ thus eliminating the convective term. At the inflow and outflow boundaries $\boldsymbol{n} \cdot \nabla \psi = 0$ thereby eliminating the electrophoretic (2nd) term (though not completely correct it is often common to also neglect the convective (3rd) term, as outlined above, leaving the much simpler $\boldsymbol{n} \cdot \nabla n_i = 0$ condition).

As mentioned, the above formulation, while general, is typically very difficult to implement in most practical cases. By decoupling the potential field from the double layer field, assuming that the dielectric constant is uniform everywhere and using a model a symmetric electrolyte where both n$^+$ and n$^-$ have the same bulk concentration of n$_o$, the system of equations above reduces to the much simpler Poisson - Boltzmann distribution which, after linearization, yields the Debye-Hückel approximation to the double layer field,

$$\nabla^2 \psi - \kappa^2 \psi = 0 \tag{14.9}$$

where κ is the Debye-Hückel parameter and is equivalent to $\kappa = (2z^2 e^2 n_o / \varepsilon_w \varepsilon_o k_b T)^{1/2}$, which can be solved directly subject to boundary conditions Eq. (14.7a) and Eq. (14.7b) and then used to calculate the net charge density via,

$$\rho_e = -2n_o |z| e \sinh\left(\frac{|z|e\psi}{k_b T}\right) \tag{14.10}$$

for use in Eq. (14.1a) (details of the Poisson-Boltzmann derivation are available through a number of sources [33, 39] and will not be discussed in detail here). As mentioned above Eq. (14.9) and Eq. (14.10) represent the linearized version of the Poisson-Boltzmann equations and thus some error is necessarily introduced in linearizing the non-linear equation (the linearized version is only considered exact for low zeta potential (e.g., $\zeta < 25$ mV) [33]). For most microfluidic and biochip situations, this error tends to be reasonably small and the additional computational expense and iterative solution required to solve the full non-linear equation is often not justified (see Erickson and Li [15] for an example). The major drawback of using such a formulation is that information regarding the convective and electrical effects on the double layer field and the resulting influence on the flow structure cannot be obtained.

As is described in detail in the aforementioned reference texts, the inverse of the Debye-Hückel parameter (i.e. $1/\kappa$) is representative of the double layer thickness. Depending on the value of n_o this thickness can vary from close to 1 μm at low ionic concentration down to a few nanometers at high ionic concentration consistent with the buffers used in most Lab-on-Chip applications. This tends to cause a variety of numerical difficulties and requires further simplification of the above formulation the details of which are outlined in Section 14.3.

14.2.3. Applied Electrical Field

The applied electric field occurs either through direct application of an external voltage (as in electroosmotic flow) or induced via an effect known as the streaming potential. A streaming potential occurs when ions from the double layer are convected along with the bulk flow, typically pressure driven, accumulating at the downstream end resulting in a potential differences between the upstream and downstream reservoirs. In either case, the potential field is most generally governed by the conservation of current condition as below,

$$\nabla \cdot \boldsymbol{j} = 0 \quad (14.11a)$$

where j is the current flux. The current flux can be obtained by summation of the flux of each individual species multiplied by the valence and elemental charge yielding,

$$\nabla \cdot \left(\sum_i z_i e \left[-D_i \nabla n_i - \frac{D_i z_i e}{k_b T} n_i \nabla \phi + n_i \boldsymbol{v} \right] \right) = 0. \quad (14.11b)$$

Though derived from it, Eq. (14.11b) is different from the Nernst-Planck equations in that here we consider the conservation or charge as opposed to the conservation of individual species. As with the Nernst-Planck equations, in many situations it is not necessary to completely solve for Eq. (14.11b). In cases of pressure driven flow where the streaming potential is of interest, the diffusive flux (1st) term can be neglected as it tends to small compared to the remaining two. A further simplification often used in such cases is to ignore the divergence operator in Eq. (14.11b), integrate the remaining flux terms over the cross sectional area, and enforce a steady state zero net current condition where the conduction current (2nd term) has equal magnitude to the convection current (3rd term). For electroosmotic or combined flow in Lab-on-Chip systems, the conduction current tends to be much larger than the other terms and, as mentioned above, the EDL region tends to be thin. Under these conditions the 1st and 3rd terms in Eq. (14.11b) can be ignored as well as the additional conduction through the double layer yielding,

$$\nabla \cdot (\lambda \nabla \phi) = 0 \quad (14.11c)$$

where λ is the bulk solution conductivity and is given by

$$\lambda = \sum_i \frac{D_i z_i^2 e^2 n_{i,b}}{k_b T} \quad (14.11d)$$

where $n_{i,b}$ is the bulk concentration of the ith species. In general it is the total solution conductivity that is used in microfluidic applications and thus there is typically no need to perform the summation as shown. While there are many examples of non-uniform conductivity solutions in on-chip processes [6, 16, 34] in most cases it is assumed that a uniform solution conductivity exists everywhere. In that case λ is a constant and can be removed from the above formulation leaving a simple Laplacian to describe the applied potential field.

In most cases it is proper to assume that the channel walls are perfectly insulating, thus a zero gradient boundary condition is applied,

$$\mathbf{n} \cdot \nabla \phi = 0 \quad \text{at solid walls} \quad (14.12a)$$

The most commonly and easily applied boundary condition at the inflow and outflow boundaries is a fixed potential, similar to that described by Eq. (14.12b),

$$\begin{aligned}\phi_{inflow} &= \phi_1 \\ \phi_{outflow} &= \phi_2\end{aligned} \quad \text{at inflow and outflow boundaries for fixed potential situations} \quad (14.12b)$$

Such an approach works quite well for fundamental studies, if an entire microfluidic chip or microchannel network is to be modeled (such that the magnitude of the externally applied voltages are well defined by the experimental conditions), or if a simple system is considered (e.g. a capillary tube). If it is desired to model only a local section of a chip however, it is often non-trivial to estimate what the magnitude of the potential field is at the various inlets and outlets. In many such cases it is easier to apply a current based boundary condition, as the current can be more easily measured externally or estimated from a circuit model. In such cases the proper boundary condition is

$$\mathbf{n} \cdot \nabla \phi - J/\lambda A \quad \text{at inflow and outflow boundaries for fixed current situations} \quad (14.12c)$$

where J is the current and A is the channel cross sectional area. It is important to note that in order for the solution to remain bounded a fixed potential condition must be applied at a minimum of one inflow/outflow boundary (typically an outflow boundary is fixed at zero). Floating reservoirs (i.e. those where no potential is applied) are represented by a zero current condition (similar to that applied at the channel walls).

14.2.4. Microtransport Analysis

In the previous section the general theory required for microscale flow field simulation has been outlined. In most situations however the fluid flow itself it not of primary interest so much as it is a mechanism which can be exploited to transport the various reactants or products from one site to another. On the microscale this species transport is accomplished by 3 mechanisms: diffusion, electrophoresis, and convection. In the most general case the superposition of these three mechanisms results in, analogous to Eq. (14.5), the following conservation equation

$$\frac{\partial c_i}{\partial t} = \nabla \cdot (D_i \nabla c_i + \mu_{ep,i} c_i \nabla \phi - c_i v) + R_i \tag{14.13}$$

where c_i is the local concentration of the ith species, μ_{ep} is the electrophoretic mobility ($\mu_{ep} = D_i z_i e / k_b T$) and R_i is a bulk phase reaction term. Species transport typically occurs on timescales much longer than those for fluid flow and thus often the transient regime is of interest. A consequence of this is that Eq. (14.13) can solved after the steady state flow field has been determined, as for most dilute solutions the species transport does not globally influence the fluid flow. An important aspect of dilute species transport analysis is that in the absence of interspecies interaction (e.g. due to either bulk or surface phase reactions), the equations are not coupled and thus can be solved separately.

The reaction term in Eq. (14.13) can take many forms but for a typical reaction a rate law type relation is often assumed which, as an example, may take the form,

$$R_C = k_{a,3} c_A^l c_B^m - k_{d,3} c_C^n \tag{14.14}$$

where $k_{a,3}$ and $k_{d,3}$ are the forward and backwards reaction rate constants (the subscript 3 is used here to emphasize that these rate constants pertain to the 3D bulk region) and the superscripts l, m and n are the order of the reaction and A, B and C represent different species [3]. If the reaction were to go to effective completion, k_d is often very small and could thus be ignored. In addition to multispecies bulk phase reactions, other examples where such a reaction term may be used is to account for photon induced uncaging or photobleaching of fluorescence labeled molecules.

Boundary conditions on species concentration equations are typically very dependent on the situation of interest. However, in most Lab-on-Chip applications species are transported into the region of interest from a particular inlet and transported out through an outlet (note that due to electrophoretic transport, what constitutes an inlet for species transport may well be an outlet in fluid flow), yielding boundary conditions of the form,

$$c_i = c_o \quad \text{at the inlet for the } i^{th} \text{ species.} \tag{14.15a}$$
$$c_i = 0 \quad \text{at all other inlets} \tag{14.15b}$$
$$\mathbf{n} \cdot \nabla c_i = 0 \quad \text{at all outlets} \tag{14.15c}$$

where c_o is a known concentration. The latter of boundary conditions is not ideal, as its proper application requires that the computational domain be extended sufficiently far such that the species concentration is no longer expected to exhibit large spatial variations. It is however typically the best approximation that can be made. Boundary conditions along solid surfaces are governed by the flux of species from the bulk solution onto the solid surface and typically takes the form shown in Eq. (14.15d)

$$D_i(\mathbf{n} \cdot \nabla c_i) = \partial c_{i,2}/\partial t \qquad \text{at solid surfaces} \qquad (14.15d)$$

where $c_{i,2}$ represents the surface concentration of the ith species and t is time. In the absence of a heterogeneous reaction or significant adsorption/sorption into the solid matrix (typically the most common case), Eq. (14.15d) reduces to a simple zero flux condition. When heterogeneous reactions or adsorption/sorption is to be considered another level of modeling (identical in principal to the above) must typically be considered for the 2D surface. For an example of such a case the reader is referred to Erickson et al. [17] who presented a model for on-chip DNA hybridization kinetics.

14.3. NUMERICAL CHALLENGES DUE TO LENGTH SCALES AND RESULTING SIMPLIFICATION

As alluded to above, numerical modeling of microscale flow, particularly electroosmotic flow, in microstructures is complicated by the simultaneous presence of three separate length scales; the channel length (mm), the channel depth or width (μm) and the double layer thickness, $1/\kappa$ (nm), which we will refer to as L_1, L_2 and L_3 respectively. In general the amount of computational time and memory required to fully capture the complete solution on all three length scales would make such a problem nearly intractable. Since the channel length and cross sectional dimensions are required to fully define the problem (L_1, L_2), most computational studies have resolved this problem by either eliminating or increasing the length scale associated with the double layer thickness (L_3). Bianchi et al. [5], Patankar and Hu [49] and Fu et al. [23] accomplished this by artificially inflating the double layer thickness to bring its length scale nearer that of the channel dimensions. This allowed them to solve for the EDL field, calculate the electroosmotic body force term, and incorporate it into the Navier-Stokes Equation without any further simplification. It did however not fully eliminate the third length scale and significant mesh refinement was still required near the channel wall. In a different approach Ermakov et al. [20] removed the double layer length scale (L_3) from the formulation by applying a slip boundary condition to Eq. (14.1a), at the edge of the double layer, given by Eq. (14.16),

$$v_{slip} = \frac{\varepsilon_w \varepsilon_o \zeta}{\eta} \nabla \phi = \mu_{eo} \nabla \phi, \qquad \text{at solid walls} \qquad (14.16)$$

where μ_{eo} is the electroosmotic mobility and is a quantity commonly quoted in microfluidic studies (though geared towards electrophoresis of colloidal spheres, one of the more complete derivations of this velocity condition at the edge of the double layer is provided by Keh and Anderson [35]) and $\nabla \phi$ is evaluated at the boundary. Note that the application of boundary condition Eq. (14.12a) necessarily implies that $\nabla \phi$ is directed parallel to the boundary and that the velocity normal to the wall is identically zero as expected. Since the slip condition is applied at the edge of the double layer and not the channel wall the net charge density in the bulk solution is by definition negligible and thus the electroosmotic body force term in Eq. (14.1a) is eliminated. The most important consequence of the implementation of this boundary condition is that a description of the double layer field is no longer required, thus greatly simplifying the problem. While such a simplification cannot be used in cases where information regarding the flow in the double layer region is desired or required, it has been used successfully by a number of authors (Stroock et. al [56], Erickson and Li [18]) when species transport or bulk fluid motion is primary interest. An example of the implementation of this condition is provided in the following section.

Before proceeding it is worthwhile to briefly mention a few other approaches to microscale flow and transport simulation, not mentioned in the above, which may be of interest to the reader. The aforementioned paper by Fu et al. [23] provides alternative approach to the boundary conditions outlined above which they have applied to a variety of on-chip transport situations [24]. Molho et al. [43] demonstrate the use of a combined numerical simulation and optimization to study turn geometries and the resulting band spreading in microfluidic systems. Fiechtner and Cummings [21] also looked at this problem using the automated "Laplace" code, developed at Sandia National Laboratories. Though not strictly computational studies the analytical and numerical work by Griffiths and Nilson [27] and the stability analysis performed by Chen et al. [6] are also of significant interest.

14.4. CASE STUDY I: ENHANCED SPECIES MIXING USING HETEROGENEOUS PATCHES

In the preceding sections we have provided details of the equations for simulation of microfluidics and transport systems and discussed their implementation. In the following sections we will present three case studies with the objective of demonstrating the appropriate numerical implementation and approximation level depending on the information of interest. For the first of these examples we consider the enhanced species mixing in electroosmotic flow due to the presence of non-uniform electrokinetic surface properties. As discussed above, most microfluidic systems, particularly electroosmotically driven ones, are limited to the low Reynolds number regime and thus species mixing is largely diffusion dominated, as opposed to convection or turbulence dominated at higher Reynolds numbers. Consequently, mixing tends to be slow and occur over relatively long distances and times. As an example, the concentration gradient generator presented by Dertinger et al. [9] required a mixing channel length on the order of 9.25 mm for a 45 μm × 45 μm cross sectional channel or approximately 200 times the channel width to achieve nearly complete mixing. Here we use microfluidic and microtransport analysis to investigate if bulk flow circulation regions induced by heterogeneous patches can enhance species mixing.

As is shown in Figure 14.1, we consider the mixing of equal portions of two buffer solutions, one of which contains a concentration, c_o, of a species of interest. In general the introduction of surface heterogeneity induces flow in all three coordinate directions, thus necessitating the use of a full 3D numerical simulation. In all simulations presented here a square cross section was used with the depth equaling the width of the channel, w, and the arm length, L_{arm}. The length of the mixing channel, L_{mix}, was dictated by that required to obtain a uniform concentration (*i.e.* a fully mixed state) at the outflow boundary (thus making the application of Eq. (14.15c) at the outlet reasonable). Depending on the simulation conditions this required L_{mix} to be on the order of 200 times the channel width.

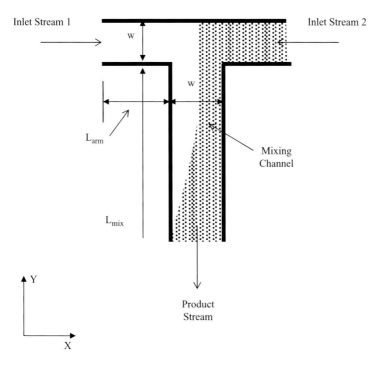

FIGURE 14.1. T-Shaped micromixer formed by the intersection of 2 microchannels, showing a schematic of the mixing/dilution process.

This is an example of a case where we are primarily interested in steady state, bulk phase fluid flow and species transport. As such we have no specific interest the double layer field and thus the electroosmotic slip condition approach, outlined in Section 14.3, is the relevant level of approximation. Therefore it was applied on all surfaces and used to simulate the hydrodynamic influence of the surface heterogeneities on the flow field and used to solve the lower Reynolds number, steady state versions of Eqs. (14.1a) and (14.1b). Additionally we assume that the transported species is dilute within a relatively highly concentrated buffer, such that the uniform conductivity assumption is met and the potential field can be determined from the Laplacian form of Eq. (14.11c) and solved subject to fixed potential conditions, Eq. (14.12b), at the inflow and outflow boundaries and insulation conditions, Eq. (14.12a), along the channel walls. The resulting species transport is modeled by the steady state, non-reactive version of Eq. (14.13).

The system of equations was solved over the computational domain via the finite element method using 27-noded triquadratic brick elements for ϕ, v and c and 8-noded trilinear brick elements for p. Through extensive numerical experimentation these higher order elements were found to be much more stable, especially when applied to the convection-diffusion-electrophoresis equation, than their lower order counterparts. In all cases the discretized systems of equations were solved using a quasi-minimal residual method solver and were preconditioned using an incomplete LU factorization. The advantages and disadvantages of using the finite element method (as opposed to other techniques such as control volume or finite difference) are well documented [26] and will not be discussed here other than to say that it is the authors' experience that the techniques outlined above can be most easily implemented using the finite element method. Details of the finite element technique used here can be found in Heinrich and Pepper [30].

14.4.1. Flow Simulation

For the purposes of this example, we consider a test case of a channel 50 μm in width and 50 μm in depth, an applied voltage of $\phi_{app} = 500$ V/cm ($\phi_{inlet,\ 1} = \phi_{inlet,\ 2} = \phi_{app}(L_{mix} + L_{arm})$, $\phi_{outlet} = 0$), and a mixing channel length of 15mm. We choose a homogeneous electroosmotic mobility of -4.0×10^{-8} m^2/Vs, corresponding to a ζ-potential of -42 mV. An electroosmotic mobility of $+4.0 \times 10^{-8}$ m^2/Vs was assumed for the heterogeneous patches.

In Figure 14.2 the mid-plane flow fields near the T-intersection for the (a) homogeneous case is compared with that generated by (b) a series of 6 symmetrically distributed heterogeneous patches on the left and right channel walls and (c) a series of offset patches also located on the left and right walls respectively. For clarity the heterogeneous patches are marked as the crosshatched regions in this and all subsequent figures. As can be seen both Figures 14.2b and 14.2c do exhibit regions of local flow circulation near these heterogeneous patches, however their respective effects on the overall flow fields are dramatically

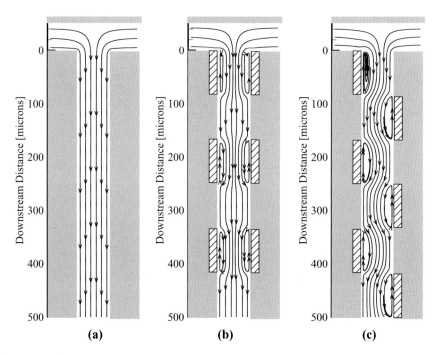

FIGURE 14.2. Electroosmotic streamlines at the midplane of a 50 μm T-shaped micromixer for the (a) homogeneous case with $\zeta = -42$ mV, (b) heterogeneous case with six symmetrically distributed heterogeneous patches on the left and right channel walls and (c) heterogeneous case with six offset patches on the left and right channel walls. All heterogeneous patches are represented by the crosshatched regions and have a $\zeta = +42$ mV. The applied voltage is 500 V/cm.

different. In Figure 14.2b it is apparent that the symmetric circulation regions force the bulk flow streamlines to converge into a narrow stream through the middle of the channel. The curved streamlines shown in Figure 14.2c show the more tortuous path through which the bulk flow passes as a result of the offset, non-symmetric circulation regions.

14.4.2. Mixing Simulation

As discussed above the goal of these simulations was to examine the effect of these circulation regions on species mixing. Figure 14.3 compares both the 3D and the channel

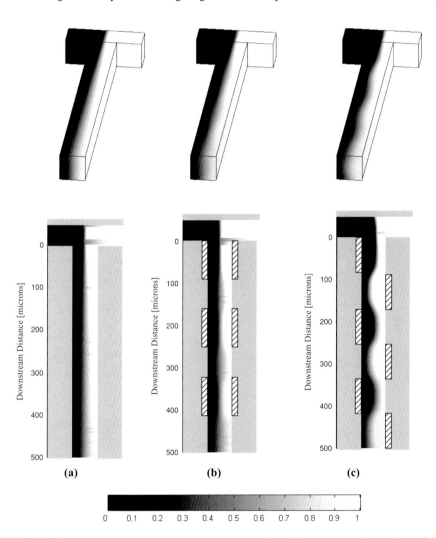

FIGURE 14.3. 3D species concentration contours (upper image) the midplane contours (lower image) for the 50 μm T-shaped micromixer resulting from the flow fields shown in Figure 14.2; (a) homogeneous case, (b) heterogeneous case with symmetrically distributed heterogeneous patches, and (c) heterogeneous case with offset patches. Species diffusivity is 3×10^{-10} m^2/s and zero electrophoretic mobility is assumed.

midplane concentration profiles for the three flow arrangements shown in Figure 14.2. In all these figures a neutral mixing species (i.e. $\mu_{ep} = 0$, thereby ignoring any electrophoretic transport) with a diffusion coefficient $D = 3 \times 10^{-10}$ m^2/s is considered. As expected both symmetric flow fields discussed above have yielded symmetric concentration profiles as shown in Figure 14.3a and 14.3b. While mixing in the homogeneous case is purely diffusive in nature, the presence of the symmetric circulation regions, Figure 14.3b, enables enhanced mixing by two mechanisms, firstly through convective means by circulating a portion of the mixed downstream fluid to the unmixed upstream region, and secondly by forcing the bulk flow through a significantly narrower region, as shown by the convergence of the streamlines in Figure 14.2b. In Figure 14.3c the convective effects on the local species concentration is apparent from the concentration contours generated for the non-symmetrical, offset patch arrangement.

14.5. CASE STUDY II: OSCILLATING AC ELECTROOSMOTIC FLOWS IN A RECTANGULAR MICROCHANNEL

In the previous example we simulated the electroosmotically driven transport of a dilute species. As discussed in Section 14.2.1, the flow transients in such a situation need not be considered since they occur on timescales much shorter than other on-chip processes (*i.e.* species transport). A situation where this condition is not met is when the applied electric field is transient or time periodic. Recently the development of a series of related applications has led to enhanced interest in such time periodic electroosmotic flows or AC electroosmosis. Oddy et al. [46] for example proposed and experimentally demonstrated a series of schemes for enhanced species mixing in microfluidic devices using AC electric fields. Green et al. [28] experimentally observed peak flow velocities on the order of hundreds of micrometers per second near a set of parallel electrodes subject to two AC fields, 180° out of phase with each other. Brown et al. [4] and Studer et al. [57] presented microfluidic devices that incorporated arrays of non-uniformly sized embedded electrodes which, when subject to an AC field, were able to generate a bulk fluid motion. Other prominent examples include the works by Selvaganapathy et al. [55], Hughes [32] and Barragán [1].

In this section we will demonstrate the modeling of an AC electroosmotic flow in a rectangular channel geometry, shown in Figure 14.4. AC electroosmotic flows are in general a good candidate for demonstrative simulation as they represent a case where the appropriate time scale (*i.e.* the period of oscillation) is significantly shorter than that required for the flow field to reach a steady state. For example if the AC field is applied at a frequency of 1000Hz the period of oscillation is 0.001s compared with 0.1s for the conservative case of $Re = 1$, discussed in Section 14.2.1. As such the transient term in Eq. (14.1a) must be considered (note that this is a simplified analysis for estimating when flow field transients should be considered, for a more detailed discussion of the relevant timescales in AC electroosmosis see Erickson and Li [15]). Further complicating matters is the work by Dutta and Beskok [12] who demonstrated that depending on the frequency and the double layer thickness there may be insufficient time for the liquid within the double layer to fully respond. As such the use of a slip condition at the edge of the double layer was shown to produce largely inaccurate results.

Thus to model this situation we consider the transient form of Eq. (14.1a), replacing the steady state $\nabla \Phi$, with $\phi_{app} sin(2\pi ft)$, where f is the frequency of oscillation. Since the flow field is uniaxial and symmetric about the center axes the computational domain is reduced

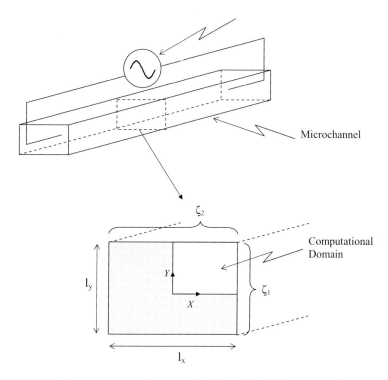

FIGURE 14.4. System geometry and computational domain for AC electroosmosis simulation.

to 2D in a single quadrant as shown in Figure 14.4. The double layer and applied potential fields can be decoupled, as is implied above, since the gradient of EDL field is perpendicular to the wall and the applied potential field parallel with it. The uniaxial condition implies that the derivative of the flow velocity in the lengthwise direction is zero, thus the momentum convection term in Eq. (14.1a) disappears, independent of the flow Reynolds number. Here we used the Debye-Hückel linearization to determine the EDL field and net charge density, Eq. (14.9) and Eq. (14.10).

14.5.1. Flow Simulation

Figure 14.5 compares the time-periodic velocity profiles in the upper left hand quadrant of a 100 μm square channel for two cases (a) $f = 500$ hz and (b) $f = 10$ khz with $\phi_{app} = 250$ V/cm. To illustrate the essential features of the velocity profile a relatively large double layer thickness has been used, $\kappa = 3 \times 10^6$ m^{-1} (corresponding to a bulk ionic concentration $n_o = 10^{-6}$ M), and a uniform surface potential of $\zeta = -25$ mV was selected. For a discussion on the effects of double layer thickness the reader is referred to Dutta and Beskok [12]. From Figure 14.5, it is apparent that the application of the electrical body force results in a rapid acceleration of the fluid within the double layer. In the case where the momentum diffusion time scale is much greater than the oscillation period (high f, Figure 14.2b) there is insufficient time for fluid momentum to diffuse far into the bulk flow and thus while the fluid within the double layer oscillates rapidly the bulk fluid

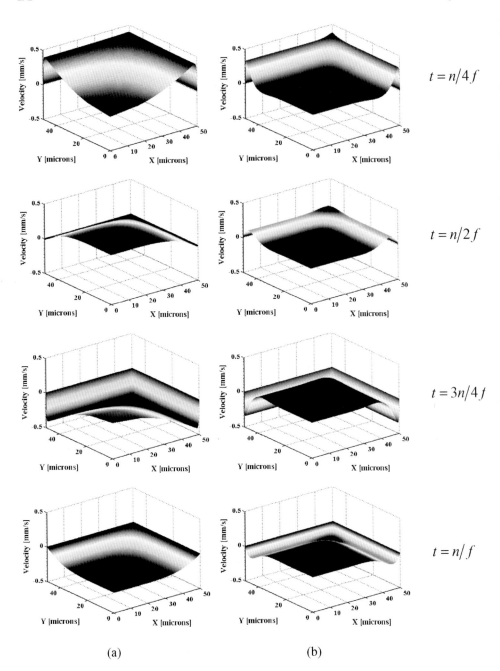

FIGURE 14.5. Steady state time periodic electroosmotic velocity profiles in upper quadrant of a 100 μm square microchannel at (a) $f = 500$ hz and (b) $f = 10$ khz with $\phi_{app} = 250$ V/cm. In each case n represents a "sufficiently large" whole number such that the initial transients have died out.

remains nearly stationary. At $f = 500$ hz there is more time for momentum diffusion from the double layer, however the bulk fluid still lags behind the flow in the double layer.

Another interesting feature of the velocity profiles shown in Figure 14.5 is the local velocity maximum observed near the corner (most clearly visible in the $f = 10$ khz case at $t = n/4f$ and $t = n/2f$). The intersection of the two walls results in a region of double layer overlap and thus an increased net charge density. This peak in the net charge density increases the ratio of the electrical body force to the viscous retardation allowing it to respond more rapidly to changes in the applied electric field and thus resulting in a local maximum in the transient electroosmotic flow velocity. Such an effect would not be observed had the slip condition approach used in Section 14.4 been implemented here.

14.6. CASE STUDY III: PRESSURE DRIVEN FLOW OVER HETEROGENEOUS SURFACES FOR ELECTROKINETIC CHARACTERIZATION

In the above two examples we limited ourselves to low Reynolds number cases where the flow field had no presumed effect on the EDL field. Here we will examine a case where such a condition is not met, specifically pressure driven slit flow over heterogeneous surfaces. Such a situation arises in a variety of electrokinetic characterization and microfluidics based applications. For example the streaming potential technique has been used to monitor the dynamic or static adsorption of proteins [13, 45, 60, 62], surface active substances [25] and other colloidal and nano-sized particles [29, 63] onto a number of surfaces. In this technique the surface's electrokinetic properties are altered by introducing a heterogeneous region, for example due to protein adsorption, which has a different ζ-potential than the original surface. The introduction of this heterogeneous region induces a change in the surface's average ζ-potential, which is monitored via a streaming potential measurement, and related back to the degree of surface coverage.

In this final example we demonstrate the simulation of pressure driven flow through a slit microchannel with an arbitrary but periodic patch-wise heterogeneous surface pattern on both the upper and lower surfaces (such that the flow field is symmetric about the midplane). Since the pattern is repeating, the computational domain is reduced to that over a single periodic cell, as shown in Figure 14.6. As mentioned above, in this example we are interested in examining the detailed flow structure and coupled effect of the flow field on the electrical double layer. As such we require a very general solution to the Poisson, Eq. (14.4), Nernst-Planck, Eq. (14.5), and Navier-Stokes, Eq. (14.1), equations in order to determine the local ionic concentration, double layer distribution and the overall flow field. Note that here in this theoretical treatment we limit ourselves to steady state flow, but we do wish to consider cases of higher Reynolds number. Thus the transient term in Eq. (14.1a) can be ignored however the higher momentum convection term should not.

14.6.1. System Geometry, Basic Assumptions and Modeling Details

In general boundary conditions on the respective equations are as described in section 2 with the exception of the inlet and outlet boundaries where periodic conditions are applied (for a general discussion on simulation in periodic domains the reader is referred to the work by Patankar et al. [50]). As mentioned above, in the absence of externally applied

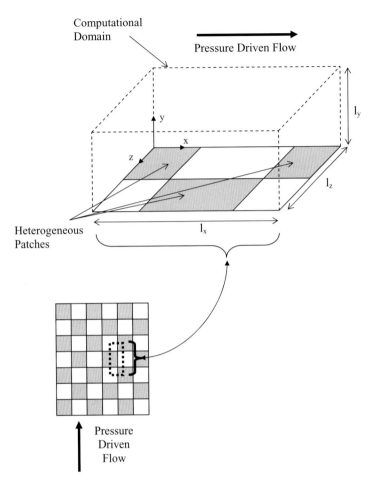

FIGURE 14.6. 3D spatially periodic computational domain for pressure driven flow through a slit microchannel with "close-packed" heterogeneous surface pattern.

electric fields the convective, streaming current components of the current conservation equation, Eq. (14.11b), is no longer dwarfed by the conduction component and thus must be considered. Additionally since we have rapid changes in the surface charge pattern, charge diffusion cannot be neglected. The highly coupled system consisting of all the general equations is difficult to solve numerically, thus several simplifying assumptions were made. Firstly it was assumed that the gradient of the induced potential field, ϕ, was only significant along the flow axis. Also, we assumed that Eq. (14.11b) was satisfied by applying a zero net current flux at every cross section along the flow axis, consistent with what has been done by others [7, 42]. In general we are interested in the steady state behavior of the system, thus the transient term in Eq. (14.1a) could be ignored, however we do wish to examine higher Reynolds number cases thus the convective terms should be included.

As discussed above, the resulting system of equations was solved using the finite element method. The computational domain was discretized using 27-noded 3D elements, which were refined within the double layer region near the surface and further refined in locations where discontinuities in the surface ζ-potential were present (*i.e.* the boundaries between patches). In all cases periodic conditions were imposed using the technique described by Sáez and Carbonell [53].

The solution procedure here began with a semi-implicit, non-linear technique in which the ionic species concentrations and the double layer potential equations were solved for simultaneously. In general it was found that a direct solver was required as the resulting matrix was neither symmetric nor well conditioned. Once a solution was obtained the forcing term in the Navier-Stokes equations was be evaluated the flow field solved using a penalty method [30] which eliminated the pressure term from the formulation. While this reduced the number of equations that had to be solved, the resulting poorly conditioned matrix dictated the use of a direct solver over potentially faster iterative solver, such as that described in Section 14.4. Further details on the solution procedure and model verification stages are available elsewhere [19].

14.6.2. Flow Simulation

This example considers the flow over the periodic surface pattern described above with a periodic length of $l_x = 50$ µm and width $l_y = 25$ µm, a channel height of 50 µm (therefore $l_y = 25$ µm in Figure 14.6). The liquid is considered as an aqueous solution of a monovalent species (physical properties of KCl from Vanysek [58] are used here) with a bulk ionic concentration of 1×10^{-5} M. Note that the selection of this reasonably low ionic concentration is equivalent to the double layer inflation techniques discussed earlier in order to bring the double layer length scale closed to that of the flow system. The Reynolds number Re = 1 and a homogeneous ζ-potential of -60 mV was selected. Figure 14.7 shows the flow streamlines for different heterogeneous ζ-potentials (a) -40 mV, (b) -20 mV and (c) 0 mV. As can be seen the streamlines along the main flow axis remain relatively straight, as would be expected for typical pressure driven flow; however, perpendicular to the main flow axis a distinct circular flow pattern can be observed. In all cases the circulation in this plane is such that the flow near the surface is always directed from the lower ζ-potential region (in these cases the heterogeneous patch) to the higher ζ-potential region. As a result the direction of circulation is constantly changing from clockwise to counterclockwise along the x-axis as the heterogeneous patch location switches from the right to the left side. It can also be observed that the center of circulation region tends to shift towards the high ζ-potential region as the difference in the magnitude of the homogeneous and heterogeneous ζ-potential is increased. The observed flow circulation perpendicular to the pressure driven flow axis is the result of an electroosmotic body force applied to the fluid continua caused by the differences in electrostatic potential between the homogeneous surface and the heterogeneous patch [19]. In general it was found that the strength of this circulation was proportional to both Reynolds number and double layer thickness.

14.6.3. Double Layer Simulation

Figure 14.8 shows the influence of the flow field on the double layer distribution along the indicated plane in the computational domain at Reynolds numbers of (a) Re = 0.1, (b) Re = 1 and (c) Re = 10 respectively. In all cases the heterogeneous ζ potential is

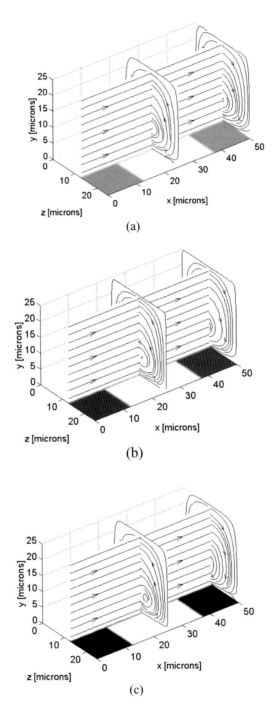

FIGURE 14.7. Influence of heterogeneous patches on flow field streamlines for close packed pattern. Homogeneous ζ-potential $= -60$ mV (white region), heterogeneous patches (dark patches) have ζ-potential (a) -40 mV, (b) -20 mV, (c) 0 mV. In all cases Re $= 1$ and $n_o = 1 \times 10^{-5}$ M KCl.

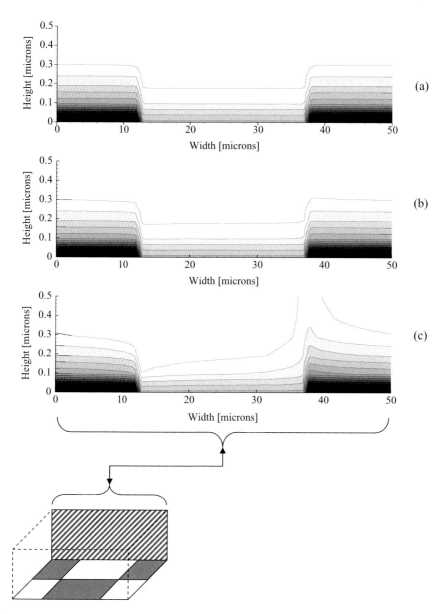

FIGURE 14.8. Electrostatic potential contours in the double layer region parallel with the flow axis on z = 0 μm (see Figure 14.6) for (a) Re = 0.1, (b) Re = 1.0 and (c) Re = 10.0. In each case the ζ-potential of the homogeneous region and heterogeneous patch are $\zeta_0 = -60$ mV and $\zeta_p = -40$ mV respectively and the bulk ionic concentration is $n_o = 1 \times 10^{-5}$ M.

−40 mV and the bulk ionic concentration is 1×10^{-5} M. As can be seen the double layer field becomes significantly distorted at higher Reynolds numbers, when compared to the diffusion-dominated field at Re = 0.1 and Re = 1. The diffusion-dominated field in these cases represents a classical Poisson-Boltzmann distribution. This distortion of the double layer field is the result of a convective effect observed by Cohen and Radke [7] in their 2D numerical work and comes about as a result of an induced y-direction velocity in the transition region between the two heterogeneous surfaces. This velocity perpendicular to the surface is a result of a weaker electroviscous effect over the lower ζ potential heterogeneous patch (see Li [38] for a discussion on the electroviscous effect) and thus the velocity in the double layer is locally faster. To maintain continuity, a negative y-direction velocity is induced when flow is directed from a higher ζ-potential region (*i.e.* more negative) to a lower and a positive y-direction velocity is induced for the opposite case.

14.7. SUMMARY AND OUTLOOK

In this chapter we have described the relevant equations, assumptions and numerical techniques involved in performing efficient microscale flow and transport simulation for microfluidic and lab-on-chip systems. There are very few "standard" problems in such systems and thus through a series of illustrative examples we have endeavored to demonstrate the range of complexity which may be required depending on the information of interest.

In this work we have limited our discussion to flow and species transport systems. These two aspects form the fundamental basis for the operation of most Lab-on-Chip devices and cover the majority of what many researchers are interested in simulating at present. In general however they do not present a complete picture of what is required to engineer a true lab-on-chip. The majority of the developmental work in Lab-on-Chip simulation is currently directed towards either the examination of more complex flow systems, particularly those with non-uniform electrical conductivity [52], those involving thermal analysis [16], or those which couple reaction kinetics to the transport problem [17]. The eventual goal is towards the integration of all these models into a single "numerical prototyping" platform enabling the coupled simulation of microfluid dynamics, microtransport, microthermal, micromechanics, microelectronics and optics with chemical and biological thermodynamics and reaction kinetics.

REFERENCES

[1] V.M. Barragán and C.R. Bauzá. Electroosmosis through a cation-exchange membrane: Effect of an ac perturbation on the electroosmotic flow. *J. Colloid Interface Sci.*, 230:359, 2000.
[2] G.K. Batchelor. *An Introduction to Fluid Dynamics*, Cambridge University Press, Cambridge, 2000.
[3] R.B. Bird, W.E. Stewart, and E.N. Lightfoot. *Transport Phenomena*, John Wiley & Sons, New York, 1960.
[4] B.D. Brown, C.G. Smith, and A.R. Rennie. Fabricating colloidal particles with photolithography and their interactions at an air-water interface. *Phys. Rev. E,* 63:016305, 2001.
[5] F. Bianchi, R. Ferrigno, and H.H. Girault. Finite element simulation of an electroosmotic-driven flow division at a T-junction of microscale dimensions. *Anal. Chem.*, 72:1987, 2000.
[6] C.-H. Chen, H. Lin, S. K. Lele, and J.G. Santiago. *Proceedings of ASME International Mechanical Engineering Congress*. ASME, Washington IMECE2003–55007, 2003.

[7] R. Cohen and C.J. Radke. Streaming potentials of nonuniformly charged surfaces. *J. Colloid Interface Sci.*, 141:338, 1991.
[8] A. de Mello. Plastic fantastic? *Lab-on-a-Chip*, 2:31N, 2002.
[9] S.K. Dertinger, D.T. Chiu, N.L. Jeon, and G.M. Whitesides. Generation of gradients having complex shapes using microfluidic networks. *Anal. Chem.*, 73:1240, 2001.
[10] V. Dolnik, S. Liu, and S. Jovanovich. Capillary electrophoresis on microchip *Electrophoresis*, 21:41, 2000.
[11] D.C. Duffy, J.C. McDonald, O.J.A. Schueller, and G.M. Whitesides. Rapid prototyping of microfluidic systems in poly(dimethylsiloxane) *Anal. Chem.*, 70:4974, 1998.
[12] P. Dutta and A. Beskok. Analytical solution of combined eloectoossmotic/pressure driven flows in two-dimensional straight channels: Finite debye layer effects. *Anal. Chem.*, 73:5097, 2001.
[13] A.V. Elgersma, R.L.J. Zsom, L. Lyklema, and W. Norde. Adsoprtion competition between albumin and monoclonal immunogammaglobulins on polystyrene lattices. *Coll. Sur.*, 65:17. 1992.
[14] D. Erickson and D. Li. Integrated microfluidic devices. *Anal. Chimica Acta*, 507:11, 2004.
[15] D. Erickson and D. Li. Analysis of alternating current electroosmotic flows in a rectangular microchannel. *Langmuir*, 19:5421, 2003.
[16] D. Erickson, D. Sinton, and D. Li. Joule heating and heat transfer in poly(dimethylsiloxane) microfluidic systems. *Lab-on-a-Chip*, 3:141, 2003a.
[17] D. Erickson, D. Li, and U.J. Krull. Modeling of DNA hybridization kinetics for spatially resolved biochips. *Anal. Biochem.*, 317:186, 2003b.
[18] D. Erickson and D. Li. Influence of surface heterogeneity on electrokinetically driven microfluidic mixing. *Langmuir*, 18:1883, 2002a.
[19] D. Erickson and D. Li. Microchannel flow with patchwise and periodic surface heterogeneity. *Langmuir*, 18:8949, 2002b.
[20] S.V. Ermakov, S.C. Jacobson, and J.M. Ramsey computer simulations of electrokinetic transport in micro-fabricated channel structures. *Anal. Chem.*, 70:4494, 1998.
[21] G.J. Fiechtner and E.B. Cummings. Faceted design of channels for low-dispersion electrokinetic flows in microfluidic systems. *Anal. Chem.*, 75:4747, 2003.
[22] D. Figeys and D. Pinto. Proteomics on a chip: Promising developments. *Electrophoresis*, 22:208, 2001.
[23] L.-M. Fu, R.-J. Yang, and G.-B. Lee. Analysis of geometry effects on band spreading of microchip electrophoresis. *Electrophoresis*, 23:602, 2002a.
[24] L.-M. Fu, R.-J. Yang, G.-B. Lee, and H.H. Liu. Electrokinetic injection techniques in microfluidic chips. *Anal. Chem.*, 74:5084, 2002b.
[25] D.W. Fuerstenau. Streaming potential studies on quartz in solutions of aminium acetates in relation to the formation of hemimicelles at the quartz-solution interface. *J. Phys. Chem.*, 60:981, 1956.
[26] V.K. Garg. In V.K. Garg (ed.). *Applied Computational Fluid Dynamics*, Marcel Dekker, New York, p. 35, 1998.
[27] S.K. Griffiths and R.H. Nilson. Band spreading in two-dimensional microchannel turns for electrokinetic species transport. *Anal. Chem.*, 72:5473, 2000.
[28] N.G. Green, A. Ramos, A. González, H. Morgan, and A. Castellanos. Fluid flow induced by nonuniform ac electric fields in electrolytes on microelectrodes. I. Experimental measurements. *Phys. Rev. E*, 61:4011, 2000.
[29] R. Hayes, M. Böhmer, and L. Fokkink. A study of silica nanoparticle adsorption using optical reflectometry and streaming potential techniques. *Langmuir*, 15:2865, 1999.
[30] J.C. Heinrich and D.W. Pepper. *Intermediate Finite Element Method*, Taylor & Francis, Philadelphia, 1999.
[31] J.D. Holladay, E.O. Jones, M. Phelps, and J. Hu. Microfuel processor for use in a miniature power supply *J. Power Sources*, 108:21, 2002.
[32] M.P. Hughes. Strategies for dielectrophoretic separation in laboratory-on-a-chip systems. *Electrophoresis*, 23:2569, 2002.
[33] R.J. Hunter. *Zeta Potential in Colloid Science: Principles and Applications*, Academic Press, London, 1981.
[34] B. Jung, R. Bharadwaj, and J.G. Santiago. Thousandfold signal increase using field-amplified sample stacking for on-chip electrophoresis. *Electrophoresis*, 24:3476, 2003.
[35] H.J. Keh and J.L. Anderson. Boundary effects on electrophoretic motion of colloidal spheres. *J. Fluid Mech.*, 153:417, 1985.
[36] N.A. Lacher, K.E. Garrison, R.S. Martin, and S.M. Lunte. Microchip capillary electrophoresis/electrochemistry. *Electrophoresis*, 22:2526, 2001.

[37] A.V. Lemoff and A.P. Lee. An AC magnetohydrodynamic micropump. *Sens. Actu. B*, 63:178, 2000.
[38] D. Li. Electro-viscous effects on pressure-driven liquid flow in microchannels. *Coll. Surf. A*, 195:35, 2001.
[39] J. Lyklema. *Fundamentals of Interface and Colloid Science, Vol. 1: Fundamentals*, Academic Press, London, 1991.
[40] J. Lyklema. *Fundamentals of Interface and Colloid Science, Vol. 2: Solid-Liquid Interfaces*, Academic Press, London, 1995.
[41] J. Masliyah. *Electrokinetic Transport Phenomena*. Alberta Oil Sands Technology and Research Authority, Edmonton, 1994a.
[42] J. Masliyah. Salt rejection in a sinusoidal capillary tube. *J. Coll. Int Sci.*, 166:383, 1994b.
[43] J.I. Molho, A.E. Herr, B.P. Mosier, J.G. Santiago, T.W. Kenny, R.A. Brennen, G.B. Gordon, and B. Mohammadi. Optimization of turn geometries for microchip electrophoresis. *Anal. Chem.*, 73:1350, 2001.
[44] J.M.K. Ng, I. Gitlin, A.D. Stroock, and G.M. Whitesides. Components for integrated poly(dimethylsiloxane) microfluidic systems. *Electrophoresis*, 23:3461, 2002.
[45] W. Norde and E. Rouwendal. Streaming potential measurements as a tool to study protein adsorption-kinetics. *J. Coll. Int. Sci.*, 139:169, 1990.
[46] M.H. Oddy, J.G. Santiago, and J.C. Mikkelsen. Electrokinetic instability micromixing. *Anal. Chem.*, 73:5822, 2001.
[47] B.M. Paegel, R.G. Blazej, and R.A. Mathies. Microfluidic devices for DNA sequencing: Sample preparation and electrophoretic analysis. *Curr. Opin. Biotechnol.*, 14:42, 2003.
[48] R. Panton. *Incompressible Flow*, John Wiley & Sons, New York, 1996.
[49] N.A. Patankar and H.H. Hu. Numerical simulation of electroosmotic flow. *Anal. Chem.*, 70:1870, 1998.
[50] S. Patankar, C. Liu, and E. Sparrow. Fully developed flow and heat-transfer in ducts having streamwise-periodic variations of cross-sectional area. *J. Heat Trans.*, 99:180, 1977.
[51] J.N. Reddy and D.K. Gartling. *The Finite Element Method in Heat Transfer and Fluid Dynamics*, CRC Press, Boca Raton, 2001.
[52] L. Ren and D. Li. Electrokinetic sample transport in a microchannel with spatial electrical conductivity gradients *J. Colloid Intefrace Sci.*, 294:482, 2003.
[53] A. Sáez and R. Carbonell. On the performance of quadrilateral finite-elements in the solution to the stokes equations in periodic structures. *Int. J. Numer. Meth. Fluids*, 5:601, 1985.
[54] D.A. Saville. Electrokinetic effects with small particles. *Ann. Rev. Fluid Mech.*, 9:321, 1977.
[55] P. Selvaganapathy, Y.-S.L. Ki, P. Renaud, and C.H. Mastrangelo. Bubble-free electrokinetic pumping. *J. Microelectromech. Syst.*, 11:448, 2002.
[56] A.D. Stroock, M. Weck, D.T. Chiu, W.T.S. Huck, P.J.A. Kenis, R.F. Ismagilov, and G.M. Whitesides. Patterning electro-osmotic flow with patterned surface charge. *Phys. Rev. Lett.*, 84:3314, 2000.
[57] V. Studer, A. Pépin, Y. Chen, and A. Ajdari. Fabrication of microfluidic devices for AC electrokinetic fluid pumping. *Microelect. Eng.*, 61:915, 2002.
[58] P. Vanysek. In D.R. Lide (ed.). *CRC Handbook of Chemistry and Physics*, CRC Press, 2001.
[59] B.H. Weigl, R.L. Bardell, and C.R. Cabrera. Lab-on-a-chip for drug development. *Advan. Drug Del. Rev.*, 55:349, 2003.
[60] C. Werner and H.J. Jacobasch. Surface characterization of hemodialysis membranes based on electrokinetic measurements. *Macromol. Symp.*, 103:43, 1996.
[61] F.M White. *Fluid Mechanics*, McGraw-Hill, New York, 1994.
[62] M. Zembala and P. Déjardin. Streaming potential measurements related to fibrinogen adsorption onto silica capillaries. *Coll. Surf. B*, 3:119, 1994.
[63] M. Zembala and Z. Adamczyk. Measurements of streaming potential for mica covered by colloid particles. *Langmuir*, 16:1593, 2000.

15

Modeling Electroosmotic Flow in Nanochannels

A.T. Conlisk[*] and Sherwin Singer[†]

[*] Department of Mechanical Engineering, The Ohio State University, Columbus, Ohio 43202
[†] The Department of Chemistry, The Ohio State University, Columbus, Ohio 43210-1107

15.1. INTRODUCTION

The determination of the nature of fluid flow at small scales is becoming increasingly important because of the emergence of new technologies. These techologies include Micro-Electro Mechanical Systems (MEMS) comprising micro-scale heat engines, micro-aerial vehicles and micro pumps and compressors and many other systems. Moreover, new ideas in the area of drug delivery and its control, in DNA and biomolecular sensing, manipulation and transport and the desire to manufacture laboratories on a microchip (lab-on-a-chip) require the analysis and computation of flows on a length scale approaching molecular dimensions. On these small scales, new flow features appear which are not seen in macro-scale flows. In this chapter we review the state-of-th-art in modeling liquid flows at nanoscale with particular attention paid to liquid mixture flows applicable to rapid molecular analysis and drug delivery and other applications in biology.

The governing equations of fluid flow on length scale orders of magnitude greater than a molecular diameter are well known to be the Navier-Stokes equations which are a statement of Newton's Law for a fluid. Along with conservation of mass and appropriate boundary and initial conditions in the case of unsteady flow, these equations form a well-posed problem from which, for an incompressible flow (constant density) the velocity field and the pressure may be obtained.

However, as the typical length scale of the flow approaches the micron level and below, several new phenomena not important at the larger scales appear. A perusal of the literature suggests that these changes may be classified into three rather general groupings:

- Fluid properties, especially transport properties (e.g. viscosity and diffusion coefficient) may deviate from their bulk values.
- Fluid, especially gases may slip at a solid surface.
- Channel/tube surface properties such as roughness, hydrophobicity/hydrophilicity and surface charge become very important.

Evidence already exists that there may be slip at hydrophobic surfaces in liquids; no slip still appears to hold at a hydrophilic surface. In liquids, slip or no-slip at the wall is a function of surface chemistry and roughness whereas in gases, slip is entirely controlled by the magnitude of the Knudesn number, the ratio of the mean free path to the characteristic length scale.

Transport at the nanoscale especially in biological applications is dominated by electrochemistry. There are a number of textbooks on this subject typified by [1–7] among many others. The term electrokinetic phenomena in general refers to three phenomena: (1) electrophoresis, which is the motion of ionic or biomolecular transport in the absence of bulk fluid motion; (2) electroosmosis, the bulk fluid motion due to an external electric field, (3) streaming potential, the potential difference which exists at the zero total current condition. In this chapter we focus primarily on electroosmosis and in the spirit of addressing biological applications we consider aqueous solutions only.

We shall see that it is impossible to pump fluid through very small channels mechanically via a pressure drop; one alternative is to pump the fluid by the imposition of an external electric field. *Electroosmosis* requires that the walls of the channel or duct be charged. Biofluids such as Phosphate Buffered Saline (PBS) in aqueous solution contain a number of ionic species. Because the mixture has a net charge balancing the wall charge, an electric field oriented in the desired direction of motion can be employed to induce bulk fluid motion. In addition, at the same time, because of the different diffusion coefficients of the different species, the ionic species will move at different velocities relative to the mass or molar averaged velocity of the mixture. This process is called *electrophoresis*.

In dissociated electrolyte mixtures, even in the absence of an imposed electric field, an electric double layer (EDL) will be present near the (charged) surfaces of a channel or tube. The nominal thickness of the EDL is given by

$$\lambda = \frac{\sqrt{\epsilon_e RT}}{F \left(\sum_i z_i^2 c_i\right)^{1/2}} \tag{15.1}$$

where F is Faraday's constant, ϵ_e is the electrical permittivity of the medium, c_i the concentrations of the electrolyte constituents, R is the gas constant, z_i is the valence of species i and T is the temperature. Here the ionic strength $I = \Sigma_i z_i^2 c_i$. The actual thickness of the electrical double layer is actually an asymptotic property much like the boundary layer thickness in classical external fluid mechanics. If we define the dimensionless parameter

$$\epsilon = \frac{\lambda}{h}$$

where h is the channel height, then for $\epsilon \ll 1$ the thickness of the EDL is normally $\sim 4 - 6\epsilon$. For $\epsilon \sim 1$ we say that the electrical double layers overlap.

Typically, the width of the electric double layer is on the order of 1 *nm* for a moderately dilute mixture; for extremely dilute mixtures, width of the electric double layer may reach several hundred nanometers. In the case where $\epsilon \ll 1$ the problem for the electric potential and the mole fraction of the ions is a singular perturbation problem and the fluid away from the electric double layers is electrically neutral. In the case where $\frac{\lambda}{h} = O(1)$ the channel height is of the order of the EDL thickness. In this case, electroneutrality need not be preserved in the core of the channel; however, the surface charge density will balance this excess of charge to keep the channel (or tube) electrically neutral. We assume that the temperature is constant and that the ionic components of the mixture are dilute.

There is a clear advantage to electroosmotic pumping versus pressure pumping in very small channels. When the EDLs are thin the flow rate is given to leading order by

$$Q_e \sim U_0 h W \tag{15.2}$$

where U_0 is independent of h (as we will see) and W is the width. Thus the flow rate is proportional to h and not h^3 as for pressure-driven flow for which the volume flow rate in a parallel plate channel is

$$Q_p = \frac{Wh^3}{12\mu L}\Delta p \tag{15.3}$$

where Δp is the pressure drop. This means that driving the flow by a pressure gradient is not feasible as depicted on Figure 15.1; note that at a channel height of 10*nm* three atmospheres of pressure drop are required to drive a flow of $Q = 10^{-6} L/min$ which is a characteristic flow rate in many applications. This is a large pressure drop in a liquid and clearly, a relatively awkward pump would be required to provide this pressure drop.

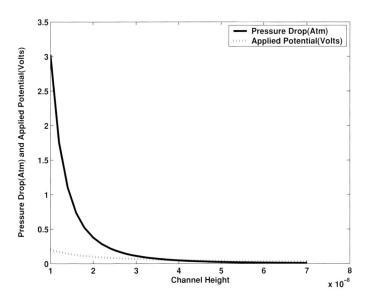

FIGURE 15.1. Pressure drop and applied voltage as a function of channel height to achieve a flowrate of $Q = 10^{-6} L/min$ in the system of Figure 15.2(b).

The plan of this chapter is as follows. We consider electroosmotic flow for the transport of ionic species; we focus on internal flow since the vast majority of the biomedical applications involve internal flow. In the next section various aspects of electrokinetic phenomena are outlined and then the governing equations are derived; the flow field, the electric field and the mass transfer problems are coupled. We then present results for several different parameters including channel height. The channels considered here are nano-constrained in one dimension. Next we discuss comparisons with experiment and conclude with a discussion of molecular dynamics simulations for probing the limits of continuum theory.

15.2. BACKGROUND

15.2.1. Micro/Nanochannel Systems

Because it is not feasible to transport fluids in nanochannels using an imposed pressure drop, electroosmosis and electrophoresis are often used. Application areas include rapid bio-molecular analysis; these devices are called lab-on-a-chip. Small drug delivery devices may employ electrokinetic flow to control the rate of flow of drug to the patient. These devices may also be used as biomolecular separators because different ionic species travel at different speeds in these channels. Natural nanochannels exist in cells for the purpose of providing nutrients and discarding waste. In general, these devices act as electroosmotic and/or diffusion pumps (Figure 15.2). The diffusion pump for drug delivery depicted on Figure 15.2 (a) has been tested on rats; the human version is expected to be orders of magnitude smaller. On Figure 15.2 (b) is a sketch of an electroosmotic nanopump whose operation has been documented by iMEDD, Inc. of Columbus, Ohio [8].

Nanochannel systems applicable to biomolecular analysis require interfacing microchannel arrays with nanochannel arrays. An example is depicted on Figure 15.3. These nanochannels are fabricated by a sacrificial layer lithography method which can produce channels with different surface properties [9]. On Figure 15.3, note that fluid enters from a bath on one side and is forced through several microchannels. The fluid is then forced to turn into a number of much smaller nanochannels which have the ability to sense and interrogate single molecules. Many biomolecules have a characteristic size on the order of 1–5 nm. On the other hand, in the iMEDD device (Figure 15.2(b)), fluid is forced directly into the nanochannels. In most systems of this type, the flow must pass through a micro scale channel into a nanochannel and then into a micro channel again.

One difficulty with modeling these systems is that the voltage drop in the nanochannels is not easily determined. This is because in systems of this type, the electrodes are placed in the baths upstream and downstream of the channels. The electric field must be determined in a complex geometry and in general must be calculated numerically. Moreover, the iMEDD device has over 47,000 nanochannels each of which may operate independently. Clearly each channel cannot be modeled independently in a single numerical simulation.

15.2.2. Previous Work on Electroosmotic Flow

Compared to the amount of work done on flow in micro-channels, there has been relatively little modeling work done on flow in channels whose smallest dimension is on the order

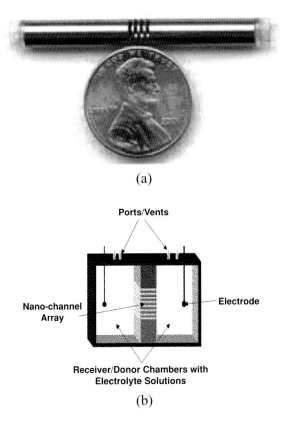

FIGURE 15.2. (a) Diffusion pump fabricated by iMEDD, Inc. of Columbus, Ohio. (b) Nanochannels fabricated by iMEDD, Inc. of Columbus, Ohio [8].

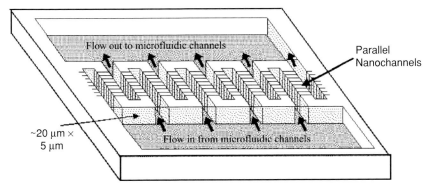

FIGURE 15.3. A micro/nanochannel system fabricated by Hansford [9].

of the electric double layer. In all of this work fully-developed flow is assumed. The problem for channel heights on the order of the electric double layer, that is, for overlapped double layers has been investigated by Verwey [10]. There the solution for the potential is based on a Boltzmann distribution for the number concentration of the ions; the potential is calculated based on a symmetry condition at the centerline. Note that the electric double layer will always be present near charged walls whether or not an external potential is used to drive a bulk motion of the liquid. Qu and Li [11] have recently produced solutions which do not require the Boltzmann distribution, but they also assume symmetry at the centerline; the results show significant differences from the results of Verwey and Overbeek [10]. The results of Qu and Li [11] are valid for low voltages since the Debye-Huckel approximation is invoked.

The first work on the electroosmotic flow problem discussed here appears to have been done by Burgeen and Nakache [12] who also considered the case of overlapped double layers. They produced results for the velocity field and potential for two equally charged ions of valence z; a Boltzmann distribution is assumed for the number of ions in solution. The convective terms in the velocity momentum equation are assumed to be negligible and the solution for the velocity and potential is assumed to be symmetric about the centerline of the channel.

Levine et al. [13] solved the same problem as Burgeen and Nakache [12] and produced results for both thin and overlapping double layers for a single pair of monovalent ions. Again symmetry of the flow with respect to the direction normal to the channel walls is assumed. The current flow is also calculated.

Rice and Whitehead [14] seem to be the first to calculate the electrokinetic flow in a circular tube; they assume a weak electrolyte and so assume the Debye-Huckel approximation holds. Levine et al. [15] also consider flow in a tube under the Debye-Huckel approximation; an analytical solution is found in this case. Stronger electrolyte solutions valid for higher potentials are also condidered using a simplified ad hoc model for the charge density.

Conlisk et al. [16] solve the problem for the ionic mole fractions and the velocity and potential for strong electrolyte solutions and consider the case where there is a potential difference in the direction normal to the channel walls corresponding in some cases to oppositely charged walls. They find that under certain conditions reversed flow may occur in the channel and this situation can significantly reduce the flow rate.

All of the work discussed so far assumes a pair of monovalent ions. However, many biofluids contain a number of other ionic species. As noted above, an example of one of these fluids is the common Phosphate Buffered Saline (PBS) solution. This solution contains five ionic species, with some being divalent. Zheng et al. [17] have solved the entire system of equations numerically; they show that the presence of divalent ions has a significant effect on the flow rate through the channel.

15.2.3. Structure of the Electric Double Layer

Consider the case of an electrolyte mixture which is bounded by a charged wall. If the wall is negatively charged then a surplus of positively charged ions, or cations can be expected to be drawn to the wall. On the other hand, if the wall is positively charged, then it would be expected that there would be a surplus of anions near the wall. The question then becomes, what is the concentration of the cations and anions near a charged surface? This question has vexed chemists for years and the concepts described below are based on qualitative, descriptive models of the region near a charged surface.

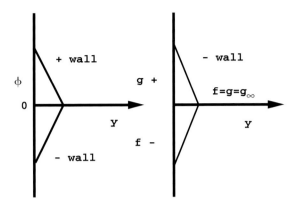

FIGURE 15.4. Potential and mole fractions near a negatively charged wall according to the ideas of Helmholtz [18]. Here g denotes the cation mole fraction and f denotes the anion mole fraction.

The electric double layer has been viewed as consisting of a single layer of counterions pinned to the wall outside of which is a layer of mobile coions and counterions. The simplest model for the EDL was originally given by Helmholtz [18] long ago and he assumed that electrical neutrality was achieved in a layer of fixed length. For a negatively charged wall he assumed the distributions of the anions and the cations are linear with distance from the wall as depicted on Figure 15.4.

In the Debye-Huckel picture of the electric double layer [19], the influence of the ionic species are equal and opposite so that the wall mole fractions of the coion (anion for a negatively charged wall), f^0 and counterion g^0 are symmetric about their asymptotic value far from the wall and the picture is as on Figure 15.5. The Gouy-Chapman [20, 21] model of the electric double layer allows for more counterions to bind to the wall charges so that the counter ions accumulate near the wall. This means that for a negatively charged wall g^0 can be much larger than its asymptotic value, whereas f^0, is not much lower than the asymptotic value in the core. This situation is depicted on Figure 15.6. Whether the

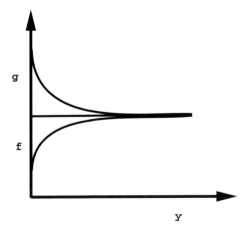

FIGURE 15.5. Debye-Huckel [19] picture of the electric double layer. Here g denotes the cation mole fraction and f denotes the anion mole fraction.

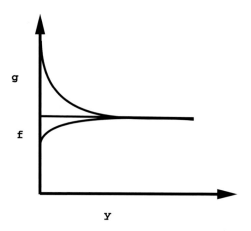

FIGURE 15.6. Gouy-Chapman model [20, 21] of the EDL.

Debye-Huckel picture or the Gouy-Chapman model of the EDL obtains depends on the surface charge density with the Debye-Huckel picture occuring at low surface charge densities and the Gouy-Chapman model occuring for higher surface charge densities.

Stern [22] recognized that there are a number of other assumptions embedded in these qualitative and simple models. In the models discussed so far the ions have been assumed to be point charges and the solvent is not modified by the presence of the charges. He proposed that the finite size of the ions affects the value of the potential at the surface. Finite-size ion effects in stagnant solutions are discussed by Bockris and Reddy [6]. Figure 15.7 shows the Stern layer consisting of a single layer of counterions at a negatively charged wall. Stern suggested that the surface potential be evaluated at the surface of shear as shown on Figure 15.7. The potential there is called the ζ potential and is a measured quantity.

15.3. GOVERNING EQUATIONS FOR ELECTROKINETIC FLOW

We consider now the case of flow of an electrolyte mixture in a slit channel for which the width and length of the channel are much bigger than its height as depicted on Figure 15.8. We have in mind the mixture consisting of water and a salt such as sodium chloride, but it is easy to see how to add additional, perhaps multivalent components. We consider the case where the salt is dissociated so that mixture consists of positively and negatively charged ions, say Na^+ and Cl^-.

In dimensional form, the molar flux of species A for a dilute mixture is a vector and given by[2]

$$\vec{n}_A = -cD_{AB}\nabla X_A + u_A z_A F X_A \vec{E}^* + c X_A \vec{u}^* \quad (15.4)$$

Here D_{AB} is the diffusion coefficient, c the total concentration, X_A is the mole fraction of species A, which can be either the anion or the cation, p is the pressure, M_A is the molecular weight, R is the gas constant, T is the temperature, u_A is the mobility, z_A is the valence, F is Faraday's constant, $\vec{E}^*$ is the total electric field and $\vec{u}^*$ is the mass average velocity of the fluid. The mobility u_A is defined by $u_A = \frac{D_{AB}}{RT}$. The mass transport equation for steady

MODELING ELECTROOSMOTIC FLOW IN NANOCHANNELS 309

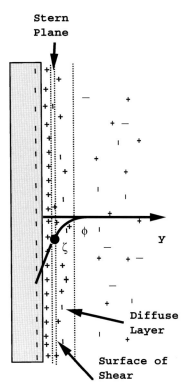

FIGURE 15.7. The Stern Layer. The potential at the edge of the plane of shear is called the ζ potential and the potential inside the Stern layer is linearly varying with distance from the wall.

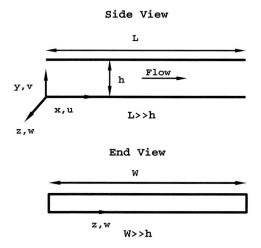

FIGURE 15.8. Geometry of the channel. Here it is only required that $h \ll W, L$ where W is the width of the channel and L its length in the primary flow direction. u, v, w are the fluid velocities in the x, y, z directions.

state is then

$$\nabla \cdot \vec{n}_A = 0 \tag{15.5}$$

In non-dimensional form the equation for the mole fraction of species A is given by

$$\frac{\partial^2 X_A}{\partial y^2} + \epsilon_1^2 \frac{\partial^2 X_A}{\partial x^2} + \epsilon_2^2 \frac{\partial^2 X_A}{\partial z^2} = ReSc\left(\epsilon_1 u \frac{\partial X_A}{\partial x} + v \frac{\partial X_A}{\partial y} + \epsilon_2 w \frac{\partial X_A}{\partial z}\right)$$
$$+ z_A \left(\epsilon_1 \frac{\partial X_A E_x}{\partial x} + \frac{\partial X_A E_y}{\partial y} + \epsilon_2 \frac{\partial X_A E_z}{\partial z}\right) \tag{15.6}$$

and we have assumed that the fluid and transport properties are constants. The externally imposed electric field, E_0, is constant in the x direction, whereas variations in the potential in y and z directions are permitted. The coordinates (x,y,z) are nondimensional; for example, $x = \frac{x^*}{L}$ and the scaling lengths in the three directions are (L, h, W) as depicted on Figure 15.8. Also (u, v, w) are the dimensionless velocities in each of the coordinate directions (x, y, z); for example $u = u^*/U_0$ where u^* is dimensional. Here $\epsilon_1 = \frac{h}{L}$ and $\epsilon_2 = \frac{h}{W}$. We assume $h \ll W, L$ so that both ϵ_1 and ϵ_2 are small. $Re = \frac{U_0 h}{\nu}$ is the Reynolds number and $Sc = \frac{\nu}{D_{AB}}$ is the Schmidt number, where ν is the kinematic viscosity. The valence $z_A = 1$ for the positive ion and $z_B = -1$ for the negative ion, although generalization to arbitrary valence is obvious. The determination of the velocity scale U_0 will be discussed below.

The mass transfer equation is subject to boundary conditions at a solid surface. Consider the wall at $y = 0$. Then if A refers to either of the ion distributions, it follows that we can specify the ion concentration or the flux at the surface. For the case of specified mole fraction,

$$X_A = X^0 \quad y = 0$$

and

$$X_A = X^1 \quad y = 1$$

where X^0 is the mole fraction at $y = 0$ and X^1 is the mole fraction at $y = 1$. Similar boundary conditions will hold at $z = 0$ and $z = 1$.

The mass transfer equation must be supplemented by an equation for the electric field. Most often in these problems, the channel is connected to large baths upstream and downstream. In this case the electric field in both baths and the channel should be calculated. To simplify the problem we have assumed that the x-component of the electric field is constant. This means that the dimensional potential is of the form

$$\phi^* = -\gamma x^* + \phi_1^*(y^*, z^*)$$

where it is seen that ϕ_1^* is the perturbation potential and γ is a constant. This form of the potential is consistent with the situation within the channel for a liquid, assuming a uniform dielectric constant across the channel. Note that a variable dielectric constant could be easily incorporated into our analysis as a function of y to accommodate changes in concentration and temperature.

MODELING ELECTROOSMOTIC FLOW IN NANOCHANNELS

Now, the potential scale is defined by taking $\phi_0 = \frac{RT}{F}$ and the equation for the electric potential is

$$\epsilon^2 \left(\frac{\partial^2 \phi}{\partial y^2} + \epsilon_1^2 \frac{\partial^2 \phi}{\partial x^2} + \epsilon_2^2 \frac{\partial^2 \phi}{\partial z^2} \right) = -\beta(X_+ - X_-) \quad (15.7)$$

where

$$\beta = 1 + \frac{c_3}{I}$$

where c_3 is the concentration of the solvent and $\epsilon = \frac{\lambda}{h}$ and I is the ionic strength. For very wide channels $\epsilon \gg \epsilon_1, \epsilon_2$; in addition, it is important to note that there will be boundary layers near the entrance and the exit of the channel and near the side walls where all of the independent variables vary rapidly. However, these regions are small, and in particular for very wide channels the influence of the side wall boundary layers will be negligible.

Since only differences in potential are important in this analysis we can specify the potential boundary conditions as

$$\phi = 0 \quad y = 0$$

and

$$\phi = \phi^1 \quad y = 1$$

The velocity field is coupled to the mass transfer equations and the equation for the potential. The governing equations of fluid flow express conservation of linear momentum along with the continuity equation which expresses conservation of mass. Conservation of mass requires

$$\epsilon_1 \frac{\partial u}{\partial x} + \frac{\partial u}{\partial y} + \epsilon_2 \frac{\partial w}{\partial z} = 0 \quad (15.8)$$

The x-direction is the primary direction of flow. Here it is noted that the velocity v is small and of $O(max(\epsilon_1, \epsilon_2))$.

We assume that the flow may be driven by an electrical body force oriented in the x-direction. The three momentum equations for an incompressible, steady flow are, in dimensionless form,

$$\epsilon^2 Re \left(\epsilon_1 u \frac{\partial u}{\partial x} + v \frac{\partial u}{\partial y} + \epsilon_2 w \frac{\partial u}{\partial z} \right) = -\epsilon_1 \epsilon^2 \frac{\partial p}{\partial x} + \beta(X_+ - X_-) + \epsilon^2 \nabla^2 u \quad (15.9)$$

$$\epsilon^2 Re \left(u \epsilon_1 \frac{\partial v}{\partial x} + v \frac{\partial v}{\partial y} + \epsilon_2 w \frac{\partial v}{\partial z} \right) = -\epsilon^2 \frac{\partial p}{\partial y} + \beta \Lambda \frac{\partial \phi}{\partial y}(X_+ - X_-) + \epsilon^2 \nabla^2 v \quad (15.10)$$

$$\epsilon^2 Re \left(\epsilon_1 u \frac{\partial w}{\partial x} + v \frac{\partial w}{\partial y} + \epsilon_2 w \frac{\partial w}{\partial z} \right) = -\epsilon_2 \epsilon^2 \frac{\partial p}{\partial z} + \epsilon_2 \Lambda \beta \frac{\partial \phi}{\partial z}(X_+ - X_-) + \epsilon^2 \nabla^2 w \quad (15.11)$$

where Re is the Reynolds number,

$$Re = \frac{\rho U_0 h}{\mu}$$

and U_0 is the velocity scale. These equations are the classical Navier-Stokes equations for constant density and viscosity which govern fluid flow of Newtonian liquids in a continuum. Here $p = \frac{p^*}{\mu U_0 / h}$ is the dimensionless pressure and

$$\nabla^2 = \frac{\partial^2}{\partial y^2} + \epsilon_1^2 \frac{\partial^2}{\partial x^2} + \epsilon_2^2 \frac{\partial^2}{\partial z^2}$$

Also $\Lambda = \frac{\phi_0}{E_0 h}$ and the velocity scale is

$$U_0 = \frac{\epsilon_e E_0 \phi_0}{\mu}$$

The momentum equations are subject to boundary conditions for which

$$u = v = w = 0 \text{ on all solid surfaces}$$

The governing equations form a set of seven equations in seven unknowns for the three velocity components, the two mole fractions of the electrolytes, the electric potential and the pressure. In electrochemistry these equations are usually not solved in their full form and several approximations may be made. First, the nonlinear terms can be neglected for small Reynolds numbers which usually is the case in biological problems. This makes the solution of the equations a lot easier. Second, if the channel is such that $h \ll W, L$ then essentially the entire problem may be said to be fully developed, that is all of the flow variables only in the y-direction. In this case the equations reduce to

$$\frac{\partial}{\partial y}\left(\frac{\partial X_+}{\partial y} + X_+ \frac{\partial \phi}{\partial y}\right) = 0 \qquad (15.12)$$

$$\frac{\partial}{\partial y}\left(\frac{\partial X_-}{\partial y} - X_- \frac{\partial \phi}{\partial y}\right) = 0 \qquad (15.13)$$

$$\epsilon^2 \frac{\partial^2 \phi}{\partial y^2} = -\beta(X_+ - X_-) \qquad (15.14)$$

and

$$\epsilon^2 \frac{\partial^2 u}{\partial y^2} = -\beta(X_+ - X_-) \qquad (15.15)$$

These are now four equations in four unknowns. Further it is easy to see how to generalize these equations to more ionic species of arbitrary valence.

For clarity and completeness we repeat the boundary conditions for the simplified problem (with $X_+ = g$ and $X_- = f$) as

$$\phi = 0 \text{ at } y = 0 \tag{15.16}$$
$$\phi = \phi^1 \text{ at } y = 1 \tag{15.17}$$
$$f = f^0 \text{ at } y = 0 \tag{15.18}$$
$$f = f^1 \text{ at } y = 1 \tag{15.19}$$
$$g = g^0 \text{ at } y = 0 \tag{15.20}$$
$$g = g^1 \text{ at } y = 1 \tag{15.21}$$
$$u = 0 \text{ at } y = 0, \quad \text{and} \quad y = 1 \tag{15.22}$$

The last equation is the no-slip condition for the velocity.

The governing equations indicate that the flow, in the symmetric case has a four-parameter family of solutions: $\varepsilon, \beta, g^0 f^0$. However, we can define

$$\delta^2 = \frac{\varepsilon^2}{\beta g^0}$$

and rescale the mole fractions g, f on the quantity g^0. In this case, for symmetry the boundary conditions become

$$f = \gamma \text{ at } y = 0, 1 \tag{15.23}$$
$$g = 1 \text{ at } y = 0, 1 \tag{15.24}$$

where $\gamma = \frac{f^0}{g^0}$ and we now have only a two-parameter family of solutions δ, γ. The only thing that needs to be done is to specify δ, γ and then to calculate the effective channel height from the definition of ϵ given a concentration. Note that β is a large number and g^0 is a small number so that in general $\delta = O(\epsilon)$.

The theory is easily extended to mixtures having an arbitrary number of species. In this case there will be N parameters corresponding to $\gamma_k, k = 1, \ldots, N-1$, in addition to δ. In the asymmetric case where there may be walls of opposite charge, for two electrolyte species, there will be two additional parameters associated with the boundary conditions at $y = 1$. Thus for N species all of which have boundary values at $y = 1$ that are different from those at $y = 0$, we will have $2N$ parameters to specify plus the value of the potential at $y = 1$, so that the total number of parameters is $2N + 1$.

The classical Poisson-Boltzmann distribution is obtained by integrating equations (15.12) and (15.13) resulting in

$$f = f^0 e^\phi \tag{15.25}$$
$$g = g^0 e^{-\phi} \tag{15.26}$$

where we have taken $\phi^1 = 0$. Note that the functional form of equations (15.25) and (15.26) are valid both inside and outside the electric double layer. This is termed the Poisson-Boltzmann distribution for the ionic species and is valid even in more than one dimension.

15.4. FULLY DEVELOPED ELECTROOSMOTIC CHANNEL FLOW

15.4.1. Asymptotic and Numerical Solutions for Arbitrary Electric Double Layer Thickness

Most often the fully developed assumption in slit pores is invoked because channels with more than one nano-constrained dimension have not been fabricated. As mentioned above we will consider only the case of two ionic species although the extension to an arbitrary number of species is easy to formulate; thus we consider a three component mixture containing the ionic species plus the solvent water; however it is easy to extend the theory to mixtures such as PBS [17]. Because the mole fractions sum to one, it is sufficient to consider only the equations governing the ionic species.

Equations (15.14) and (15.15) suggest clearly the nature of the solution as a function of ϵ. For $\epsilon \ll 1$, the electric double layers are thin and equations (15.14) and (15.15) reduce to

$$g - f = 0$$

and so the core region away from the walls the fluid is electrically neutral. Near the walls, the velocity and potential vary rapidly and so we define the new variable appropriate in the region near $y = 0$, $\hat{y} = \frac{y}{\epsilon}$ and, for example equation, (15.15) becomes

$$\frac{\partial^2 u}{\partial \hat{y}^2} = -\beta(g - f)$$

This is expected to be the situation in the case of a micro-channel. On the other hand when $\epsilon = O(1)$ the fluid is not electrically neutral and the whole problem must be solved in the entire domain $0 \leq y \leq 1$. This is the case for a nano-channel.

The case of $\epsilon \ll 1$ has been considered in detail by Conlisk et al. [16] and we now outline the solution. Electroneutrality requires

$$f_o = g_o \tag{15.27}$$

where the subscript o denotes the outer solution outside of the double layers.

Now since the outer solutions for f and g are equal, we can subtract the equations for g and f to find

$$\phi_0 = C \int_y^1 \frac{dy}{g_o} + D \tag{15.28}$$

Now the dimensionless velocity in the outer region satisfies

$$u_o = \phi_o + Ey + F \tag{15.29}$$

and in the inner region near $y = 1$ for example,

$$u_i = \phi_i + G \tag{15.30}$$

Near $y = 0$ we have

$$u_i = \phi_i + K \qquad (15.31)$$

The unknown constants are obtained by matching and the solution in the core of the channel is

$$u_o = \phi_o = \frac{1}{2}\ln\left(\frac{g^0}{f^0}\right) \qquad (15.32)$$

$$f_o = \sqrt{g^0 f^0} = g_o \qquad (15.33)$$

Note that both positive and negative velocities may occur based on the relative values of the mole fractions f^0 and g^0. If the two electrolytes have valences z_g and z_f, the solution is

$$u_o = \phi_o = \frac{1}{z_g - z_f}\ln\frac{-Z_g g^0}{z_f f^0} \qquad (15.34)$$

$$f_o = \sqrt{-\frac{z_g}{z_f} g^0 f^0 e^{-(z_g+z_f)\phi_o}} = g_o \qquad (15.35)$$

Now consider the inner solution within the electric double layer. Substituting expressions (25) and (26) in the potential equation (15.14) we can integrate once to obtain

$$-\frac{1}{2}(\phi_i')^2 = -f^0(e^{\phi_i} - 1) - g^0(e^{-\phi_i} - 1) \qquad (15.36)$$

leading to the solution for ϕ in integral form as

$$\phi_i(y) = \sqrt{2(g^0 + f^0)} \int_0^\phi \sqrt{ae^{-\phi} + be^\phi - 1}\, d\phi \qquad (15.37)$$

where

$$a = \frac{g^0}{g^0 + f^0}$$

$$b = \frac{f^0}{g^0 + f^0}$$

The integral is over the domain $(0, \phi_o)$ which is given above.

Much information about the solution in the inner region may be obtained without performing the integration explicitly. The uniformly valid solution for the velocity is given by [23]

$$u_{uv} = u_i^0 + -\phi^1 y + u_i^1 \qquad (15.38)$$

where u_i^0 is the inner solution near $y = 0$ for example. Note that near $y = 0$, $u_{uv} = \phi_i^0$, and $u_{uv} = \phi_i^1 - \phi^1$ near $y = 1$ where ϕ_i^0 is the inner solution for the potential near $y = 0$.

Once the solution for the potential and velocity is computed, the flow rate can be calculated in the asymptotic case as

$$Q == U_0 A \int_0^1 u\, dy = U_0 A \left(\frac{g^0 - f^0}{g^0 + f^0} - 2\epsilon \right)$$

where $A = Wh$ is the cross-sectional area of the channel. The corresponding surface charge density is given by

$$\sigma_0 = \frac{\epsilon_e RT}{\epsilon h F} \int_0^\infty \frac{\partial^2 \phi}{\partial \hat{y}^2} d\hat{y} = -\frac{\epsilon_e RT}{\epsilon h F} \frac{\partial \phi}{\partial \hat{y}}\bigg|_0 \tag{15.39}$$

In classical electrochemistry it is argued that for the case of thin, non-overlapping double layers, the appropriate reference concentration or mole fraction to use is the value in the bulk. Thus $\phi^* = \frac{\partial \phi^*}{\partial y^*} = 0$ in the bulk and [10]

$$\phi^*(y^*) = \frac{2RT}{zF\epsilon_e} \ln \left(\frac{1 + \gamma e^{-\kappa y^*}}{1 - \gamma e^{-\kappa y^*}} \right) \tag{15.40}$$

where

$$\gamma = \frac{e^{\frac{zF\phi_0^*}{2RT}} - 1}{e^{\frac{zF\phi_0^*}{2RT}} + 1}$$

and all variables are dimensional. The quantity $\phi_0^* = \zeta^*$ the ζ-potential and $\kappa = \frac{1}{\lambda}$. This approach may be useful for non-overlapping bouble layers but is not appropriate for overlapping EDLs.

We will present results for an aqueous solution at a temperature $T = 300°K$, imposed electric potential is six volts applied over a length of $3.5 \mu m$ with a dielectric constant of water, 78.54. We assume two monovalent ions, say Na+ and Cl−; the baseline mixture wall mole fractions correspond to a concentration of Na equal to .154 M, Cl .141 M and water 55.6 M for which $\beta \sim 190$. Results are also produced for other concentrations as noted.

On Figure 15.9 are the results for a number of values of δ corresponding to different channel heights from overlapping double layers to vary thin double layers. In the overlapped case, a numerical solution using classical finite-difference techniques [16] is computed for the potential. Note the plug flow nature of the velocity for small δ; this is the classical "top hat" electroosmotic velocity as depicted in textbooks. Note that we compute the potential as a perturbation from the ζ potential at the side walls and thus the velocity is equal to the potential. Moreover, the ζ potential is not needed in the simulation.

Figure 15.10 demonstrates the concentration, velocity, and potential profiles across channels of different heights and oppositely charged surfaces. That is, $y = 0$ corresponds to a negatively charged surface and $y = 1$ corresponds to a positively charged surface.

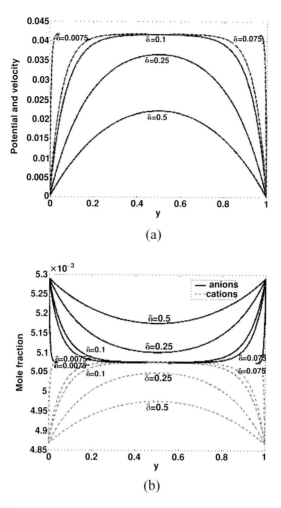

FIGURE 15.9. Results for the dimensionless velocity and potential along with mole fractions for the symmetric case of a pair of monovalent ionic species for an electric field of 6 volts over a channel of length $L = 3.5\mu m$; the channel height varies and the value of δ is shown; here $\gamma = 0.92$. (a) Velocity and potential. (b) Mole fractions for $f^0 = f^1 = 0.00252$ and $g^0 = g^1 = .00276$.

Accordingly, the ionic wall mole concentrations show an increase in the counterion for each surface, leading to increased concentrations of opposing ions at the opposite walls. For this simulation, the same mole fractions were used at $y = 0$ as in Figure 15.9 with the cation concentration at $y = 1$ equal to 0.5 that of the concentration at $y = 0$. The anion concentration at $y = 1$ is dictated by the condition of equal electrochemical potential at each wall. Note that within the channel there is flow in opposing directions due to competing electroosmotic effects at the two surfaces. This gives the considerably lower flowrate of $Q = 6.5 \times 10^{-13} \frac{L}{min}$ when compared with the symmetric case for which the velocity is always positive.

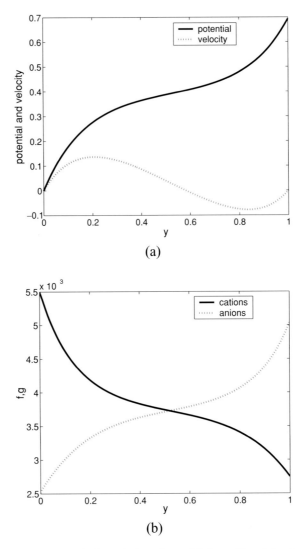

FIGURE 15.10. Results for the dimensionless velocity and potential along with mole fractions for the asymmetric case of an NaCl-water mixture. Here the electric field corresponds to 6 volts over a channel of length $L = 3.5\ \mu m$; the channel height $h = 4nm$. (a) Velocity and potential. (b) Mole fractions for $f^0 = 0.00252$, $f^1 = 0.00503$ and $g^0 = .00550$, $g^1 = 0.00275$.

15.4.2. Equilibrium Considerations

The results of the previous section are for arbitrarily specified wall boundary conditions and they do not take into account the presence of upstream and downstream reservoirs present in actual systems; the situation has been decpicted on Figure 15.2(b). In experiments, the molarities in the reservoirs are known and the mixture is electrically neutral there. In this situation, incorporation of the reservoirs without calculating the solutions there

MODELING ELECTROOSMOTIC FLOW IN NANOCHANNELS

can be achieved by calculating the wall mole fractions using the requirement that the electrochemical potential in the reservoirs far upstream be the same as the average value at any channel cross section as described by Zheng, Hansford and Conlisk [17]. This requirement leads to the Nernst equation [3] which is given by

$$\Delta\Psi = \frac{RT}{z_i F} \ln \frac{c_{iR}}{\bar{c}_{iC}} \quad i = 1, \ldots, N \tag{15.41}$$

where c_{iR} and $\bar{c}_{iC}$ are the average values of the concentration of species i in the reservoir and the fully developed region within the channel respectively.

The channel must be electrically neutral as well so that

$$z_f c_f + \sum_i z_i \bar{c}_{iC} = 0 \tag{15.42}$$

where z_f and c_f are the valence and the concentration of the fixed charges on the wall. Electroneutrality in the reservoir requires

$$\sum_i z_i c_{iR} = 0 \tag{15.43}$$

Equations (15.41) and (15.42) are $N+1$ equations in $N+1$ unknowns for the average concentrations in the channel and the Nernst Potential $\Delta\Psi$.

If the volume of the reservoir is much larger than the volume of the channels as is the case in practice, then the concentration in the reservoir will be fixed at the value of the concentration prior to the initiation of flow into the channel. To obtain each of the average concentrations in the channel, we equate the Nernst potential for each of the species and substituting into equation (15.42) we have

$$z_f c_f + z_1 \bar{c}_{1C} + \sum_{i=2}^{N} \frac{z_i c_{iR}}{(C_{1R})^{\frac{z_i}{z_1}}} (\bar{c}_{1C})^{\frac{z_i}{z_1}} = 0 \tag{15.44}$$

where 1 is the most populous species, for example. This is a single equation for the average concentration of species 1 in the chaannel $\bar{c}_{1C}$, assuming the surface charge density is known.

To be dimensionally consistent in equation (15.44) the surface charge density should be converted to Moles/liter(M) and

$$c_f = \frac{A\sigma}{1000 V_0 F} \tag{15.45}$$

where A is the surface area and V_0 is the volume. The factor 1000 converts m^3 to liter.

The average concentration in the channel is defined in dimensional form as

$$\bar{c}_i = \frac{1}{h} \int_0^h c_i dy^* \tag{15.46}$$

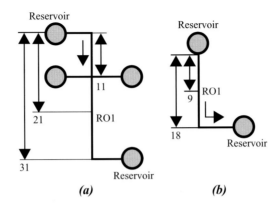

FIGURE 15.11. Oak Ridge (left) and Georgia Tech (right) configurations for the experimental measurement of the mobility.

and in nondimensional form, dividing by the total concentration of the mixture we have

$$\bar{X}_i = \int_0^1 X_i dy \qquad (15.47)$$

where X_i is the mole fraction. Using equations (15.44) and (15.47) we can iterate until the wall mole fractions are consistent with the reservoir concentrations for given surface charge density defined by

$$\sigma = \frac{\epsilon_e RT}{\epsilon h F} \int_0^1 (g - f) dy \qquad (15.48)$$

The precise procedure is described by Zheng et al. [17] It is useful to point out that for small surface charge densities ($\sim 0.01 \frac{C}{m^2}$) the Debye-Huckel picture of the EDL is recovered, whereas for $\sigma \sim 0.1 \frac{C}{m^2}$ the Gouy-Chapman picture of the EDL is recovered.

15.5. COMPARISON OF CONTINUUM MODELS WITH EXPERIMENT

In this section we compare the continuum model of EOF with several sets of experimental data. The first set of data is for channels fabricated by a group at Oak Ridge National Laboratory[1] as depicted on the left side of Figure 15.11. The mixture is 50% sodium tetraborate, 50% methanol solution. The wall mole fractions are calculated by the procedure described in the previous section and the geometry of the channels is described on Table 15.1 [24]. These channels are essentially slit pores as we have considered in the analysis. We have compared the model with data for over fifteen operating conditions and a small portion of those data comparisons are shown here. Figure 15.12 shows the results for the case of $2mM$ and $20mM$ in the upstream reservoir. Here the results are very good; clearly the basic

---

[1] The authors appreciate the use of the data received from Dr. J. Michael Ramsey and his group at Oak Ridge National Laboratoriy. This group included at the time of receipt of the data Dr. Steve Jacobson and Dr. J. P. Alarie.

MODELING ELECTROOSMOTIC FLOW IN NANOCHANNELS

TABLE 15.1. ORNL channel dimensions.

Channel height (nm)	Channel width μm
83	20.3
98	18.4
290	18.3
300	20.2
1080	20.1

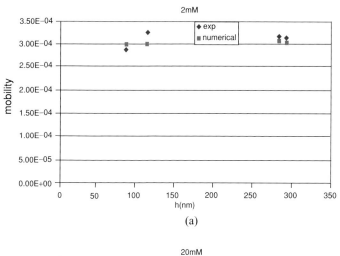

(a)

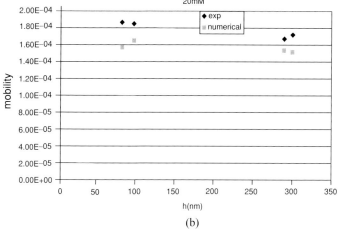

(b)

FIGURE 15.12. Comparison of the experimental data with the theoretical model for the mobility. The converged surface charge density is held fixed at $-0.0154 C/m^2$. (a) Ionic strength 2 mM; (b) 20 mM.

TABLE 15.2. Results for the mobility and dimensional ζ potential for the indicated in the $83nm$ channels of ORNL. Both numerical and asymptotic values of the mobility and ζ potential are shown. The subscript e stands for experiemental, a for asymptotic and the units of the mobility μ are $\frac{cm^2}{Vsec}$. The units of surface charge density are $\frac{C}{m^2}$ and ζ is in volts.

mM	$\mu_e \times 10^4$	$\mu \times 10^4$	$\mu_a \times 10^4$	σ	ζ_e	ζ	ζ_a
0.2	3.23	3.52	7.53	−0.0154	−0.138	−0.141	−0.185
2	2.85	2.97	3.28	−0.0154	−0.098	−0.114	−0.125
20	1.89	1.44	1.21	−0.0154	−0.054	−0.049	−0.050
150	0.80	0.58	0.59	−0.0154	−0.017	−0.019	−0.019

electrochemical effects are present in the model. Note that the model yields a true prediction; there are no adjustable parameters with which to match the data.

It is useful to point out that the ζ potential can be defined by setting the potential

$$\psi = \phi + \zeta$$

equal to zero in the core. This yields in dimensionless form

$$\zeta = -\frac{1}{2}\ln\frac{g^0}{f^0}$$

and note that this implies a theoretical relationship between the ζ potential and the surface charge density. On Table 15.2 are the results at various molarities for the $83nm$ channel. Note the good agreement between the experimental data and the model. Clearly the prediction of the ζ potential by the model is extremely good. The differences in the numerically computed ζ potential(ζ) and the asymptotic value (ζ_a) valid for $\epsilon \ll 1$ on Table 15.2 indicates that the double layers on the channel walls may be overlapping.

A similar set of results have been produced by Yoda's group at Georgia Institute of Technology [25]. Figure 15.13 compares the experimentally measured mobility μ_{ex} with the dimensional electroosmotic mobility μ_{eo} predicted by an asymptotic model described in Section 15.4 which is

$$\mu_{eo} = \frac{\varepsilon_e \phi_o}{2\mu} \ln\left\{\frac{g^0}{f^0}\right\}. \tag{15.49}$$

Here, ε_e and μ are the electrical permittivity and absolute viscosity, respectively, of the fluid $a\phi_o \equiv RT/F$ (R is the universal ideal gas constant, T the absolute temperature of the fluid and F is Faraday's constant) is the characteristic scale for the potential, and (g^0, f^0) are the wall mole fractions of the cationic and anionic species, respectively. The wall mole fractions are predicted by the method described in the previous section. The experimental and model values for mobility agree within 10% over a 200-fold change in molar concentration, suggesting that the particle tracers follow the flow with good fidelity over the range of E studied.

The linear result on Figure 15.13 suggests that the mobility scales as a power of the concentration:

$$\mu = \mu_o \left(\frac{C}{1mM}\right)^{-N} \tag{15.50}$$

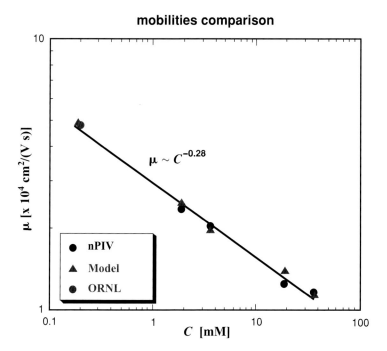

FIGURE 15.13. Graph of mobility values $\mu_{ex}()$ and $\mu_{eo}(\times)$ as a function of molar concentration of EOF of sodium tetraborate buffer. The solid line shows Equation 50 with $\mu_0 = 3.00 \times 10^{-4}$ cm^2/(V · sec) and $N = 0.277$.

A curve-fit of the experimental data points gives $\mu_o = 2.94 \times 10^{-4}$ cm^2/(V · sec) and $N = 0.278$. These are the first results suggesting this correlation.

We have compared the model with another data set taken by iMEDD, Inc. of Columbus, Ohio. Using techniques developed by Hansford. et al. [21], they fabricate a variety of silicon membranes consisting of a series of nanochannels and measure electroosmotic flow rate across those membranes. The membranes are 3.5mm long, 1.5mm wide, and 3.5 (m thick. In their configuration, there are 47,500 nanochannels on each of these membranes. The nanochannels themselves are 3.5 (m in length and 44 (m in width. iMEDD inc. fabricate and made tests to channels with different heights: 4nm, 7nm, 13nm, 20nm, 27nm and 49nm in particular. The geometry of the iMEDD nanochannel membrane and the testing apparatus are shown in Figure 15.2(b).

The two reservoirs are filled with buffer solution. The buffer used in iMEDD's experiments are PBS, 10 times diluted PBS, and 100 times diluted PBS. 17 Volts DC is applied on the two electrodes at the far ends of the two reservoirs. It should be noted that the voltage drop across the membrane is much less than 17 Volts, because most of the 17 Volts is dissipated outside the nanopore membrane, due to the resistance of the buffer in the reservoir. In addition to those experimental parameters mentioned in the ORNL comparison section, the voltage drop across the membrane is required and this value is calculated from a resistance model for each channel height.

Using membrane resistance data determined by iMEDD, we can calculate the volume flow rates based on the calculated voltage drops. The comparison of the calculated flow

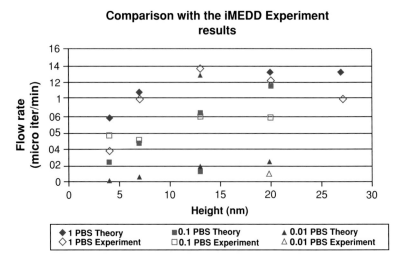

FIGURE 15.14. Comparison of the experimental data with the model for the flow rate through the iMEDD membrane.

rates with iMEDD's experimental results is shown in Figure 15.14. They agree very well except at 4nm. It is possible that at this channel height, some non-continuum effects are present which cannot be captured by the purely continuum theory described here. Also there is some disagreement at larger heights; we believe these discrepancies are due to the estimate of the voltage drop.

Note the interesting result that there appears to be a maximum in the flowrate at the 13 nm location. In the iMEDD experimental configuration, as the channel height increases, the total cross-sectional area of the channels increases, and the resistance of the nanopore membrane decreases. Therefore the ratio of the cross membrane voltage to the total voltage decreases, so the cross membrane voltage decreases. The rapid drop in the cross membrane voltage leads to a maximum in the volume flow rate curve. This phenomenon is unusual and does not occur in pressure-driven flow.

From all of these results it is clear that continuum models compare well with the experimental data. There may be a lower limit down around channel heights of 4–5nm where continuum theory may begin to fail, but this can only be confirmed by comparisons with a molecular dynamics simulation. This is considered next.

15.6. MOLECULAR DYNAMICS SIMULATIONS

15.6.1. Introduction

Non-equilibrium molecular dynamics (NEMD) offers a completely molecular description of electroosmotic flow (EOF) in nanochannels, and some studies have recently appeared [26–28]. Implementation of NEMD requires specification of an interaction model for the fluid and channel walls. To date, the chemical composition and roughness of the channel walls has not been characterized well experimentally. Therefore, currently MD studies of EOF have used, to varying extent, schematic models, particularly with respect to the channel walls. A further consideration is the model used to describe how frictional heating is dissipated. Despite the simplifying assumptions that, at this stage of our understanding, have to be made, simulation studies have helped determine the extent to which continuum theories will apply at the nanoscale and have identified important issues that need further study.

The issues involved in comparing with continuum theory can be divided into two groups. First, the driving force for EOF is determined by the charge distribution within the fluid. In the linear response regime, the charge distribution governing EOF is the equilibrium ionic distribution within the electrolyte. Therefore, one important issue concerning EOF is essentially static in nature: how well continuum methods like Poisson-Boltzmann theory predict the equilibrium ion distribution. Secondly, there is a group of issues in setting up continuum equations for fluid flow. The type of boundary conditions, i.e. no-slip or partial slip, and where to place the boundary surface are crucial issues. There is also a question of whether a linear constitutive relation between shear stress and strain exists, and whether this relation can be taken as fixed throughout the channel.

15.6.2. Statics: the Charge Distribution in a Nanochannel

In nanochannels, the width of the channel is typically between 10 and 100 solvent molecular diameters. Under many experimental conditions this is less than a Debye length. Since Poisson- Boltzmann theory (PBT) neglects the molecular nature of solvent and ions, it is not surprising that simulation studies of nanochannel flow indicate that PBT is a poor starting point for description of the charge distribution in very small channels [26–28]. Here we identify and discuss several shortcomings of Poisson-Boltzmann theory—neglect of specific ion-wall interactions, molecular nature of the solvent, ion correlations, and finite-size corrections to the solvent dielectric response.

PBT theory is a mean field theory in which the instantaneous, fluctuating interactions of ions with other fluid particles and the walls are replaced by an average potential. PBT has been solved analytically for ions between uniform charged surfaces [29, 30]. Classical PBT neglects any specific interactions between ions and the wall, other than incorporating very strong binding of some ions in the Stern layer through a modified wall charge. Simulations show that specific ion-wall interactions significantly affect the ion density and need to be incorporated to predict EOF flow characteristics.

Liquids tend to order near walls, an effect which arises form the molecular nature of the solvent. The effect has been elegantly confirmed by the oscillatory dependence of the force on distance as solvent is squeezed out between mica sheets. To illustrate this effect, the density profile of a fluid of spherical particles interacting via a Lennard-Jones interaction is shown in Fig. 15.15, where layering is clearly visible in the snapshot and density profile. The ion distribution is consequently modified by the structured environment near the channel walls in a complex fashion. Spherical ions in a solvent of spherical particles of the same size tend to lie within the solvent layers [28]. However, in simulations of ions in water using the SPC [31] or SPC/E [32] water models, the ions tend to fit between solvent layers [26, 27].

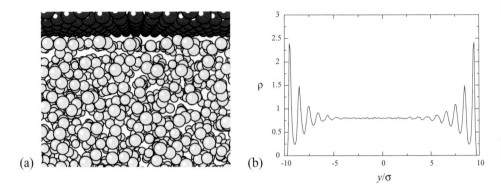

FIGURE 15.15. An illustration of fluid layering near a wall. Panel (a) is a snapshot of the region near the wall from a simulation of particles interacting via pairwise Lennard-Jones potentials, $4\epsilon \left[\left(\frac{\sigma}{r}\right)^{12} - \left(\frac{\sigma}{r}\right)^6 \right]$. The overall density of the fluid is $\rho\sigma^3 = 0.8$. Fluid layers are visible next to the wall, near the top of the figure. Panel (b) is the equilibrium density profile across the channel.

Since PBT is a mean field theory, it neglects correlations between ions. Recently, the problem of ion correlation has received much attention. While PBT predicts that the effective interaction between like-charged walls, rods or spheres will always be repulsion, correcting for ion correlations predicts an effective attraction at high ionic strength. It remains to be seen whether correlation effects are significant under typical EOF conditions.

Finally, in classical PBT the solvent dielectric response is captured by modifying Coulomb interactions with a dielectric constant. This procedure is only correct for a dielectric medium of infinite size. The well-known correction when ions approach an interface with a medium of differing dielectric constant takes the form of image charges. For example, Onsager and Samaras showed that ions in a medium with a high dielectric constant, such as water, will be repelled from an interface with a medium of low dielectric constant [33]. What the fictitious image charge repulsion describes is the actual lessened dielectric solvation when the ion is near the interface. This is the likely situation for aqueous electrolyte flow in nanochannel fabricated in low silicon or polymeric materials. Image charge effects are included in molecular dynamics simulations. In simulations of a model electrolyte, decreased solvation near the wall results in complete exclusion of ions from the solvent layer adjacent to the wall [28].

15.6.3. Fluid Dynamics in Nanochannels

As discussed in the previous section, we know that classical PBT does not give sufficient accuracy to understand EOF in nanochannels [26–28]. Now we consider whether, with an accurate ion distribution, linear hydrodynamics can predict EOF in nanochannels. There are two studies of this issue, one by Qiao and Aluru [27] (QA) and the other by Zhu, Singer, Zheng and Conlisk [28] (ZSZC). In both studies simulations are compared with continuum fluid mechanics using an ion density that agreed with simulations (and not from PBT). The conclusions reached concerning the applicability of linear hydrodynamics are somewhat different.

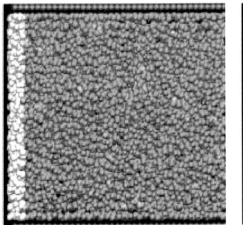
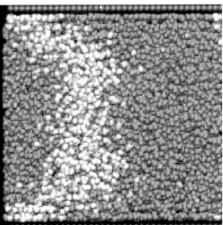

FIGURE 15.16. Snapshots from a simulation of EOF by Zhu, Singer, Zheng and Conlisk [28]. Cations are red, anions are blue, wall particles are green. In the left-hand frame the solvent particles within a slab are colored white, and the other solvent particles are left as gray. The right-hand frame, from a later time, shows the evolution of the slab of white-colored particles.

QA compared continuum theory with simulations in simulations of aqueous chloride ions confined by walls of Lennard-Jones particles. The width of their nanochannel is $3.49 nm$. They enforce no-slip boundary conditions at the positions of the outermost wall particles, and find that continuum theory predicts a velocity profile more than double that of simulations. They attribute this disagreement to a steep rise of solvent viscosity near the channel walls, and advocate incorporation of position-dependent viscosity into the Navier-Stokes equations.

ZSZC simulate EOF in a model fluid composed of spherical Lennard-Jones solvent and ions, as pictured in Fig. 15.16. Reflecting the exclusion of ions from the wall layer because of solvation forces discussed in the previous region, ZSZC find that solving the PBT equations within a region excluding the fluid near the walls gives a qualitatively accurate description of the charge density, and also yields analytic expressions for the velocity profile. In their comparison with continuum theory, ZSZC enforce no-slip boundary conditions at a surface that represents the furthest penetration of solvent against the walls. This surface is closer to the channel center than that of QA. In contrast to QA, ZSZC find that continuum hydrodynamics gives a good account of the velocity profile.

Differing points of view about the placement of the no-slip surface led QA and ZSZC to different conclusion about the applicability of standard hydrodynamics with a constant viscosity throughout the channel. By placing their continuum fluid up to the position of wall particles, QA neglect excluded volume interactions which keep solvent and walls atoms separated by the sum of their radii. To correct this picture, they invoke position dependent viscosity. ZSZC estimates a surface of closest penetration of solvent against the wall, which can be done in several ways. One method ZSZC uses is to fit the velocity profile from Poiseuille flow to the well-known analytical formula and set the no-slip surface at the point where the Poiseuille flow profile went to zero. They find that this procedure set the no-slip surface very close to the simple estimate of Travis and Gubbins [34], which is half a Lennard-Jones particle diameter from the centers of the wall particles. Using a better estimate of the no-slip surface for the system simulated by QA, ZSZC recalculate the velocity profile and found much better agreement between continuum theory and MD simulation without having to invoke position-dependent viscosity.

To obtain a completely analytical expression that matches simulation results, ZSZC solve PBT and corresponding hydrodynamic equations for fluid flow with adjustable volume available to the ions. The original intent is to describe the exclusion of ions from the wall region found in their simulations. However, the theory also brings to light a qualitative relationship between ion distribution and velocity profile: When ions are excluded from the wall region, the flow increases. Conversely, attraction of the ions to the walls decreases the flow rate. This trend exists because the greatest flow response occurs when the driving force is in the center of the channel, away from the no-slip boundaries.

15.7. SUMMARY

In this chapter we have described modeling of electroosmotic flow in micro- and nano-channels. We have reviewed the structure of the electric double layer and its role in the transport of ions and solvent in channels. The flow is driven by the difference in cation/anion concentrations at the walls of the channel. At small cation, anion concentration differences the Debye-Huckel picture of the electric double layer is appropriate; at larger concentration differences, the Gouy-Chapman picture of the electric double emerges naturally. We show how analytical solutions for the velocity and concentrations or mole fractions may be derived for electric double layer thicknesses small compared with the channel height.

Analytical results are compared with several sets of experimental data and the model compares well. In addition, the comparisons with experiments suggest that the continuum approximation may break down for channel hieghts on the order of $5nm$ and so the mechanics and results of molecular dynamics simulations are also described. It is shown that MD simulations allow preferential solvation near the wall in contrast to the continuum Poisson-Boltzmann approach, resulting in significant differences in the magnitude of bulk solution and ionic transport.

The persistence of the validity of continuum theory down to under about $5nm$ in height, at first glance seems surprising. However, on further reflection, since liquid molecules are so close together an effective Knudsen number defined as the ratio of a molecular diameter to the smallest dimension of the channel remains small down to under $10nm$. For water, the molecular diameter is about $0.3nm$ and so using the well-developed gas phase theory, an effective Knudsen number of 0.1 corresponds to a channel height of $3nm$. Indeed it has often been mentioned that continuum theory in liquids persists to about 10 molecular diameters.

Most of the devices being fabricated today have channels greater than 2 or 3 nanometers. Continuum or near continuum behavior should obtain down to these scales. Thus despite the well-known lack of a well-defined theory of liquids, modeling of liquid flows at nanoscale seems to be relatively straightforward when compared to gases, even though there is a well-developed kinetic theory for gases. Indeed, for air a Knudsen number of around 1 corresponds to a channel height of about $60nm$ which is in the transition regime where continuum theory breaks down and expensive and time consuming molecular simulations are required.

ACKNOWLEDGEMENTS

Much of the work described in this chapter was begun under funding from DARPA under agreement number F30602–00–2–0613. ATC is grateful to the contract monitors Dr. Anantha Krishnan (DARPA), Mr. Clare Thiem and Mr. Duane Gilmour of Air Force Research Lab (IFTC) for their support. ATC is also grateful to Dr. Mauro Ferrari and Professor Derek Hansford for getting him involved in this work. SJS gratefully acknowledges support from NSF grant CHE-0109243.

REFERENCES

[1] R.A. Robinson and R.H. Stokes. *Electrolyte Solutions*, Academic Press: New York, p. 284, 1959.
[2] J.S. Newman. *Electrochemical Systems*, Prentice-Hall, Englewood Cliffs, NJ, p. 138, 1973.
[3] R.J. Hunter. *Zeta Potential in Colloid Science*, Academic Press: London, p. 59, 1981.
[4] R.F. Probstein. *Physicochemical Hydrodynamics*, Butterworths: Boston, p. 161, 1989.
[5] Paul Delahay. *Double Layer and Electrode Kinetics*, Wiley Interscience, New York, 1965.
[6] John O'M. Bockris and Amulya K.N. Reddy. *Modern Electrochemistry, Volume 1 Ionics*, (2 Ed.), Plenum Press, New York, London, pp. 273f, 1998.
[7] J. Israelachvili. *Intermolecular and Surface Forces*, (2 Ed.), Academic Press, London, 1991.
[8] Private communication by Tony Boiarski, 2002.
[9] D. Hansford, T. Desai, and M. Ferrari. Nanoscale size-based biomoleculart separation technology, In J. Cheng, and L.J. Kricka (eds.), *Biochip Technology*, Harwood Academic Publishers, 341, 2001.
[10] E.J.W. Verwey and J.Th. G. Overbeek. *Theory of Stability of Lyophobic Colloids*, Elsevier: Amsterdam, 1948.
[11] W. Qu and D. Li. A model for overlapped EDL fields. *J. Coll. Interface Sci.*, 224:397, 2000.
[12] D. Burgeen and F.R. Nakache. Electrokinetic flow in ultrafine capillary slits. *J. Phys. Chem.*, 68:1084, 1964.
[13] S. Levine, John R. Marriott, and Kenneth Robinson. Theory of electrokinetic flow in a narrow parallel-plate channel. *Farad. Trans., II*, 71:1, 1975.
[14] C.L. Rice and R. Whitehead. Electrokinetic flow in a narrow capillary. *J. Phys. Chem.*, 69(11):4017, 1965.
[15] S. Levine, J.R. Marriott, G. Neale, and N. Epstein, N. Theory of electrokinetic flow in fine cylindrical capillaries at high zeta potentials. *J. Coll. Int. Sci.*, 52(1):136, 1975.
[16] A.T. Conlisk, Jennifer McFerran, Zhi Zheng, and Derek Hansford. Mass transfer and flow in electrically charged micro- and nano-channels. *Anal. Chem.*, 74(9):2139, 2002.
[17] Zhi Zheng, Derek J, Hansford, and A.T. Conlisk. Effect of multivalent ions on electroosmotic flow in micro and nanochannels. *Electrophoresis*, 24:3006, August 2003.
[18] H.L.F. Helmholtz. *Ann. Physik.*, 7(3):337, 1879.
[19] P. Debye, and E. Huckel. The interionic attraction theory of deviations from ideal behavior in solution. *Z. Phys.*, 24:185, 1923.
[20] G. Gouy. About the electric charge on the surface of an electrolyte. *J. Physics A*, 9:457, 1910.
[21] D.L. Chapman. A contribution to the theory of electrocapillarity. *Phil. Mag.*, 25:475, 1913.
[22] O. Stern. The theory of the electrolytic double layer. *Z. Elektrochem.*, 30:508, 1924.
[23] J. Kevorkian, and J.D. Cole. *Perturbation Methods in Applied Mathematics*, Springer-Verlag, New York, 1981.
[24] S.C. Jacobson, S.V. Ermakov, and J.M. Ramsey. Minimizing the number of voltage sources and fluid reservoirs for electrokinetic valving in microfluidic devices. *Anal. Chem.*, 71:3273, 1999.
[25] R. Sadr, M. Yoda, Z. Zheng, and A.T. Conlisk. An experimental study of electroosmotic flow in rectangular microchannels. *J. Fluid Mech.*, 506:357, 2004.
[26] J.B. Freund. Electro-osmosis in a Nanometer-scale Channel Studied by Atomistic Simulation. *J. Chem. Phys.*, 116(5):2194, 2002.
[27] R. Qiao and N.R. Aluru. Ion Concentrations and Velocity Profiles in Nanochannel Electroosmotic Flows. *J. Chem. Phys.*, 118(10):4692, 2003.

[28] W. Zhu, S.J. Singer, Z. Zheng, and A.T. Conlisk. Electroosmotic Flow of a Model Electrolyte. *Phys. Rev.*, E 71(4):41501, 2005.

[29] A.J. Corkhill and L. Rosenhead. Distribution of Charge and Potential in an Electrolyte Bounded by Two Infinite Parallel Plates. *Proc. Royal Soc.*, 172 (950):410, 1939.

[30] S. Levine and A. Suddaby. Simplified Forms for Free Energy of the Double Layers of Two Plates in a Symmetrical Electrolyte. *Proc. Phys. Soc.*, A 64(3):287, 1951.

[31] H.J.C. Berendsen, J.P.M. Postma, W.F. von Gunsteren, and J. Hermans. Interaction Models for Water in Relation to Protein Hydration. In B. Pullman (ed.), *Intermolecular Forces*. Reidel, Dordrecht, Holland, p. 331, 1981.

[32] H.J.C. Berendsen, J.R. Grigera, and T.P. Straatsma. The Missing Term in Effective Pair Potentials. *J. Phys. Chem.*, 91(24):6269, 1987.

[33] L. Onsager and N.N.T. Samaras. The Surface Tension of Debye-Hückel Electrolytes. *J. Chem. Phys.*, 2(8):1934.

[34] K.P. Travis and K.E. Gubbins. Poiseuille Flow of Lennard-Jones Fluids in Narrow Slit Pores. *J. Chem. Phys.*, 112(4):1984, 2000.

16

Nano-Particle Image Velocimetry: A Near-Wall Velocimetry Technique with Submicron Spatial Resolution [1]

Minami Yoda [2]

G.W. Woodruff School of Mechanical Engineering, Georgia Institute of Technology, Atlanta, GA 30332-0405

16.1. INTRODUCTION

Over the last decade, characterizing and understanding fluid flow and transport at spatial scales of 100 µm or less has become a major area of research in fluid mechanics because of the rapid development of microscale devices based upon microelectromechanical systems (MEMS) fabrication techniques. Examples of such microfluidic devices include Labs-on-a-Chip for biochemical separation and analysis, inkjet printer heads, various types of microelectronic cooling devices, microscale fuel cells, microthrusters, and genomic and proteomic "chips" capable of sequencing and identifying various proteins including RNA and DNA. More recently, nanotechnology and the promise of engineering new devices at the molecular scale has sparked interest in understanding flow at spatial scales below 1µm. Optimizing and controlling the flow of liquids and gases in micro- and nanoscale devices requires both well-validated models of transport down to the molecular scale and diagnostic techniques with spatial resolution well below 1µm.

A major "bottleneck" in developing such flow models is our limited fundamental understanding of microscale flows and limitations in our ability to measure flow parameters such as velocity, pressure, concentration and temperature at these spatial scales. At the microscale, surface (*vs.* bulk) effects become significant; the new physical phenomena that may therefore be significant in micro- and nanoscale flows include apparent slip [3] (*i.e.*, nonzero flow velocity at the wall due possibly to a "soft" dissociated fluid or gas layer [4, 5]); the effects of surface properties such as chemistry (*e.g.* hydrophobic *vs.* hydrophilic [6, 7]), molecular-scale roughness [8], and charge; and discrepancies in fluid properties (*e.g.*

viscosity) between the interfacial and bulk flow regions. Diagnostic techniques capable of interrogating the near-wall region to measure flow parameters within the first 100 nm of the wall are therefore important in understanding how surface effects impact micro- and nanoscale transport, and validating our current models of transport at these spatial scales.

This chapter first briefly reviews current experimental techniques in microfluidics, with an emphasis on velocimetry techniques and their spatial resolution and near-wall capabilities. The chapter then describes a new technique, nano-particle image velocimetry (nPIV), based upon evanescent-wave illumination generated by the total internal reflection of light at a refractive index interface. The applications of nPIV are illustrated by velocity and mobility results obtained in steady, fully-developed two-dimensional electroosmotic flow (EOF) of liquids.

16.2. DIAGNOSTIC TECHNIQUES IN MICROFLUIDICS

The majority of diagnostic techniques in fluid mechanics were originally developed for turbulent flows, where flow parameters such as velocity and pressure have strong stochastic fluctuations. "Classical" flow measurement diagnostic techniques such as hot-wire anemometry and laser-Doppler velocimetry (LDV) therefore emphasized measurements at a single spatial location with high temporal resolution or sampling rate. Over the last twenty years, advances in lasers and imaging technology have led to the development of a variety of nonintrusive optical diagnostic techniques for obtaining spatially and temporally resolved measurements of velocity and such as particle-image velocimetry (PIV) and molecular-tagging velocimetry (MTV).

Given that a microchannel typically has a cross-section dimension of $O(100 \, \mu m)$ and speeds of $O(1 \, mm/s)$ or less, nearly all microchannel flows are laminar, and, in most applications, steady. Spatial (vs. temporal) resolution is therefore the major limiting issue in microfluidic measurements. Given the limited amount of available space and access in most microdevices, diagnostic techniques must also be nonintrusive methods. Yet most of the nonintrusive optical diagnostic techniques have minimum spatial resolutions that are comparable to or exceed the entire dimension of the microchannel—perhaps 50 μm for LDV and 500 μm for PIV.

Nevertheless, both PIV and MTV have been extended to microscale flows. The most common microfluidic diagnostic technique, micro-particle image velocimetry (μPIV), has been used extensively over the last five years to obtain velocity fields in steady flows in microchannels with dimensions of $O(1-100 \, \mu m)$ [9]. In μPIV, water is seeded with 200–500 nm diameter colloidal fluorescently dyed polystyrene spheres (density $\rho \approx 1.05$ g/cm^3) at volume fractions of $O(10^{-4}-10^{-3})$. The region of interest in the flow is then volumetrically illuminated by light of the wavelength required to excite the fluorescent dye (e.g. from an argon-ion or frequency-doubled Nd:YAG), and the longer-wavelength fluorescence from the particle tracers is imaged through a high-magnification microscope objective and epi-fluorescent filter cube onto a CCD camera. The small depth of field of the microscope objective ensures that only tracer particles in a "slice" of the flow centered about the focal plane of the imaging optics are in focus, and hence have relatively high grayscale values and small particle diameters, on the CCD.

Assuming that the flow component along the optical axis is negligible, the depth of field of the imaging system therefore determines the "out-of-plane" spatial resolution (*i.e.*, the spatial resolution along the optical axis). Santiago *et al.* [9] cited a depth of field of 1.5 µm for a 100×1.4 oil immersion objective based upon a criterion that particle images more than 25% larger than "in-focus" particle images were considered out of focus. More recently, Tretheway and Meinhart [6] reported a depth of field of 1.8 µm for a 60×1.4 oil immersion objective. The minimum out-of-plane spatial resolution of µPIV therefore exceeds 1 µm at present.

In most µPIV measurements, "interrogation windows", or portions of two images (the "image pair") separated by a given time interval Δt are cross-correlated. The location of the cross-correlation peak then gives the average displacement for the tracer particles in both interrogation windows, and this displacement divided by Δt gives the average velocity of the group of tracers, and presumably the flow velocity. The minimum "in-plane" spatial resolution (*i.e.*, the spatial resolution in the plane normal to the optical axis) of µPIV is then determined by the largest of the following factors:

- the dimensions of the interrogation window;
- the displacement over the time interval;
- the minimum spatial resolution of the image;
- the size of the tracer particle.

For most µPIV applications, the minimum in-plane spatial resolution is determined by the interrogation window. Santiago *et al.* [9] used 32×32 pixel interrogation windows, corresponding to an in-plane spatial resolution of 6.9 µm; Tretheway and Meinhart [6] used 128×8 pixel rectangular interrogation regions, corresponding to an in-plane spatial resolution of 14.7×0.9 µm.

MTV using both caged fluorophores and phosphorescent supramolecules has also been used to measure velocity fields in microchannels. In molecular tagging velocimetry, molecular tracers (*e.g.* phosphorescent supramolecules and caged fluorophores) are excited or uncaged by light at UV wavelengths and emit light (after excitation by a visible laser for caged fluorophores) at visible wavelengths [10, 11]. The velocity field can then be obtained by exciting or "tagging" a material line or grid of material lines in the flow of a liquid seeded with molecular tracers and measuring the convection of the tagged lines over a known time interval. Paul *et al.* [12] used a relatively high molecular weight caged fluorophore, caged fluorescein dextran, to visualize Poiseuille and electroosmotic flows through 100 µm open capillaries. Sinton and Li [13] used 5-carboxymethoxy-2-nitrobenzyl (CMNB)-caged fluorescein to obtain electroosmotic flow velocity profiles through 20–200 µm square and circular microchannels. Lum and Koochesfahani [14] recently reported MTV measurements in Poiseuille flow through a 300 µm channel using the phosphorescent supramolecule based upon the lumophore 1-Bromonaphthalene (1-BrNp).

Assuming negligible flow along the optical axis, the minimum out-of-plane spatial resolution of MTV data is usually determined by the dimensions of the tagged line normal to the object plane. The minimum in-plane spatial resolution for MTV is determined by the largest of the following three factors: 1) the dimension of the tagged line; 2) the displacement over the time interval; and 3) the spatial resolution of the image. It is therefore usually determined by the dimension of the tagged line. To date, the smallest dimension reported for a tagged line is about 25 µm [14], but the diffraction-limited spot size at UV wavelengths for high magnification and numerical aperture optics can be as small as a few micrometers, suggesting that the minimum spatial resolution for MTV should be comparable to the values reported for µPIV.

Nevertheless, both μPIV and MTV have minimum spatial resolutions exceeding 1 μm, suggesting that neither technique is capable of spatially resolving velocity fields for geometries with dimensions below 10 μm. Moreover, the near-wall capabilities of both techniques are limited. The best μPIV measurements to date were obtained within 450 nm of the wall, and are actually the velocity averaged over the first 900 nm (based upon interrogation window size) next to the wall [7]. The near-wall capabilities of MTV are yet largely unexplored. Nevertheless, assuming that velocity data can be obtained within a few pixels of the wall for MTV, the current minimum pixel dimension for intensified CCD cameras is just below 10 μm. For a magnification of 100, MTV may ultimately be able to obtain velocity data averaged over the first 300–400 nm next to the wall.

Current microfluidics diagnostic techniques therefore appear to be limited to measuring velocities averaged over about the first 400 nm next to the wall. This near-wall capability is inadequate to investigate many interfacial phenomena of interest such slip length, with reported values ranging from about 8 nm to 1 μm [7, 15]. Over the last three years, we have therefore developed an extension of μPIV based upon evanescent-wave illumination generated by total internal reflection (TIR) of a laser beam at the fluid-solid interface between the flow and the wall [16]. This nano-particle image velocimetry (nPIV) technique is inherently limited to interrogating the near-wall region, and as such, complements bulk flow velocimetry techniques such as μPIV and MTV.

16.3. NANO-PARTICLE IMAGE VELOCIMETRY BACKGROUND

Nano-PIV, like μPIV, obtains measurements of two components of the flow velocity field by tracking the motion of groups of fluorescent colloidal tracer particles over time as they are convected by the flow. Nano-PIV differs from μPIV, however, in its use of evanescent-wave (*vs.* volumetric) illumination.

16.3.1. Theory of Evanescent Waves

Consider a beam of light propagating through a dense transparent medium of refractive index n_2 incident upon a interface with a less dense transparent medium of refractive index $n_1 < n_2$. If the angle of incidence θ_i measured with respect to the y-axis, or the normal to the interface, is greater than the critical angle $\theta_c = \sin^{-1}(n_{12})$, where the relative index of refraction $n_{12} = n_1/n_2$, then the light will undergo total internal reflection (TIR) at the interface (Figure 16.1). Snell's Law gives:

$$n_2 \sin\theta_i = n_1 \sin\theta_t \quad \Rightarrow \quad \begin{cases} \sin\theta_t = \sin\theta_i/n_{12} > 1 \\ \cos\theta_t = \sqrt{1-\sin^2\theta_t} = \dfrac{\sqrt{n_{12}^2 - \sin^2\theta_i}}{n_{12}} = \pm i\beta \end{cases} \quad (16.1)$$

where the angle of transmission θ_t has a complex cosine. The wavefunction of the transmitted electric field is therefore [17]:

$$\begin{aligned}\mathbf{E}_t &= \mathbf{E}_{0t} \exp\{ik_t[\sin\theta_t x + \cos\theta_t y - \omega t]\} \\ &= \mathbf{E}_{0t} \exp\{\mp k_t \beta y\} \exp\left\{-i\left[\omega t - \frac{k_t}{n_{12}}(\sin\theta_i)x\right]\right\}\end{aligned} \quad (16.2)$$

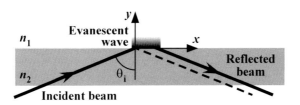

FIGURE 16.1. TIR of an incident beam of light at a refractive index interface and the generation of an evanescent wave in the less dense medium of index n_1. Note that the actual reflected beam is slightly offset from the geometrically reflected light ray due to the Goos-Hänchen effect [18].

where k_t is the wavenumber. The solution with the positive exponential is physically impossible. The resultant wave therefore has an amplitude that decays exponentially with distance normal to the interface y and propagates along the direction parallel to the interface x. The intensity of the evanescent wave is proportional to the square of the transmitted electric field magnitude, and it can then be shown that this intensity

$$I(y) = I_o \exp\left\{-\frac{y}{y_p}\right\} \quad \text{where} \quad y_p = \frac{1}{2k_t\beta} = \frac{\lambda_o}{4\pi n_2}\frac{1}{\sqrt{\sin^2\theta_i - n_{12}^2}} \quad (16.3)$$

Here, I_o is the intensity at the interface (i.e., $y = 0$), y_p is the penetration depth of the evanescent wave, and λ_o is the wavelength of the light in vacuum.

The components of the electric field tangential to the interface must be continuous across the interface. For light polarized perpendicular to the plane of incidence (i.e., the x–y plane) with an electric field magnitude incident upon the interface of $E_i^\perp$, the Fresnel Equations for the amplitude transmission coefficient gives the y-component of the electric field of the evanescent-wave at the interface as [17, 19]:

$$E_z^o = E_i^\perp \frac{2\cos\theta_i}{\cos\theta_i + n_{12}\cos\theta_t}$$

$$= E_i^\perp \frac{2\cos\theta_i}{\sqrt{1 - n_{12}^2}} \exp\{-i\delta_\perp\} \quad \text{where} \quad \delta_\perp = \tan^{-1}\left\{\frac{\sqrt{\sin^2\theta_i - n_{12}^2}}{\cos\theta_i}\right\} \quad (16.4)$$

For parallel polarized light with an electric field magnitude incident upon the interface of $E_i^\|$, the Fresnel Equations give the x- and y-components of the evanescent-wave electric field at the interface as:

$$E_x^o = E_i^\| \frac{2\cos\theta_i \cos\theta_t}{\cos\theta_t + n_{12}\cos\theta_i} = E_i^\| \frac{2\cos\theta_i\sqrt{\sin^2\theta_i - n_{12}^2}}{\sqrt{n_{12}^4\cos^2\theta_i + \sin^2\theta_i - n_{12}^2}} \exp\left\{-i\left(\delta_\| + \frac{\pi}{2}\right)\right\}$$

$$E_y^o = E_i^\| \frac{2\cos\theta_i \sin\theta_t}{\cos\theta_t + n_{12}\cos\theta_i} = E_i^\| \frac{2\cos\theta_i \sin\theta_i}{\sqrt{n_{12}^4\cos^2\theta_i + \sin^2\theta_i - n_{12}^2}} \exp\{-i\delta_\|\}$$

where

$$\delta_\| = \tan^{-1}\left\{\frac{\sqrt{\sin^2\theta_i - n_{12}^2}}{n_{12}^2\cos\theta_i}\right\} \quad (16.5)$$

Note that these two components are exactly 90° out of phase with each other.

The evanescent-wave intensity at the interface I_o depends on θ_i and the polarization of the incident beam, with I_o proportional to square of the magnitude of the electric field $\mathbf{E}$ at the refractive-index interface. The evanescent-wave intensities for parallel and perpendicular polarized light, or $I_o^{\parallel}$ and $I_o^{\perp}$, respectively, are then:

$$\frac{I_o^{\parallel}}{J^{\parallel}} = \frac{4\cos^2\theta_i}{1-n_{12}^2} \frac{2\sin^2\theta_i - n_{12}^2}{(1+n_{12}^2)\sin^2\theta_i - n_{12}^2}$$

$$\frac{I_o^{\perp}}{J^{\perp}} = \frac{4\cos^2\theta_i}{1-n_{12}^2} \tag{16.6}$$

where $J^{\parallel}$ and $J^{\perp}$ are the intensities of light incident upon the interface with polarizations parallel and perpendicular to the plane of incidence, respectively. Surprisingly, the evanescent-wave intensity at the interface often exceeds that of the incident light. For $\theta_i = \theta_c$, Eq. (16.6) reduces to $I_o^{\parallel}/J^{\parallel} = 4/n_{12}^2$ and $I_o^{\perp}/J^{\perp} = 4$; note that higher evanescent-wave intensities can be achieved with parallel polarized light since $n_{12} < 1$. Even for TIR at a glass-water interface with $n_2 = 1.5$ and $n_1 = 1.33$, and an angle of incidence slightly above the critical angle $\theta_i = 63°$, for example, the evanescent-wave intensity is several times that of the incident light, with $I_o^{\parallel}/J^{\parallel} = 4.89$ and $I_o^{\perp}/J^{\perp} = 3.86$.

Since $y_p < \lambda_o$ for most cases and the intensity of the evanescent wave decays exponentially with y, this illumination is confined to a narrow region immediately adjacent to the interface. For TIR of blue light with $\lambda_o = 488$ nm at a glass-water interface, $y_p < 120$ nm. Evanescent waves are hence the basis of the total internal reflection fluorescence (TIRF) microscopy used to visualize cell-substrate contact and measure binding kinetics of proteins to cell surface receptors (for example) in biophysics [20] and the total internal reflection microscopy (TIRM) technique used to measure colloidal forces in surface science [21]. Based upon the properties of evanescent-wave illumination, nPIV should therefore be able to obtain velocity data averaged over the first $O(100$ nm$)$ thick layer next to the wall with an out-of-plane spatial resolution based upon the penetration depth of $O(100$ nm$)$.

Nano-PIV tracks the displacement of fluorescent particle tracers illuminated by evanescent waves. To model the resultant particle image intensity, consider a fluorescent sphere of radius a illuminated by an evanescent wave with penetration depth y_p where $a \sim y_p$ whose center and edge are at distances z' and h from the interface, respectively (Figure 16.2). Under the assumptions that:

a) disturbances due to the presence of the sphere have a negligible effect on the exponential decay of the evanescent wave-intensity in the vicinity of the sphere (note that essentially a single "layer" of spheres are illuminated by the evanescent wave since $y_p \sim 2a$ at these low tracer seeding densities),
b) only the fluorophores on the surface of the sphere are excited by the evanescent wave, and
c) the intensity at the interface I_o is constant,

the net power emitted by this sphere when illuminated by the evanescent wave is directly proportional to the product of $I(y)$ and the surface area of the sphere. Moreover, since this fluorescence is typically detected by a microscope objective placed below the sphere

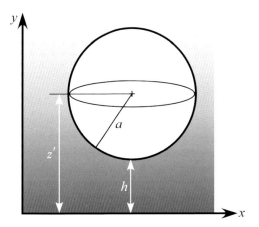

FIGURE 16.2. Sphere illuminated by an evanescent wave.

through the dense medium, only light emitted by the bottom half of the sphere facing the interface will be detected by the imaging system, and the amount of light imaged, or the "brightness" B of the particle, will be proportional to the net power emitted by the bottom half of the sphere. It can then be shown that the image brightness of a spherical particle illuminated by an evanescent wave also decays exponentially, albeit in h, or the distance from its nearest edge to the interface, as follows:

$$B \propto B_o \exp\left\{-\frac{h}{y_p}\right\} \tag{16.7}$$

where B_o is the brightness of the particle when it is in contact with the wall [22].

The size of such a particle imaged through a microscope objective and recorded on a CCD camera is essentially the convolution of the geometric and diffraction-limited images since $a < \lambda_o$ in nPIV. If both these images can be approximated as Gaussian functions, this convolution is then also a Gaussian function for a spherical particle with its center at the focal plane with an effective diameter [23, 24]

$$d_e = \sqrt{4M^2 a^2 + d_{s\infty}^2} \quad \text{where} \quad d_{s\infty} = 1.22 M \lambda \sqrt{\left(\frac{n_o}{\text{NA}}\right)^2 - 1} \tag{16.8}$$

Here, M and NA are the magnification and numerical aperture, respectively, of the microscope objective and n_o is the refractive index of the immersion medium (*e.g.*, oil) between the objective and the object.

16.3.2. Generation of Evanescent Waves

Most TIRM experiments use one of two methods to generate evanescent-wave illumination, namely the prism method based upon a prism optically coupled to the sample, or the "prismless" method, where the evanescent wave is introduced through a high numerical aperture (NA) objective (Figure 16.3) [19]. In general, the prism method gives more flexibility in terms of optical setup, since the optics for generating the evanescent wave can be

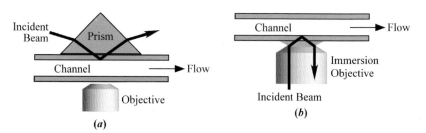

FIGURE 16.3. Examples of TIRM configurations where the evanescent wave is generated using the (a) prism or (b) "prismless" methods.

aligned independently of the imaging system. Two more advantages of the prism method are that it usually gives cleaner illumination, with better control over autofluorescence and/or autoluminescence, and, given the high cost of high NA microscope objectives, the prism method is usually more economical. Note that neither the prism nor the optical coupling fluid between the prism and the microchannel wall are required to have the same refractive index as the wall.

The prismless method requires a high NA immersion objective. For such an objective, Snell's Law gives:

$$\text{NA} = n_2 \sin \theta_i > n_1 \qquad (16.9)$$

For water, $n_1 = 1.33$, suggesting that NA must be well above this value; the few microscope objectives with magnifications of 60 or greater and numerical apertures of 1.45 or 1.65 are quite expensive. The major advantages of the prismless method are that this method is generally easier to implement in most microscopes, with commercial microscope vendors offering "off-the-shelf" TIRF configurations, and that arc-lamp illumination (vs. a laser) can be used to generate the evanescent wave.

16.3.3. Brownian Diffusion Effects in nPIV

Brownian diffusion of the 100–500 nm diameter colloidal particle tracers used in micro- and nano-PIV can be a major source of error in both techniques. For a convective timescale $\tau_c = 1$s (based upon a velocity scale of 100 µm/s and a length scale of 100 µm), the Peclet number, which compares the time required for a particle to diffuse its own radius to the convective timescale, is

$$Pe \equiv \frac{6\pi\mu a^3}{kT\tau_c} \sim 10^{-3}$$

for a particle of radius $a = 50$ nm suspended in water at absolute temperature $T = 300$ K. In this expression, µ is the fluid viscosity and k is the Boltzmann constant. Early µPIV studies noted that the error due to Brownian diffusion imposes a practical lower limit on both the time interval between images within an image pair and the tracer particle size for a given accuracy [9]. Olsen and Adrian [25] showed that Brownian motion reduces the signal-to-noise ratio (SNR) of the cross-correlation. In steady flows, the effects of "in-plane" Brownian diffusion can be greatly reduced by temporal averaging, with the best results achieved by averaging the cross-correlation (vs. the actual displacement) [26]. "Out-of-plane" Brownian diffusion is usually not an issue, given the out-of-plane spatial resolution of μPIV of at least 1.5 μm.

In contrast with μPIV, Brownian diffusion of nPIV tracers is hindered by the presence of the wall since particles experience increased hydrodynamic drag as they approach the wall. One of the consequences of hindered Brownian diffusion is that the diffusion coefficients are now functions of the particle distance from the wall; the hindered diffusion coefficients normal and parallel to the wall $D_\perp$ and $D_\parallel$ are [27, 28]:

$$D_\perp \approx \frac{6h^2 + 2ah}{6h^2 + 9ah + 2a^2} D_\infty$$

$$D_\parallel = \left[1 - \frac{9}{16}\left(\frac{a}{z'}\right) + \frac{1}{8}\left(\frac{a}{z'}\right) - \frac{45}{256}\left(\frac{a}{z'}\right)^4 - \frac{1}{16}\left(\frac{a}{z'}\right)^5\right] D_\infty \quad (16.10)$$

where h and z' are the distances of the particle edge and center from the wall, respectively, and $D_\infty = kT/(6\pi\mu a)$ is the Brownian diffusion coefficient for unconfined flow [29]. Note that the expression for $D_\perp$ is an approximation of the infinite series solution given by Goldman et al. [30].

If the tracers sample all values of h with equal probability within the y-extent illuminated by the evanescent wave, in-plane hindered Brownian diffusion can be greatly reduced for steady flows by time-averaging the nPIV data. Out-of-plane Brownian diffusion is, however, a major issue for nPIV (unlike μPIV) due to its much smaller out-of-plane spatial resolution. Diffusion normal to the wall leads to "particle mismatch, where the tracers leave ("drop out") or enter ("drop in") the region illuminated by the evanescent wave within the image pair, affecting the SNR of the cross-correlation.

To quantify the error due to this Brownian diffusion-induced particle mismatch, simulations were carried out of colloidal particles with initial locations generated using a Monte Carlo approach [22]. The particles were then convected over a given time interval Δt by a uniform velocity along x of magnitude U subject to hindered Brownian diffusion using the Langevin Equation. If the particles interact with the wall through some type of collision (instead of sticking to the wall), it was assumed that the particles interacted with the wall via perfectly elastic collisions as $z'/a \to 1$. This "wall interaction model" has a marked effect on the hindered diffusion normal to the wall, biasing the displacements due to diffusion towards lower values. Table 16.1 compares $D_\perp$, the diffusion coefficients predicted by Eq. (16.10) from the y-position of each particle, with $\mathcal{D}_\perp$, the diffusion coefficient calculated directly from the actual particle displacement based upon the wall interaction model used here, averaged over the motion of 10^5 particles over 6 ms with an initial position

TABLE 16.1. Comparison of hindered Brownian diffusion coefficients for motion normal to the wall.

Π	$D_\perp/D_\infty$	$\mathcal{D}_\perp/D_\infty$
3	0.582	0.261
5	0.637	0.353
7	0.683	0.440
9	0.720	0.516
11	0.750	0.578

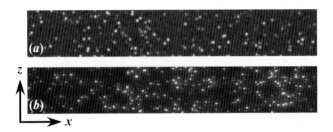

FIGURE 16.4. Representative (a) artificial and (b) experimental particle tracer images.

$z'/a = [1 : \Pi]$. For all values of Π, $\mathcal{D}_\perp < D_\perp$, and, as expected, the discrepancy between the two values decreases as Π increases. This result demonstrates that experimental measurements of hindered Brownian diffusion normal to the wall where the particle interacts with the wall via some type of elastic collision will underestimate the Brownian diffusion coefficient and that any such measurement will be very sensitive to particle-wall interaction effects due, for example, to surface charge or roughness.

An artificial image pair is then generated based upon the initial locations of these particle tracers and their locations after a time interval Δt. The particle images are modeled as a Gaussian intensity distribution with an effective diameter given by Equation (16.8) and a peak grayscale value calculated using Equation (16.7). The particle number density, image shape and image background noise characteristics are adjusted so that they match those for a typical experimental nPIV image following the guidelines suggested by Westerweel [31]. Figure 16.4 shows representative artificial and experimental images.

Datasets consisting of 5000 such artificial image pairs were processed using FFT–based cross-correlation (cf. Sec. 16.4.1) and a Gaussian (correlation) peak-finding algorithm using a Gaussian surface fit over at least 11 neighboring points for 118×98 pixel interrogation windows with 50% overlap. These relatively large interrogation windows were chosen to match experimental data (cf. Section 16.4); in all cases, at least 30 particle images were present in each window. Figure 16.5 shows the fraction of mismatched particles (as a percentage of the total number of particles) within an image pair ξ as a function of the time interval within the image pair normalized by the time required for a particle to diffuse its radius due to Brownian motion $\tau = \Delta t/(a^2/D_\infty)$. The results averaged over 5000 image pairs for both unconfined (□) and hindered (■) Brownian diffusion show that the percentage of mismatched particles increases logarithmically with τ for $\tau > 5$, and that mismatch is reduced by about 5% for hindered diffusion.

Figure 16.6 shows the displacement error ε between the results obtained by processing an artificial image pair with the actual displacement of $U(\Delta t)$ as a function of particle mismatch averaged over 5000 image pairs. This error is well below the typical minimum accuracy of cross-correlation-based PIV results of about 0.05 pixels [32] and essentially constant for $\xi < 45\%$. Results for both unconfined (□) and hindered (■) Brownian diffusion show a sharp increase in the displacement error once ξ exceeds 55%. This rapid increase in ε for larger values of ξ is due to a decrease in SNR of the cross-correlation peak from both a decrease in the height and an increase in the width of the peak [22]. Note that for a minimum of 30 particles per interrogation window, $\xi = 55\%$ implies that there are at least 13 "matched" particles in each image pair, more than twice the minimum value of

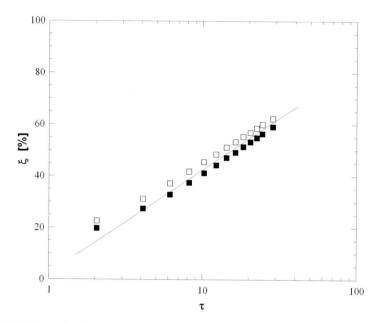

FIGURE 16.5. Plot of particle mismatch as a function of normalized time interval within the image pair.

5 matched particles recommended for cross-correlation-based PIV based upon Monte Carlo simulations [33]. The consequences of Brownian diffusion-induced particle mismatch on the accuracy of nPIV data are illustrated in the next Section, which describes the use of nPIV to measure velocity fields in steady, fully-developed electroosmotic flow.

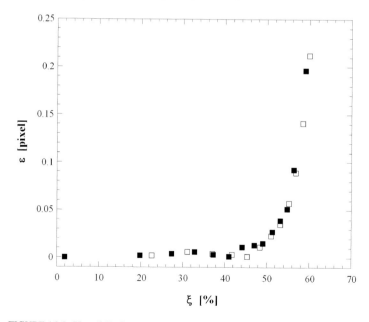

FIGURE 16.6. Plot of displacement error in pixels as a function of particle mismatch.

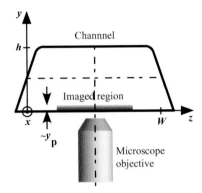

FIGURE 16.7. Definition sketch of channel cross-section and summary of channel dimensions. Here, x is the flow direction, y is the direction normal to the wall (i.e., along the optical axis) and z is along the wall normal to x.

16.4. NANO-PIV RESULTS IN ELECTROOSMOTIC FLOW

16.4.1. Experimental Details

Electroosmotic flow (EOF) of dilute aqueous solutions of the weak base sodium tetraborate ($Na_2B_4O_7$), or borax, at molar concentrations $C = 0.19$–36 mM was studied in rectangular fused silica microchannels with cross-sectional dimensions $h \times W$ (Figure 16.7) [34]. At these concentrations, the salt dissociates in aqueous solution to give anion and cation species with a valence of unity:

$$Na_2B_4O_7 + 7H_2O \rightleftharpoons 2Na^+ + 2B(OH)_3 + 2B(OH)_4^-$$

The ionic strength of the solution $I = 2C (I \equiv \sum_{i=1}^{N} z_i^2 c_i$, where z_i and c_i are the valence and molar concentration of the ith ionic species, respectively).

The microchannels were fabricated using chemical wet etching to produce an open trapezoidal channel, which was then sealed with a cover slip. The region of the microchannel imaged in these experiments was at least $360h$ downstream of any bends or intersections to ensure fully-developed flow, next to the cover slip at the bottom of the channel to minimize surface roughness and in the center of the channel to ensure uniform flow along the z-direction.

The working fluid was seeded with 100 nm diameter fluorescent polystyrene spheres (Estapor XC, Bangs Laboratories, Inc.) at volume fractions of 0.07–0.1%. The spheres have excitation and emission maxima at 480 nm and 520 nm, respectively, and a density $\rho = 1.05$ g/cm^3. The microchannel was allowed to sit for several minutes after filling the reservoirs at both ends of the channel with the working fluid to ensure that any Poiseuille flow was eliminated before imposing an electric field using platinum electrodes in the reservoirs.

The evanescent-wave illumination was generated using the prism method (cf. Section 16.3.2) from a continuous argon-ion laser (Coherent Innova 90) beam at 488 nm with an output power of approximately 0.1 W. The beam is refracted through an isosceles

C [mM]	(h, W) [μm]
0.19	(4.9, 17.3)
1.9	(10.2, 26.1)
3.6	(24.7, 51.2)
18.4	(4.8, 17.9)
36	(24.7, 51.2)

right-angle triangle prism and undergoes TIR with an angle of incidence $\theta \approx 70°$ at the interface between the fused silica cover slip with a refractive index $n_2 = 1.46$ and the water with refractive index $n_1 = 1.33$. The penetration depth of the evanescent wave y_p, which determines the out-of-plane spatial resolution of these measurements, is 90–110 nm.

The fluorescent particle tracers are imaged with an inverted epi-fluorescent microscope (Leica DMIRE2) equipped with a 63 × 0.5 objective (Leica PL Fluotar L) and a dichroic beamsplitter cube (Leica I3) to ensure that only the longer-wavelength fluorescence from the particles is imaged as 100 row × 653 column images (nominal dimensions) on a CCD camera with on-chip gain (Photometrics Cascade 650) at typical framing rates of 160 Hz with 1 ms exposure.

In-house software, implemented in MATLAB version 6.5, was used to calculate flow velocities from the particle images. This processing software uses a combined correlation-based interrogation and tracking algorithm [35] to analyze an interrogation window from the first (*i.e.*, earlier) image within the pair with dimensions (X, Z) and a larger interrogation window from the second image with dimensions of $(X + 2\Delta, Z + 2\Delta)$. The interrogation windows are cross-correlated using an FFT-based algorithm that can be applied to windows of arbitrary dimensions. The overlap between adjacent windows was 50%, and no window shift was used since the displacement in all cases was 5 pixels or less. In processing these data, the window dimensions varied with dataset, with $X = 120-200$ pixels, $Z = 58-88$ pixels and $\Delta = 10-16$ pixels. Note that typical window dimension of 120 × 80 pixels ($x \times z$) corresponding to an in-plane spatial resolution of 18 × 13 μm, with 50% overlap, gives 18 particle displacement vectors for each image. The range of window sizes were chosen to ensure that the number of particle images in each window is at least 20 to ensure a cross-correlation peak with a good signal to noise ratio, even with significant particle mismatch between two successive exposures due to Brownian diffusion normal to the wall. Cross-correlation averaging [26] or "single-pixel interrogation" [36], originally used to improve the spatial resolution of μPIV data, can also be used to improve the spatial resolution of nPIV data. Although not shown here, cross-correlation averaging of a subset of these images with windows as small as 16 × 10 pixels ($x \times z$), corresponding to an in-plane spatial resolution of 2.5 × 1.5 μm, give average velocity results within 2% of those reported here.

16.4.2. Results and Discussion

The EOF studied here is essentially uniform normal to the wall, since the electric double layer is very thin compared with y_p, with a Debye length $\lambda < 9$ nm. Figure 16.8 shows a typical raw particle image and the nPIV result obtained by averaging velocity data

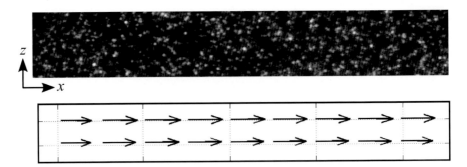

FIGURE 16.8. Representative image of particle tracers over a 110 × 16 μm ($x \times z$) field of view [top] and corresponding nPIV result [bottom] averaged over 999 image pairs. The velocity vectors represent an average velocity $U = 17.6$ μm/s.

obtained over 1000 such consecutive images (= 999 image pairs) for $C = 3.6$ mM and $E = 1.0$ kV/m. The measured velocity field verifies that the flow is essentially uniform over this x–z plane.

The nPIV results were therefore spatially and temporally averaged to obtain the bulk flow velocity U. Figure 16.9 shows U as a function of the external electric field E for various $Na_2B_4O_7$ concentrations C. The average velocity is a linear function of the driving electric field, corresponding to constant mobility. The error bars represent conservative

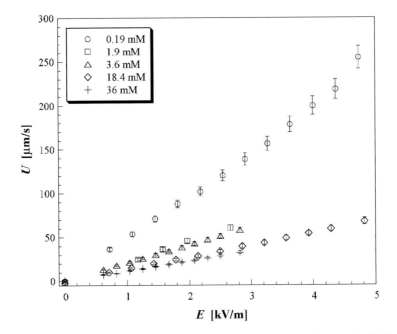

FIGURE 16.9. Graph of average velocity U as a function of external electric field E for $C = 0.19 - 36$ mM sodium tetraborate buffer.

95% confidence intervals; these error estimates include both the uncertainty in determining particle displacement in each set of images and other experimental uncertainties associated with converting this displacement into actual velocities, such as the estimated physical dimension of one pixel in each image and the CCD camera framing rate.

As discussed in Section 16.3.3, Brownian diffusion of the 100 nm tracers can be a significant issue in nPIV data. Errors due to out-of-plane Brownian diffusion can be estimated using the results from the previous Section. For a time interval between images within the image pair $\Delta t = 6.5$ ms, $\tau = 13.3$, corresponding in Figure 16.5 to a fraction of mismatched particles $\xi = 45\%$. The results shown in Figure 16.6 suggest that the displacement error due to hindered Brownian diffusion for this level of particle mismatch is negligible.

In-plane Brownian diffusion can be estimated using calculations similar to those used to generate Table 16.1 [22]. A tracer with an initial position $z'/a \leq 3$ has a mean diffusion coefficient for motion parallel to the wall $D_{||}/D_\infty = 0.774$ averaged over 10^5 particles; the rms Brownian displacement parallel to the wall for a particle using this value is then $\Delta x_{||} \approx 220$ nm, corresponding to a displacement of about one pixel, vs. the maximum displacement due to the flow of three pixels, in these experiments. Errors due to in-plane Brownian diffusion can be greatly reduced by averaging the flowfield for this steady flow over time. A typical value for the standard deviation for a single velocity vector at a given spatial location over 999 realizations—which should give an estimate of the in-plane Brownian diffusion averaged over all the tracer particles within the interrogation window—is 6%. After temporal averaging, this standard deviation in turn accounts for about 2% of the 95% confidence intervals given in Figure 16.9. Errors associated with the imaging system (due, for example, to errors in determining the magnification of the images and the limiting spatial resolution of the CCD camera) are therefore much more significant sources of error in these experimental results.

To validate the nPIV data, the mobility μ_{ex} was calculated using linear regression from the results of Figure 16.9 and compared with the electroosmotic mobility μ_{eo} predicted by the asymptotic model of Conlisk et al. [37].

$$\mu_{eo} = \frac{\varepsilon_e \phi_o}{2\mu} \ln\left\{\frac{g^0}{f^0}\right\} \tag{16.11}$$

In this equation, ε_e and μ are the electrical permittivity and absolute viscosity, respectively, of the fluid, $\phi_o \equiv RT/F$ is the characteristic scale for the potential formed from R the universal ideal gas constant, T the fluid absolute temperature and F Faraday's constant, and (g^0, f^0) are the wall mole fractions of the cationic and anionic species, respectively. As shown in Figure 16.10, the experimental and model values for mobility agree within 4.5% over a 200-fold change in molar concentration, suggesting that the particle tracers follow the flow with good fidelity over this range of electric fields. The mobility values are also compared with a single independent measurement using neutral fluorescent molecular tracers for $C = 0.2$ mM [38]. These independently obtained analytical and experimental results suggest that particle slip with respect to the fluid due to electrophoresis has a negligible effect on mobility for the EOF cases studied here. Furthermore, the mobility obtained from the nPIV data appears to be unaffected by tracer particle properties such as size and charge.

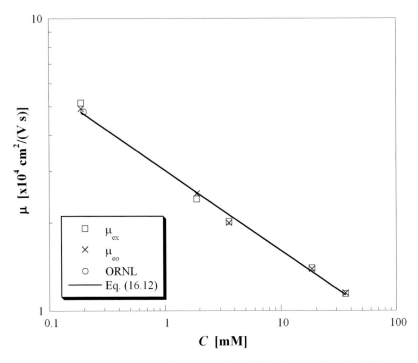

FIGURE 16.10. Log-log plot of electroosmotic mobility values as a function of C.

Finally, these results suggest that the mobility μ has a novel power-law scaling with concentration:

$$\frac{\mu}{3.03 \times 10^{-4} \text{ cm}^2/(\text{V} \cdot \text{s})} = \left(\frac{C}{1 \text{ mM}}\right)^{-0.275} \quad (16.12)$$

Note that this scaling cannot be directly derived from the analysis because of the iterative method used to derive (g^0, f^0).

16.5. SUMMARY

Over the last three years, nano-particle image velocimetry—based upon tracking the motion of colloidal fluorescent particles illuminated by evanescent waves—has been developed as a new method for measuring near-wall velocity fields with submicron spatial resolution. Initial results from this novel technique have been validated for steady fully-developed electroosmotic flows in microchannels, and suggest a novel scaling of electroosmotic mobility with electrolyte concentration.

Nano-PIV can measure velocities averaged over the first 100 nm based upon penetration depth next to the wall with an out-of-plane spatial resolution of 100 nm based upon tracer particle diameter. In comparison, the near-wall capability of μPIV is about 900 nm based upon interrogation window size and its out-of-plane spatial resolution based upon depth of field is at least 1.5 μm. The method therefore represents about an order of magnitude improvement in near-wall access and out-of-plane spatial resolution over current microfluidic diagnostic techniques. The in-plane spatial resolution of these nPIV data of 18 μm × 13 μm obtained here without cross-correlation averaging is significantly larger than that of the best reported μPIV measurements using cross-correlation averaging. Nevertheless, initial results using cross-correlation averaging of nPIV data show that the minimum in-plane spatial resolution of nPIV should be comparable to that for μPIV, namely 1–2 μm.

Brownian diffusion is potentially a significant source of error in nPIV, given typical Peclet numbers for the tracers used in this technique of $O(10^{-3})$. Results from artificial and experimental images suggest, however, that hindered Brownian diffusion—and more specifically, particle mismatch within the image pair due to Brownian diffusion—has relatively little impact on the accuracy of nPIV data for the steady uniform flows studied here provided that the interrogation window size is chosen to ensure enough (here, at least 11) matched tracers within an image pair.

These initial results demonstrate the potential of this technique. In terms of extending nPIV, the exponential decay of particle image intensity with distance from the wall given by Equation (16.3) suggests that it may be possible, for tracer images with sufficient contrast and tracers with consistent fluorescence intensity, to obtain not just the x- and z-positions of the particles, but also their y-position. If so, it should be possible to measure all three velocity components using nPIV without any additional imaging equipment. Particle tracking studies are already underway to measure hindered Brownian diffusion normal to the wall using this concept [39].

Finally, issues associated with particle tracers such as Brownian diffusion and, for EOF, electrophoresis, may be avoided altogether by using molecular tracers illuminated by evanescent waves to obtain near-wall velocimetry measurements. Current intensified CCD technology may not yet be sufficient for nanoscale MTV, given that the quantum efficiency of molecular tracers such as phosphorescent supramolecules is 3.5% [11], more than an order of magnitude lower than that of the fluorophores used in nPIV tracers of at least 90%. Nevertheless, Pit et al. [8] have already used a related technique, fluorescence recovery after photobleaching, with evanescent waves at a single point to show evidence of slip in hexadecane. The rapid development of single-photon and molecule imaging technology suggests that molecular tracers illuminated by evanescent waves will ultimately be the most accurate method for measuring near-wall velocity fields and studying interfacial phenomena such as apparent slip at the submicron scale.

ACKNOWLEDGEMENTS

This work would not have been possible without the contributions of Dr. Reza Sadr, Haifeng Li and Claudia M. Zettner in my group at Georgia Tech. We also wish to thank Dr. Terry Conlisk and Zhi Zheng at Ohio State University and Dr. Mike Ramsey and J.-P. Alarie at Oak Ridge National Laboratories for sharing their results and numerous suggestions.

REFERENCES

[1] Supported by DARPA DSO (F30602-00-2-0613) and NSF (CMS-9977314).
[2] G.W.Woodruff School of Mechanical Engineering; Georgia Institute of Technology; Atlanta,GA30332–0405 USA E-mail: minami.yoda@me.gatech.edu.
[3] S. Granick, Y. Zhu, and H. Lee. *Nat. Mat.*, 2:221, 2003.
[4] P.G. De Gennes. *Langmuir*, 18:3413, 2002.
[5] P. Attard. *Advan. Coll. Int. Sci.*, 104:75, 2003.
[6] O.I. Vinogradova. *Int. J. Min. Proc.*, 56:31, 1999.
[7] D.C. Tretheway and C.D. Meinhart. *Phys. Fluids*, 14:L9, 2002.
[8] R. Pit, H. Hervet, and L. Léger. *Phys. Rev. Lett.*, 85:980, 2000.
[9] J.G. Santiago, S.T. Wereley, C.D. Meinhart, D.J. Beebe, and R.J. Adrian. *Exp. Fluids*, 25:316, 1998.
[10] W.R. Lempert, K. Magee, P. Ronney, K.R. Gee, and R.P. Haughland. *Exp. Fluids*, 18:249, 1995.
[11] C.P. Gendrich, M.M. Koochesfahani, and D.G. Nocera. *Exp. Fluids*, 23:361, 1997.
[12] P.H. Paul, M.G. Garguilo, and D.J. Rakestraw. *Anal. Chem.*, 70:2459, 1998.
[13] D. Sinton and D. Li. *Coll. Surf. A: Physicochem. Eng. Asp.*, 222:273, 2003.
[14] C. Lum and M. Koochesfahani. *Bull. Am. Phys. Soc.*, 48(10):58, 2003.
[15] E. Bonaccurso, M. Kappl, and H.-J. Butt. *Phys. Rev. Lett.*, 88:076103, 2002.
[16] C.M. Zettner and M. Yoda. *Exp. Fluids*, 34:115, 2003.
[17] E. Hecht. *Optics* (3rd Ed.), Addison-Wesley, Reading, p. 124, 1998.
[18] H.K.V. Lotsch. *Optik* 32:189, 1970.
[19] D. Axelrod, E.H. Hellen, and R.M. Fulbright. In J.R. Lakowicz (ed.), *Topics in Fluorescence Spectroscopy, Vol. 3: Biochemical Applications*. Plenum Press, New York, p. 289, 1992.
[20] D. Axelrod, T.P. Burghardt, and N.L. Thompson. *Ann. Rev. Biophys. Bioeng.*, 13:247, 1984.
[21] D.C. Prieve and N.A. Frej. *Langmuir*, 6:396, 1990.
[22] R. Sadr, H. Li, and M. Yoda. *Exp. Fluids*, 38:90, 2005.
[23] R.J. Adrian. *Ann. Rev. Fluid Mech.*, 23:261, 1991.
[24] C.D. Meinhart and S.T. Wereley. *Meas. Sci. Technol.*, 14:1047, 2003.
[25] M.G. Olsen and R.J. Adrian. *Opt. Laser Technol.*, 32:621, 2000.
[26] C.D. Meinhart, S.T. Wereley, and J.G. Santiago. *J. Fluids Eng.*, 122:285, 2000.
[27] M.A. Bevan and D.C. Prieve. *J. Chem. Phys.*, 113:1228, 2000.
[28] H. Faxén. *Ann. Phys.*, 4(68):89, 1922.
[29] A. Einstein. *Ann. Phys.*, 17:549, 1905.
[30] A.J. Goldman, R.G. Cox, and H. Brenner. *Chem. Eng. Sci.*, 22:637, 1967.
[31] J. Westerweel. *Exp. Fluids*, 29:S3, 2000.
[32] M. Raffel, C. Willert, and J. Kompenhans. *Particle Image Velocimetry: A Practical Guide*, Springer, Berlin, p. 129, 1998.
[33] R.D. Keane and R.J. Adrian. *Appl. Sci. Res.*, 49:191, 1992.
[34] R. Sadr, M. Yoda, Z. Zheng, and A.T. Conlisk. *J. Fluid Mech.*, 506:357, 2004.
[35] D.F. Liang, C.B. Jiang, and Y.L. Li. *Exp. Fluids*, 33:684, 2002.
[36] J. Westerweel, P.F. Gelhoed, and R. Lindken. *Exp. Fluids*, 37:375, 2004.
[37] A.T. Conlisk, J. McFerran, Z. Zheng, and D.J. Hansford. *Anal. Chem.*, 74:2139, 2002.
[38] J.M. Ramsey, J.P. Alarie, S.C. Jacobson, and N.J. Peterson. In Y. Baba, S. Shoji, and A.J. van den Berg (eds.), *Micro Total Analysis Systems 2002*, Kluwer Academic Publishers, Dordrecht, p. 314, 2002.
[39] K.D. Kihm, A. Banerjee, C.K. Choi, and T. Takagi. *Exp. Fluids*, 37:811, 2004.

17

Optical MEMS-Based Sensor Development with Applications to Microfluidics

D. Fourguette*, E. Arik*, and D. Wilson[#]
*VioSense Corporation, 36 S. Chester Ave., Pasadena, CA 91106
[#]Jet Propulsion Laboratory, 4800 Oak Grove Drive, Pasadena, CA 91109

17.1. INTRODUCTION

Despite challenges associated with development and fabrication, meso-scale and MEMS devices offer many advantages such as small size and lighter weight for space applications, higher reliability, smaller and better controlled sampling volumes in particular for biological diagnostics. Traditional optical diagnostics have been used to determine parameters such as velocity, concentration, particle size and temperature in laboratory-scale and full scale flow systems over the past three decades. A great deal of efforts has been invested in these diagnostics techniques, but the size of MEMS prohibits the use of these traditional diagnostics. Spatial resolution, vibrations, and optical accessibility make traditional optical setups difficult to use for microfluidics.

Recent progress in the design and fabrication of optical MEMS and beam-shaping diffractive optical elements has enabled the development of meso-scale and microsensors whose size is comparable to the scales encountered in microfluidics applications [14]. Moreover, optical MEMS can be integrated into microfluidic devices during fabrication and as such, are also suitable for on-board process monitoring.

17.2. CHALLENGES ASSOCIATED WITH OPTICAL DIAGNOSTICS IN MICROFLUIDICS

Optical diagnostics for measurements in microfluidics present inherent challenges [7]. The most challenging aspect for optical diagnostics is the spatial resolution required to obtain meaningful results. The typical wall dimension of a microchannel is on the order of 100 micrometers in diameter, which is comparable to a typical optical probe volume dimension of a laser Doppler velocimeter.

Most optical diagnostics techniques require the presence of scatterers (particles, emulsions, or cells) in the flow to scatter the incident light. Most microfluidics applications do not allow seeding the flow with particles and prefer to rely on the presence of natural scatterers such as cell or emulsions to scatter the light. Unfortunately, the index of refraction of these scatterers is very close to that of the surrounding fluid, thus generating a weak scattered light intensity. In addition, microchannel walls often cause a lensing effect that distorts the incident and scattered light path. This path distortion, often combined with a complex optical access, causes an additional reduction in the signal-to-noise ratio of the measurements.

17.3. ENABLING TECHNOLOGY FOR MICROSENSORS: COMPUTER GENERATED HOLOGRAM DIFFRACTIVE OPTICAL ELEMENTS

The enabling optical elements in these microsensors are the diffractive optical elements (DOEs) that produce the desired optical pattern within such small space. The DOE takes the form of a shallow surface relief pattern that is etched into a transparent substrate. The surface relief pattern is designed to produce phase variation in the laser beam such that the light will form an interference pattern that approximates the desired image. DOEs of this type are frequently referred to as computer-generated holograms (CGHs).

A CGH can be designed using the iterative Fourier transform (IFT) algorithm, also known as the Gerchberg-Saxton algorithm [9]. This algorithm utilizes the well-known Fourier transform relationship between the near field (just past the DOE) and the far field (focal plane of DOE). For high efficiency and ease of fabrication, it is best if the DOE implements a *phase-only* transmission function, i.e. no spatial magnitude variation is required. This constraint makes it impossible to simply inverse the Fourier transform of the desired far-field image to find the near field (and hence the DOE function) because the inverse transform will generate magnitude variation as well as phase in the near field. The IFT design algorithm overcomes this problem by iterating between the near and far field planes many times, constraining the far-field intensity to the desired pattern and constraining the near field to have phase-only variation. After a sufficient number of iterations, the far field intensity approximates the desired image and the CGH is phase only. Over the last two decades, CGH design methodology had been advanced by a wide-variety of contributions [18].

Being able to perform the design is only the first step in realizing the required DOE for a MOEMS sensor. While there are a variety of techniques for fabricating DOEs [18], the sensors being described here generally require high numerical aperture (low f/#) optics that have micron-scale Fresnel zone spacing. Furthermore, the DOEs must be fabricated with 'blazed' analog-relief sawtooth groove shapes to have high diffraction efficiency for measuring the weakly scattering particles. Hence the DOEs must be fabricated using sub-micron features. Such high-efficiency analog-depth DOEs are fabricated at NASA's Jet Propulsion Laboratory using direct-write electron beam lithography [12, 13, 19, 22]. The DOE surface relief profiles are fabricated in a thin film of E-beam resist (PMGI, polymethylglutarimide, or PMMA, polymethylmethacrylate) that is spin-coated on a fused silica or other substrate to a thickness of approximately two microns. The electron-beam (100 or 50 kV accelerating voltage) breaks bonds in the resist making it more soluble to a development solution. Analog depth control is achieved by varying the E-beam dose (constant current, variable dwell time) in proportion to the desired depth. During pattern preparation, special deconvolution techniques are used to compensate the DOE depth profile for the E-beam proximity effect (backscattered dose) and the non-linear depth vs. dose response of the resist. After exposure, the optic is developed using an iterative procedure in which we measure the depth after each step until the desired depth is achieved. This procedure yields DOEs with depth profiles that are typically accurate to better than 5–10% ($\lambda/20$–$\lambda/10$) and have diffraction efficiencies in the range 80–90%. Generally the PMGI resist is durable enough for most applications, but if needed, techniques exist to transfer etch the E-beam resist profile into the more durable quartz substrate. If high volume fabrication is desired, a variety of replication techniques have been demonstrated [18].

DOEs are capable of shaping light into a wide range of geometries. For example, a DOE was designed and fabricated to shape light into two elongated light spots. This DOE was used in the fabrication of a time-of-flight velocimeter described in Section 17.5.1 [21]. The design was accomplished in two steps. First, the beam-shaping computer-generated hologram (CGH) phase function shown in Fig. 17.1(a) was designed using a variation of the soft-operator IFT algorithm [10] assuming a plane-wave input field and an image at infinity. Second, a commercial ray-tracing program was used to design a lens phase function to focus the light from a single-mode fiber to an image plane at a desired distance beyond the DOE. The lens phase function was optimized to correct for the aberrations introduced by focusing through the substrate. Figure 17.1(b) shows the DOE depth after the lens phase function has been added to the beam-shaping CGH phase function. After E-beam fabrication in PMGI resist on a quartz substrate, the DOE was illuminated with the single-mode output of a 660 nm fiber-pigtailed diode laser, and the intensity at focus was recorded using a charge-coupled device (CCD) camera. The agreement between the simulated (Fig. 17.1c) and the measured (Fig. 17.1d) intensity patterns at the focus is excellent, showing diffraction-limited performance due to the precision E-beam fabrication.

Another example of the capabilities afforded by these diffractive elements is provided by the dual-line-focus laser lens shown in the photograph and atomic force microscope scan of Fig. 17.2. This DOE was designed for the diverging-fringe shear stress sensor described in Section 17.2.2 [20]. When illuminated by the diverging field of a single-mode fiber, this DOE produces two line-foci that are coincident with slits in the chrome mask pattern on the other side of the 500 μm thick quartz substrate shown in Fig 17.2(c). The phase function of the DOE is comprised of a lens phase function that collimates along the vertical axis (in Fig. 17.2) and performs off-axis focusing along the horizontal axis, plus a binary π-phase grating that splits the focus into two equal-intensity orders that form the two line foci (low efficiency higher-orders are also created, but are blocked by the mask). This two-sided DOE was again fabricated by direct-write E-beam lithography in PMGI resist, and front-to-back

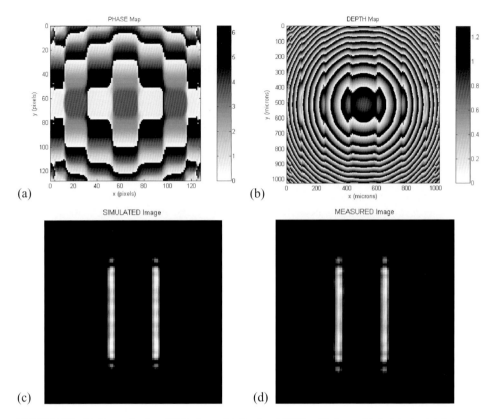

FIGURE 17.1. (a) Beam-shaping CGH phase, (b) Depth of CGH with lens added, (c) simulated image, (d) measured image.

alignment marks were used to ensure alignment of the DOEs with the chrome mask pattern to an accuracy of ~ 2 μm. Figure 17.2(d) shows photographs of the diverging fringe pattern produced when the dual-line-focus laser lens was illuminated with a single-mode output of a 660 nm fiber-pigtailed laser.

17.4. THE MINIATURE LASER DOPPLER VELOCIMETER

The miniature laser Doppler velocimeter, the MiniLDV^{TM}, is the first step toward the miniaturization of flow sensors using a combination of refractive and diffractive optical elements. Laser Doppler velocimetry (LDV) has been in use as a research tool for over three decades. Since its inception by Yeh and Cummins [23], the LDV technique has matured to the point of being well understood and, when used carefully, providing accurate experimental data. Laser Doppler Velocimetry's unique attributes of linear response, high frequency response (defined by the quality of the seed particles) and non-intrusiveness have made it the technique of choice for the study of complex multi-dimensional and turbulent flows [11, 15]. It has also been modified for use in multi-phase flows [16], particle sizing [2], and remote

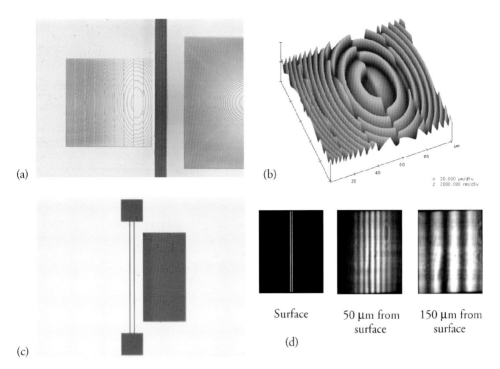

FIGURE 17.2. (a) Photograph, (b) atomic-force microscope scan of the center of the dual-line-focus laser lwns, (c) metal (chrome) mask pattern on back side of substrate, (d) diverging fringe pattern.

sensing [3], to cite a few examples. Advancements in the areas of laser performance, fiber optics, and signal processing have enhanced the utility of the technique in fluid mechanics and related research areas.

17.4.1. Principle of Measurement

Laser Doppler velocimetry utilizes the Doppler effect to measure instantaneous particle velocities. When particles suspended in a flow are illuminated with a laser beam, the frequency of the light scattered (and/or refracted) from the particles is different from that of the incident beam. This difference in frequency, called the Doppler shift, is linearly proportional to the particle velocity. With the particle velocity being the only unknown parameter, then in principle the particle velocity can be determined from measurements of the Doppler shift Δf. In the commonly employed fringe mode, LDV is implemented by splitting a laser beam to have two beams intersect at a common point so that light scattered from two intersecting laser beams is mixed, as illustrated in Figure 17.3. In this way both incoming laser beams are scattered towards the receiver, but with slightly different frequencies due to the different angles of the two laser beams [1].

When two wave trains of slightly different frequency are super-imposed we get the well-known phenomenon of a beat frequency due to the two waves intermittently interfering with each other constructively and destructively. The beat frequency, also called the

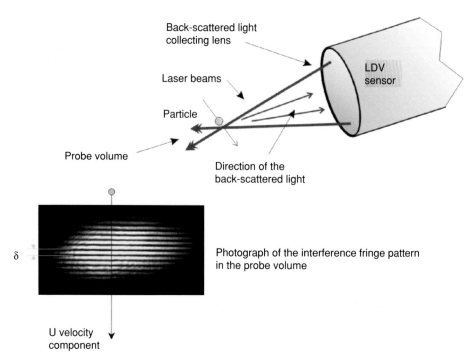

FIGURE 17.3. Principle of Laser Doppler Velocimetry.

Doppler-frequency f_D, corresponds to the difference between the two wave-frequencies:

$$f_D = \frac{1}{\lambda} \cdot 2\sin(\theta/2) \cdot u_x = \frac{2\sin(\theta/2)}{\lambda} u_x$$

where θ is the angle between the incoming laser beams and λ is the wavelength of the laser. The beat-frequency, is much lower than the frequency of the light itself, and it can be measured as fluctuations in the intensity of the light reflected from the seeding particle. As shown in the previous equation the Doppler-frequency is directly proportional to the x-component of the particle velocity, and thus can be calculated directly from f_D:

$$u_x = \frac{\lambda}{2\sin(\theta/2)} f_D$$

Further discussion on LDA theory and different modes of operation may be found in the classic text of Durst et al. [4].

17.4.2. Signal Processing

The processing is based on the Fast Fourier Transform to extract the frequency from the burst. Figure 17.4 shows the sequence involved burst processing. The raw signal contains a DC part, low frequency component, and an AC part, high frequency component. The DC-part, which is removed by a high-pass filter, is known as the Doppler pedestal, and

OPTICAL MEMS-BASED SENSOR DEVELOPMENT WITH APPLICATIONS TO MICROFLUIDICS 355

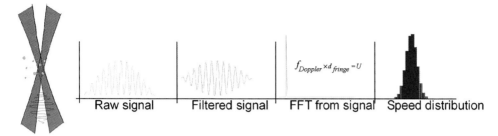

FIGURE 17.4. Signal processing for the LDV.

it is often used as a trigger-signal, which starts sampling of an assumed burst-signal. The envelope of the Doppler modulated signal reflects the Gaussian intensity distribution in the measuring volume. The AC part represents the beat frequency, and is processed using a Fast Fourier Transform to calculate the frequency of the signal, f_D.

A photograph of the miniature sensor is shown in Figure 17.5. The sensor generates two coherent beams that intersect (in this case) 80 mm from the sensor face. The two coherent beams form an interference fringe pattern. The probe is configured in back-scatter mode; the scattered light is collected using a conventional lens and imaged onto a receiving fiber. The light burst is converted into a signal using a photodetector.

17.4.3. A Laser Doppler Velocimeter for Microfluidics

A photograph of the MiniLDV^{TM} is shown in Figure 17.5. The diameter of the probe is 40 mm and the length 150 mm. Typical probe volume distance is 80 mm to 150 mm. The light source, a diode laser in this case, is contained in the body of the sensor. The light scattered off the particles is collected with a 30 mm diameter lens and imaged onto a multimode fiber. The sensor is connected to a driver containing the power supply for the laser and a photodetector to convert the light signal into an electric signal. The signal is

FIGURE 17.5. Photograph of the miniature laser Doppler velocimeter.

TABLE 17.1. LDV specifications

Working distance	50 mm
Laser wavelength	785 nm
Probe volume dimensions	$15 \times 20 \times 90 \ \mu m^3$
Fringe separation	1.27 μm
Number of fringes in probe volume	14
Frequency shifting	2 MHz

then digitized using a PC-based data acquisition and processed using software developed in LabVIEW™.

A version of the MiniLDV™ was specially designed for applications to microfluidics. Measurements in microfluidic environments generally require high spatial resolution. The probe volume size of specially designed version of the MiniLDV™ was less than 100 μm in length and 15 μm in height. The specifications for the miniature LDV sensor used to measure velocity in the microchannel are given in Table 17.1.

17.5. MICRO SENSOR DEVELOPMENT

These recent advances in micro fabrication techniques and micro-optics have enabled the development of optical microsensors for pressure and temperature, and for fluid flow measurements such as velocity and wall shear stress [6]. These microsensors offer the advantage of being non-intrusive, embeddable in small wind or water tunnel models, and capable of remote in-situ monitoring. In this Section, we describe a series of optical MEMS based microsensors for fluid flow measurements that are currently used in Research & Development settings as well as in industrial applications. These sensors are particle based, i.e., these sensors require particles to be present in the flow to scatter the light off the probe volume. The commonality for these sensors is the use of diffractive elements to shape the optical probe volume. In most cases, the generation of such optical probe volumes with conventional optics requires large optical setups prone to misalignment. Another commonality is the use of fibers to bring the light to the sensor and retrieve the backscattered signal. However, recent advances in opto-electronics will soon enable the integration of the light source and detector into the sensor head, thus bypassing the use of optical fibers.

In addition to the significant size reduction, optical MEMS provide excellent control over the shape of the probe volume, as demonstrated in the previous Section, and excellent control over the probe volume location relative to the sensor surface. Optical probe volumes generated with DOEs are diffraction limited, and most DOE designs can produce optical patterns whose characteristic length is as small as 10 micrometers. With proper optical design, a diffraction limited probe volume can be precisely and permanently positioned inside a microfluidic channel.

17.5.1. Micro Velocimeter

This sensor is based on time-of-flight velocity measurement technique. A photograph of the sensor, The MicroV™, is given in Figure 17.6. The sensor optical probe volume is

OPTICAL MEMS-BASED SENSOR DEVELOPMENT WITH APPLICATIONS TO MICROFLUIDICS 357

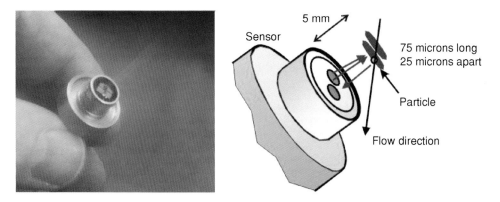

FIGURE 17.6. Photograph of the MicroV^{TM} (left), and schematic (right).

composed of two elongated light spots a few millimeters above the sensor surface [8]. The sensor uses one DOE to shape the optical probe volume as two elongated light spots (75 μm long and 25 μm apart) 5 mm off the sensor surface, and a second DOE to collect the backscattered light off the particles intersecting the light spots, as depicted in Figure 17.7. The height of the probe volume above the sensor surface can be designed to be anywhere between 800 μm to 20 mm.

Particles present in the flow intersect the two light spots and thus generate two light bursts on the sensor detector. The resulting electric signal from the detector is two pulses that are digitized using a computer-controlled data acquisition system. Autocorrelation is

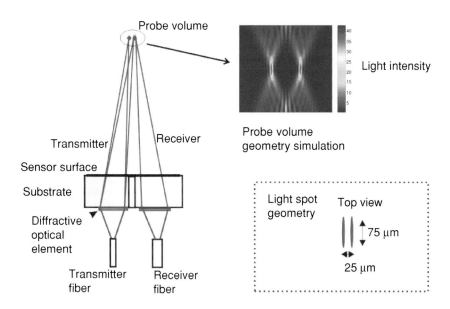

FIGURE 17.7. Schematic of the MicroV^{TM}, results of the ray tracing simulation and schematic of the probe volume.

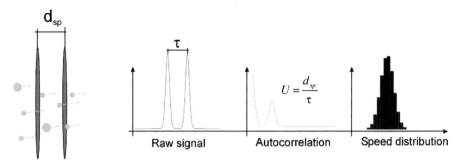

FIGURE 17.8. Schematic of the autocorrelation-based signal processing for the MicroV.

used to calculate the time elapsed between the two pulses, and therefore the particle velocity. A schematic of the processing algorithm is shown in Figure 17.8.

17.5.2. Micro Sensors for Flow Shear Stress Measurements

Flow shear stress measurements in biological applications have received a lot of interest in the last few years. Shear is associated with turbulence, and therefore molecular mixing. Many bio-MEMS applications involve mixing between two or more fluid channels, hence the need to understand microchannel flows in detail. In addition, regions of high shear in blood vessels exhibit high correlation with plaque deposition, and a better understanding of flows in small vessels will further the understanding of arteriosclerosis.

The wall shear stress σ is directly proportional to the velocity gradient at the wall. In the expression of the wall shear stress below,

$$\sigma = \mu \frac{\partial u}{\partial y}\bigg|_{y=0}$$

μ is the fluid viscosity, u is the flow velocity and y is the normal direction to the wall. The wall shear stress can be calculated by measuring the velocity gradient at the wall. In the region of the boundary layer called sub-layer, usually in the first tens of micrometers above the wall, the velocity profile is linear. Beyond the sub-layer region, the velocity profile becomes parabolic. As a result, unless the velocity is measured in the sub-layer, the velocity must be measured at a minimum of two points to calculate the velocity gradient at the wall. An iterative algorithm was successfully developed to estimate the velocity gradient at the wall using only one point velocity measurement beyond the sub-layer[1]. The following two Sections describe two optical MEMS based microsensors capable of measuring the fluid velocity in the boundary layer to estimate the wall shear stress.

17.5.2.1. Dual Velocity Sensor A MEMS-based optical sensor was designed to measure flow velocity at two locations above the sensor surface [5]. This sensor is essentially the combination of two micro velocimeters as described in Section 17.5.1. The optical chip includes one transmitter DOE (diffractive optic element) and two receiver DOEs, as shown in Figure 17.9. As for the Micro Velocimeter, the transmitter DOE is illuminated by the output of a singe mode fiber pigtailed to a diode laser. However, the transmitter is designed

OPTICAL MEMS-BASED SENSOR DEVELOPMENT WITH APPLICATIONS TO MICROFLUIDICS 359

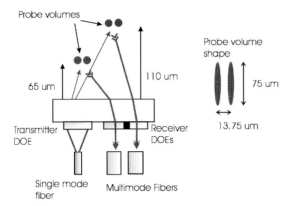

FIGURE 17.9. Design of the Dual Microvelocimeter with two probe volumes, one probe volume located at 65 μm and a second probe volume located at 110 μm above the surface of the sensor.

to generate two probe volumes, one located approximately at 65 μm and a second located approximately at 110 μm above the sensor surface. These probe locations are measured in air. Each probe volume is composed of two elongated light spots, 75 μm long and 13.7 μm apart. The scattered light generated by particles intersecting each probe volume, generating two intensity spikes, is focused onto a fiber by one receiver DOE and brought to an avalanche photodetector. An autocorrelation performed on the digitized output of each photodetector yields the time elapsed between the spikes. Again, as for the Micro Velocimeter, the distance between the light spots divided by the elapsed time yields the instantaneous flow velocity.

A photograph of the transmitter DOE and the resulting probe volumes generated by the transmitter are shown in Figure 17.10. The DOE was fabricated by direct-write electron

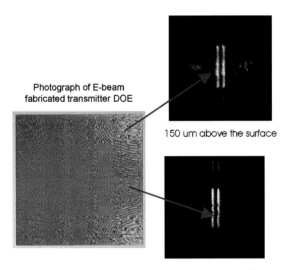

FIGURE 17.10. Photographs of the transmitter DOE and of the two probe volumes. The locations of the probe volumes are given for use in water.

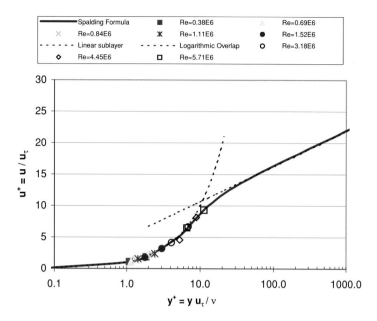

FIGURE 17.11. Spalding fit to dual velocity measurements using the Dual Velocity sensor.

beam lithography using 0.25 mm square pixels and 64 depth levels. As a result, the two peaks in the signal are well defined and the processing yields a data rate 4 times higher than in previous velocity sensors.

The Dual Micro Velocimeter was tested in the flat plate boundary layer in a water tunnel. The wall shear stress was calculated from the dual velocity measurements using a Spalding fit through the linear sub-layer and the logarithmic overlap. Figure 17.11 shows the Spalding fit to five sets of dual velocity measurements in the sub-layer.

The wall shear stress measurements obtained using the Dual Velocity sensor are shown in Figure 17.12, combined with the results from the boundary layer survey obtained with the Mini LDV. These results show that the dual velocity approach to measure wall shear stress performs very well at Reynolds number in excess of 5.5 million and is a viable extension to the Diverging Doppler sensor.

17.5.2.2. Diverging Fringe Doppler Sensor The Doppler shear stress sensor, the MicroS³™ developed for this experiment is based on a technique first presented by Naqwi and Reynolds [17] using conventional optics. They projected a set of diverging fringe pattern and used the Doppler shift to measure the gradient of the flow velocity at the wall.

This sensor measures the flow velocity within the first 100 micrometers of the sensor surface. The optical probe volume extends over 30 μm. In this region of steep velocity gradient, the probe volume is designed to measure velocity over the 30 μm span by using diverging Doppler fringes (as opposed to parallel fringes in conventional Doppler sensors).

Figure 17.13 shows a schematic of the optical MEMS sensor principle. Diverging interference fringes originate at the surface and extend into the flow. The scattered light from the particle passing through the fringes is collected through a window at the surface

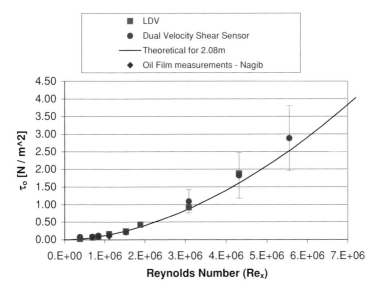

FIGURE 17.12. Wall shear stress measurements with the Mini LDV, the Diverging Doppler sensor and oil film interferometry.

of the sensor. The region defined by the intersection of the transmitter and receiver fields was centered at approximately 66 μm above the surface and measured about 30 μm high.

The local fringe separation, δ, designed to be linear with the distance from the sensor, y, is given by $\delta = k \times y$, where k is the fringe divergence rate. As particles in the fluid flow through the linearly diverging fringes, they scatter light with a frequency f that is proportional to the instantaneous velocity and inversely proportional to the fringe separation at the location of particle trajectory as shown in Figure 17.13. The velocity of the particle is therefore $u = f \times \delta$. The Doppler frequency simply multiplied by the fringe divergence yields the velocity gradient

$$\frac{u}{y} = f \times k \qquad (17.1)$$

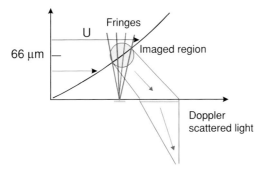

FIGURE 17.13. Schematic of the optical shear stress sensor principle of measurement.

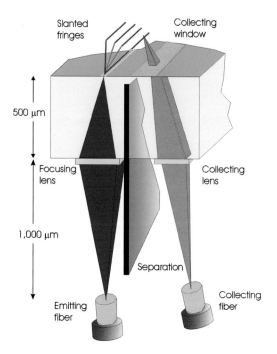

FIGURE 17.14. Schematic of the Diverging Fringe Doppler sensor.

which is equal to the wall shear,

$$\sigma = \mu \frac{\partial u}{\partial y}\bigg|_w \equiv \mu \frac{u}{y} \qquad (17.2)$$

in the linear sub-layer region of the boundary layer.

A schematic of the Diverging Fringe Doppler sensor is shown in Figure 17.14. A single mode emitting fiber illuminates a focusing lens (DOE shown in Figure 17.2) that shapes the light into two elongated beams of light. These two elongated beams of light are passed through a pair of slits to create two diffractive patterns. The interference between these diffractive patterns generates a set of diverging fringes originating at the sensor surface. The scattered light collected from the particles intersecting the probe volume is collected through window at the sensor surface and imaged into a collecting fiber with a second DOE. A detector pigtailed to the collecting fiber converts the optical burst into a signal.

Figure 17.15 shows a plot of a burst obtained with the Diverging Fringe Doppler sensor (white trace) and the frequency spectrum of the burst obtained with an FFT (yellow filled trace). The signal conditioning and processing required for the shear stress sensor is identical to those used for a laser Doppler velocimeter. The burst was previously high-pass filtered to eliminate the low frequency pedestal and low-pass filtered to eliminate high frequency noise.

A photograph of the shear stress sensor is shown in Figure 17.16. The sensor chip, 4 mm × 4 mm, is mounted into the sensor element location shown on the front face of the assembly. The diverging fringe pattern is illuminated with the help of a fogger.

OPTICAL MEMS-BASED SENSOR DEVELOPMENT WITH APPLICATIONS TO MICROFLUIDICS 363

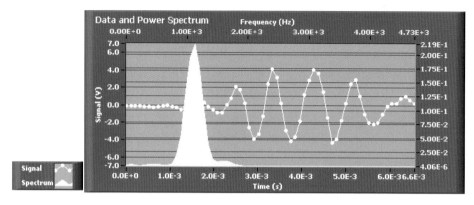

FIGURE 17.15. Plot of a filtered burst obtained with the Diverging Fringe Doppler sensor (white trace) and the frequency spectrum of the burst obtained with an FFT (yellow filled trace).

Again, the sensor was tested in the boundary layer of a flat plate and the results compared with oil film interferometry. The sensor shows excellent agreement with the other technique up to Reynolds numbers of one million.

17.6. APPLICATION TO MICROFLUIDICS: VELOCITY MEASUREMENTS IN MICROCHANNELS

17.6.1. Velocity Measurements in Polymer Microchannels

17.6.1.1. Measurements Using the MiniLDV The velocity profile obtained with the MiniLDV™ is shown in Figure 17.18. This microchannel was 200 μm × 100 μm.

The MiniLDV™ was used to measure the flow velocity in the center of the microchannel for different fluid delivery speeds. The results, shown in Figure 17.19, show that the

FIGURE 17.16. Photograph of the shear stress sensor.

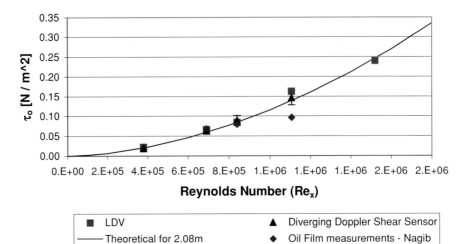

FIGURE 17.17. Shear stress measurements obtained with the shear stress sensor compared to that obtained with oil film interferometry.

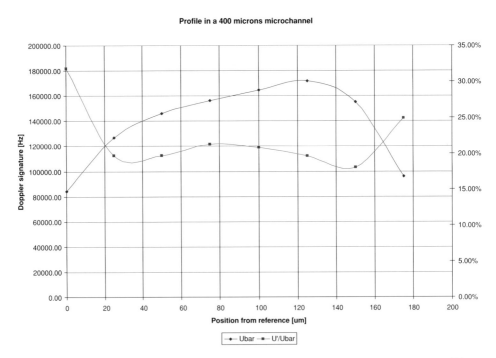

FIGURE 17.18. Velocity profile obtained in a 400 micrometer wide microchannel with the MiniLDV™.

OPTICAL MEMS-BASED SENSOR DEVELOPMENT WITH APPLICATIONS TO MICROFLUIDICS 365

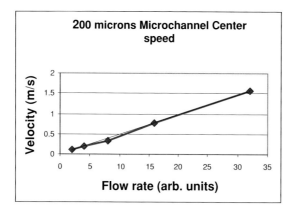

FIGURE 17.19. Linearity of the average velocity measurements in the center of the microchannel.

MiniLDV^{TM} can accurately measure the flow velocity even for very small velocities and that the measurements reflect the different flow speeds accurately.

17.6.1.2. Measurements using the Micro Velocimeter An optical MEMS based time-of-flight velocimeter, MicroV^{TM} (photograph shown in Figure 17.6, was used to measure the velocity profile inside the microchannel describe above. The following plot in Figure 17.20 shows a velocity profile obtained using the MicroV in a 400 μm × 100 μm channel.

For this data acquisition sequence, calculation of the flow velocity using volumetric quantities yielded 0.41 m/s and this value compares well with the velocity measurement using the MicroV^{TM}.

17.6.2. Test on a Caliper Life Science Microfluidic Chip using the MicroV

This test was performed by VioSense Corporation to study the feasibility of using Optical MEMS sensors for cell size measurement and sorting in microchannels. Caliper Technologies Corporation is exploring the possibility to use optical diagnostics to differentiate

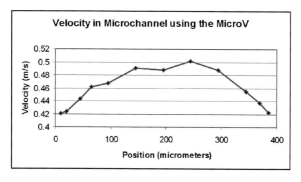

FIGURE 17.20. Velocity profile obtained with the MicroV.

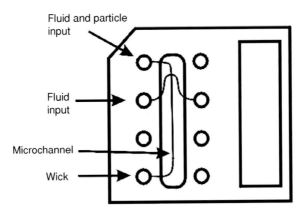

FIGURE 17.21. Schematic of the Caliper microfluidic chip.

between cells 10 μm and 20 μm in size, and also between single and paired cells 10 um in size.

Tests conducted at Caliper showed that cells 10 um in size in a buffer solution could be detected with VioSense's MicroV^{TM}, albeit with a low signal to noise ratio. The low signal to noise ratio was attributed in part to a large DC offset caused by light reflection at the surface of the chip. Additional measurements were needed to determine the origin of the light reflections and also the optical distortion sustained by the probe volume within the microchannel.

17.6.2.1. Experimental Setup A schematic of the microchannel, supplied by Caliper, is provided in Figure 17.21. The microchannel is 75 um wide and 25 um deep with a gutter type profile at the bottom. The chip material is glass. Distilled water containing particles 6.3 um in size was dropped in the upper left well while distilled water was dropped in the fluid input ports. Lens paper was used to wick the fluid through the channel.

A schematic of the experimental setup is shown in Figure 17.22. A microscope objective and a camera were placed under the chip and an OEM version of the MicroV^{TM} was placed above the chip. The camera was used to position the MicroV probe volume into the microchannel. A broadband light source was used to illuminate the chip. The light refracted by the channel walls helped identify the location of the microchannel.

Figure 17.23 shows a photograph of the microchannel with the MicroV^{TM} probe volume position. The light source was positioned such that the microchannel walls are seen as the two bright lines across the photograph. The probe volume is visible as the two slightly curved vertical lines but is still clearly defined. The streak across the probe volume is caused by a particle traveling through the probe volume at the time the frame was acquired. Some distortion and optical reflections of the probe volume by the microchannel walls are visible in the photograph.

17.6.2.2. Results Time traces of the light intensity were acquired using a digital scope and one of the traces is displayed in Figure 17.24. The scattered light from the surfaces generates a DC contribution of 7.5 volts on the detector (out of 10 volts, the saturation

OPTICAL MEMS-BASED SENSOR DEVELOPMENT WITH APPLICATIONS TO MICROFLUIDICS   367

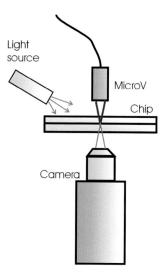

FIGURE 17.22. Schematic of the experimental setup.

limit of the detector). As a particle travels through the probe volume, the DC signal dips, therefore indicating that a lower light intensity reaches the detector. This trace behavior is explained by the fact that the particles obscure the scattered light reflected off the surfaces, in particular the back surface. As a result, the light scattered off the surface is temporarily

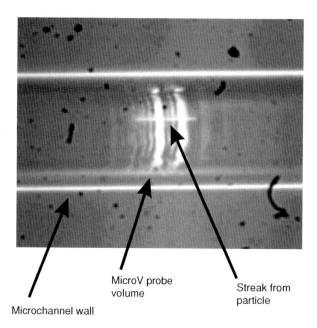

FIGURE 17.23. Streak from a particle intersecting the MicroV probe volume positioned in the microchannel.

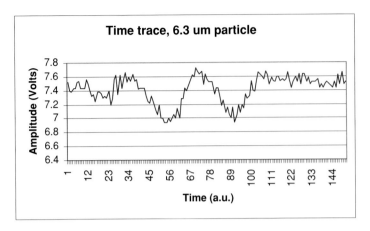

FIGURE 17.24. Time trace of the scattered light intensity as a particle passes through the probe volume.

blocked thus causing a drop in the scattered light intensity level. Another explanation can be a destructive interference between the scattered light off the surface and the scattered light off the particle. The particle velocity can be measured simply by measuring the time elapsed between the signal intensity drops.

The large contribution of the scattered light from the glass surfaces of the microchannel limits the ability of the sensor to measure small particles or particles whose index of refraction is close to that of the surrounding fluid. It is possible to design an optical sensor configuration in which the working distance of the sensor is tailored to the microchannel depth in the microfluidic chip. A shorter working distance will place the top surface far from the probe volume, thus reducing the scattered light contribution. A schematic of such sensor is shown in Figure 17.25.

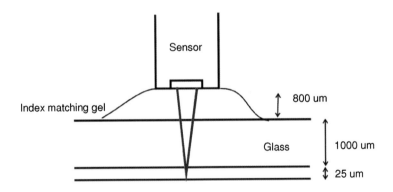

Optical path: 800 microns in index matching gel (n ~ 1.5) and 1000 microns in glass
Probe volume size: spot: 75 microns long, 20 microns apart

FIGURE 17.25. Schematic of a specially design sensor for microfluidic application.

The sensor would be designed with a working distance of 1.8 mm in glass or in a material with equivalent index of refraction. The 0.8 mm layer of index matching gel is used to minimize the reflections off the upper glass surface while still allowing vertical adjustment for probe positioning. Coupling gel reduces unwanted scattering off the glass surfaces by up to 75%.

17.7. CONCLUSIONS

Recent advances in optical MEMS technology have enabled the development of diffractive optical elements capable of generating complex optical probe volume configurations. This optical MEMS technology is the enabling factor in the development and fabrication of optical microsensors. These microsensors provide the ability to precisely control the shape and location of the optical probe volume and provide reliable optical alignment, requirements that are necessary for optical diagnostics in microfluidic environments. Preliminary tests have yielded encouraging results towards the possibility of using these optical microsensors for fluid flow characterization for microfluidics applications.

BIBLIOGRAPHY

[1] E. Arik. Current status of particle image velocimetry and laser doppler anemometry instrumentation. In R.J. Donnelly and K.R. Sreenivasan (eds.), *Flow at Ultra-High Reynolds and Rayleigh Numbers.* Springer-Verlag, NY, 1998.

[2] W. Bachalo. Method for measuring the size and velocity of spheres by Dual-beam light-scatter interferometry. *Appl. Optics,* 3:363, 1980.

[3] D. Dopheide, M. Faber, Y. Bing, and G. Taux. Semiconductor long-range anemometer using a 5 mW diode laser and a pin photodiode. In R.J Adrian (ed.), *Applications of Laser Techniques to Fluid Mechanics,* Springer-Verlag, pp. 385–399, 1990.

[4] F. Durst, A. Melling, and J.H. Whitelaw. *Principles and Practice of Laser Doppler Anemometry.* Academic Press 1976.

[5] D. Fourguette, D. Modarress, D. Wilson, M. Koochesfahani, and M. Gharib. An Optical MEMS-based Shear Stress Sensor for High Reynolds Number Applications. *AIAA Paper 2003–0742, 41st AIAA Aerospace Sciences Meeting & Exhibit,* Reno, NV, 2003.

[6] D. Fourguette, D. Modarress, F. Taugwalder, D. Wilson, M. Koochesfahani, and M. Gharib. Miniature and MOEMS Flow Sensors. *Paper AIAA-2001-2982, 31st Fluid Dynamics Conference & Exhibit Anaheim,* CA, 2001.

[7] D. Fourguette and C. Suarez. Integrated Optical Diagnostics for Miniature Devices. *AIAA Paper 99-0514, 37th AIAA Aerospace Sciences Meeting & Exhibit,* Reno, NV, 1999.

[8] M. Gharib, D. Modarress, D. Fourguette, and D. Wilson. Optical Microsensors for Fluid Flow Diagnostics. *AIAA Paper 2002–0252, 40th AIAA Aerospace Sciences Meeting & Exhibit,* Reno, NV, 2002.

[9] R.W. Gerchberg and W.O. Saxton. A practical algorithm for the determination of phase from images and diffraction plane pictures, *Optik,* 35:237–246, 1972.

[10] M. Johansson and J. Bengtsson. Robust design method for highly efficient beam-shaping diffractive optical elements using an iterative-fourier-transform algorithm with soft operations. *J. Mod. Opt.,* 47:1385–1398, 2000.

[11] B. Lehmann, C. Hassa, and J. Helbig. Three-component laser-doppler measurements of the confined model flow behind a swirl nozzle. *Developments if Laser Techniques and Fluid Mechanics, Selected Papers from the 8th International Symposium,* Lisbon, Portugal, Springer Press, pp. 383–398, 1996.

[12] P.D. Maker and R.E. Muller. Phase holograms in polymethylmethacrylate. *J. Vac. Sci. Tech. B,* 10:2516–2519, 1992.

[13] P.D Maker, D.W.Wilson, and R.E. Muller. Fabrication and Performance of Optical Interconnect Analog Phase Holograms made by E-beam Lithography. In R.T. Chen and P.S. Guilfoyle (eds.), *Optoelectronic Interconnects and Packaging, Proc. SPIE CR62*, pp. 415–430, 1996.

[14] D. Modarress, D. Fourguette, F. Taugwalder, M. Gharib, S. Forouhar, D.Wilson, and J. Scalf. Design and Development of Miniature and Micro Doppler Sensors. *10th International Symposium on Application of Laser Techniques to Fluid Mechanics*, Lisbon, Portugal, 2000.

[15] D. Modarress and D.A. Johnson. Investigation of Turbulent Boundary Layer Separation Using Laser Velocimetry. *AIAA J.*, 17(7):1979.

[16] D. Modarress and D. Tan. Application of LDA to Two-phase Flows. *Exp. Fluids*, 1:1983.

[17] A.A. Naqwi and W.C. Reynolds. Dual Cylindrical Wave Laser Doppler Method for Measurement of Skin Frictionin Fluid Flow, Report No. TF-28, Stanford University, 1987.

[18] J. Turunen and F.Wyrowski (Eds.). *Diffractive Optics for Industrial and Commercial Applications*, John Wiley & Sons, 1998.

[19] D.W. Wilson, P.D. Maker, and R.E. Muller. Binary Optic Reflection Grating for an Imaging Spectrometer. *Diffractive and Holographic Optics Technology III, SPIE Proceedings*, Vol. 2689, Jan. 1996.

[20] D.W.Wilson, J.A. Scalf, S. Forouhar, R.E. Muller, F. Taugwalder, M. Gharib, D. Fourguette, and D. Modarress. Diffractive optic fluid shear stress sensor. *Diffractive Optics and Micro Optics*, OSA Technical Digest Optical Society of America, Washington DC, pp. 306–308, 2000.

[21] D.W. Wilson, P.K. Gogna, R.J. Chacon, R.E. Muller, D. Fourguette, D. Modarress, F. Taugwalder, P. Svitek, and M. Gharib. Diffractive Optics for Particle Velocimetry and Sizing. *Diffractive Optics and Micro Optics, OSA Technical Digest*, Optical Society of America, Washington DC, pp. 11–13, 2002.

[22] D.W. Wilson, P.D. Maker, R.E. Muller, P. Mouroulis, and J. Backlund. Recent Advances in Blazed Grating Fabrication by Electron-beam Lithography. Paper 5173–16, *Proc. SPIE*, 2003.

[23] Y. Yeh and H.Z. Cummins. Localized Fluid Flow Measurements with a He-Ne Laser Spectrometer. *Appl. Phys. Lett.*, 4:176–178, 1964.

18

Vascular Cell Responses to Fluid Shear Stress

Jennifer A. McCann, Thomas J. Webster, and Karen M. Haberstroh
Weldon School of Biomedical Engineering, Purdue University, West Lafayette, IN, 47907

ABSTRACT

The development of several vascular diseases is linked to both blood flow properties and cellular behavior in the arterial and venous systems. For instance, atherosclerosis development is dependent on the blood flow profile, shear stress rate, and resulting cellular responses in the arteries. Specifically, in regions of disturbed flow behavior, cells demonstrate both altered morphology and phenotype. Based on this clinical knowledge, *in vitro* fluid flow studies have been performed on vascular endothelial and smooth muscle cells to understand the process of disease initiation and development. Ultimately, results of such studies will provide knowledge regarding key pathways involved in disease progression. Moreover, this information will be critical when designing effective drug therapies in the clinical setting.

18.1. INTRODUCTION

Cardiovascular diseases (coronary heart disease, stroke, hypertension, valvular heart disease, and diseases of the vessels) affect approximately 62,000,000 Americans annually (American Heart Association). One of the most predominant forms of vascular disease and the leading cause of death in the developing world [47] is atherosclerosis, or hardening of the arteries, which results in a narrowed vessel lumen due to an accumulation of lipids, fibrous elements, and smooth muscle cells [45, 68, 91]. In this pathological condition, the cells and mechanical properties of the affected vessel, as well as the blood flow properties through the affected region, are altered as a direct result of plaque formation.

Two hypotheses regarding the relationship between blood flow and atherosclerosis initiation have received much attention. The first states that higher than normal shear stresses

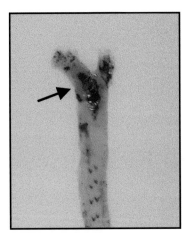

FIGURE 18.1. Oil-Red-O staining shows lipid-rich atherosclerotic lesions in the thoracic aorta arch of a Watanabe heritable hyperlipidemic rabbit. (Reprinted from Molecular Medicine Today, Vol 5, Topper and Gimbrone, Blood flow and vascular gene expression: fluid shear stress as modulator of endothelial phenotype, 40–46, Copyright 1999, with permission from Elsevier)

cause an injury to the endothelial cell layer and induce smooth muscle cell proliferation [69], while the second states that regions with lower than normal shear stresses and disturbed flows (e.g., regions with flow separation, non-laminar flow patterns, *etc.*) are more prone to plaque initiation [7, 15, 47].

Most research supports the second theory; it has been observed that flow streamlines separate at branches, curvatures, and bifurcations of large arteries, resulting in complicated flow patterns. For example, Figure 18.1 demonstrates the formation of lesions at a bifurcation in the thoracic aorta of a Watanabe heritable hyperlipidemic rabbit. In addition, the abdominal aorta develops atherosclerotic lesions that are located mainly along the posterior wall of the aortic bifurcation and extend proximally towards the renal arteries [7]. A flow study in an anatomically correct model revealed that flow patterns within the abdominal aorta distal to the renal arteries had regions of low shear stress and flow disturbances, with maximal velocities at the anterior wall and flow reversal occurring at the posterior walls [51]. Moreover, an *in vivo* study (performed with human autopsy subjects) showed a direct relationship between areas of low wall shear stress and atherosclerotic plaques in the abdominal aorta [61].

Corresponding with these defined flow patterns is the fact that endothelial cells within these specific vessel regions experience very different shear exposure than cells just microns away; these cells become proatherogenic, inducing a cascade of pathological events. The effects of atherosclerosis on endothelial and smooth muscle cell functions are discussed in further detail below.

18.1.1. Vessel Physiology

Arteries are thick, muscular, tubular structures, which serve primarily as nonthrombogenic conduits for oxygenated blood [68, 91]. The arterial wall can be divided into three circumferential layers: the intima, the media, and the adventitia. The intima comprises the innermost layer of arterial walls; it lies inside the internal elastic lamina and is directly exposed to blood flow [91]. It contains a lining of longitudinally oriented endothelial cells (which lay on the internal elastic lamina), connective tissue, and a few subendothelial smooth muscle cells [68, 91]. The media is the middle layer of arterial walls and is bounded by the internal and external elastic lamina. It is composed primarily of circumferentially oriented smooth muscle cells, which control vessel diameter and are embedded in an extracellular matrix of elastin and collagen [68, 91]. In larger arteries, elastic lamellae are also present. The amount of each component as well as the arrangement of this layer varies along the arterial tree, thus providing the necessary elasticity and mechanical properties required by the specific vessel. The adventitia is the outermost layer of arterial walls and lies outside the external elastic lamina [91]. It is made of fibrous connective tissue, with elastic fibers and fibroblasts as well. Most of its support and strength is provided by collagen. Furthermore, this layer is penetrated by small blood vessels (the vasa vasorum), which provide nutrients to the inner layers of the vessel wall [68, 91]. Finally, blood flow occurs in the lumen, or the hollow center of the blood vessel.

As was previously mentioned, endothelial cells, smooth muscle cells, fibroblasts, and macrophages are found within the various layers of the arterial wall. While each participates in homeostatic mechanisms, endothelial cells and smooth muscle cells are most actively involved in these processes.

Endothelial cells form the principal layer between blood elements and the arterial wall and actively participate in vessel homeostasis. They are thin, flat, and elongated cells, with prominent nuclei, many mitochondria, extensive endoplasmic reticulum, and Golgi apparatus [68, 91]. These cells rest on a basement membrane made primarily of collagen type IV [91]. Endothelial cells form a selective permeability barrier between the blood and underlying areas [68]. These cells produce nitric oxide (NO), prostacyclin (PGI_2), and other compounds which promote vasodilation and inhibit platelet aggregation, smooth muscle cell proliferation, and monocyte adhesion [32, 56]. Furthermore, these substances control the growth, differentiation, and function of smooth muscle cells and macrophages [56, 68].

Smooth muscle cells are primarily located in the media; they modulate vessel function by controlling vessel wall tone in response to mechanical and neural stimuli [64, 91]. These cells are spindle shaped with elongated nuclei and are characterized by a basal lamina, which surrounds individual cells. Smooth muscle cells can take on two forms: contractile or synthetic. The contractile phenotype contains many cytoplasmic filaments with less mitochondria and endoplasmic reticulum. The synthetic phenotype is characteristic of proliferating smooth muscle cells; proliferating cells have fewer contractile filaments, an increased amount of synthetic organelles, and an increased production of secretory proteins [64, 91]. Furthermore, in this phenotype, smooth muscle cells are much more responsive to mitogens. Current research supports the idea that multiple subpopulations of smooth muscle cells exist within the media, each with distinct developmental lineages and functional characteristics [27]. This is not surprising given the fact that smooth muscle cells must perform a wide range of functions, often at the same time. For instance, one group of cells may focus on maintaining vessel wall tone while another group is involved in damage repair. In this capacity, smooth muscle cells within different vessel areas may respond appropriately to local mechanical or chemical factors.

18.1.2. Vessel Pathology

In healthy vessels, cells function to provide a smooth, nonthrombogenic environment for blood flow. In contrast, endothelial dysfunction is characteristic of diseased vessels; this is an important factor in the development of atherosclerosis and its resulting disturbed flows. Specifically, injury to the endothelial cell layer (ranging from cell denudation to alterations in normal cell functions like expression of new surface factors, release of growth factors, *etc.* [91]) contributes to the initiation of lesion formation. For instance, when injured, endothelial cells express leukocyte adhesion molecules such as vascular cell adhesion molecule (VCAM)-1 and intercellular adhesion molecule (ICAM)-1 [24, 32, 43]; their involvement in recruiting monocytes and leukocytes suggest a potential role for these molecules in the development of atherosclerosis. While endothelial cells normally participate in lipid uptake, this process is enhanced in the diseased state and results in fatty streak (and ultimately plaque) formation. Finally, one of the most important endothelial cell responses to injury is decreased production of nitric oxide, which leads to impaired vasodilation and increased systemic blood pressure.

In the atherosclerotic plaque, smooth muscle cells become activated by growth factors and cytokines (e.g., platelet derived growth factor (PDGF)-β, interleukin (IL)-1) released by macrophages, platelets, and other cells. Subsequently, smooth muscle cells produce and secrete extracellular matrix components, proliferate at increased rates, migrate into the vessel lumen, and have altered gene expression [32]. The extracellular matrix components these cells release (e.g., collagen, elastin) make up the fibrous cap of the plaque. While this was first thought to be part of the natural progression of atherosclerosis, some now believe this to be a defense mechanism against the progression of the disease since the cap serves to cover the more dangerous lipid core [24]. It seems that plaques with more smooth muscle cells compared to the number of inflammatory cells rupture less often than those with a larger number of inflammatory cells within the lipid core.

While endothelial and smooth muscle cells are key components of disease development and progression, these cells do not act alone in the vessel, rather, atherosclerosis is often thought of as an inflammatory disease critically affected by the role of monocytes and other inflammatory cell types. For instance, injured endothelial cells express E and P selectins on their surface in the atherosclerotic state; expression of these adhesion molecules leads to a 'sticky' vessel wall and mediates rolling of inflammatory cells along the endothelial cell layer [13, 32, 68]. Once attached, these inflammatory cells are able to migrate between the endothelial cells and into the intima, where they differentiate into macrophages.

Monocyte-derived macrophages are present in large numbers in atherosclerotic lesions [32]; these cells are the major source of foam cells and the most predominant cell in the fatty streak [32, 68]. Furthermore, they secrete biologically active products which affect endothelial and smooth muscle cell function [32]. For instance, activated macrophages secrete IL-1, which acts on smooth muscle cells to increase their rate of proliferation.

As a result of this cellular cascade, atherosclerotic vessels become hardened and possess narrowed vessel lumens; each of these factors ultimately affects blood flow. For example, the vessel surface becomes rough as plaques protrude into the blood, which leads to thrombus formation and further blocks blood flow. In addition, a reduced luminal area results in increased blood velocity through the diseased vessel. Furthermore, atherosclerotic vessels lose their distensibility and are easily ruptured due to the hemodynamic forces associated with blood flow.

18.2. HEMODYNAMICS OF BLOOD FLOW

Before discussing studies aimed at understanding cellular responses to flow, a brief discussion of fluid flow through the circulatory system is required. The circulatory system is a closed loop flow system, with two circulations (pulmonary and systemic) arranged in series. The pulmonary circulation delivers deoxygenated blood from the right side of the heart to the lungs and then returns it to the left side of the heart, and the systemic arteries and arterioles deliver oxygenated blood from the left side of the heart to the body tissues under high pressures. Nutrient and gas exchange occurs at the tissue level in the capillary beds. Finally, the systemic veins transport deoxygenated blood back to the heart under low pressures.

The circulatory system itself is quite complex, thus precise mathematical representation of the system is difficult. The most obvious considerations are: the heart which is a non-simple pump, the vessels which are different sizes and possess diverse mechanical properties, and the blood which is a non-homogeneous fluid. Furthermore, the vessels of the circulatory system are not inert tubes, rather they are constantly remodeling to accommodate altered pressures and flow rates. Despite these problems, there are some simple relationships that are used to describe blood flow throughout the body. The primary considerations in the rate of blood flow through a vessel are the pressure between the two points of flow, the resistance which the vessel provides to flow, and the vessel area.

Based on these important parameters, three equations relating blood flow rate to vessel area, pressure, and resistance are:

$$Q = vA \tag{18.1}$$

$$Q = \frac{\Delta P}{R} \tag{18.2}$$

$$Q = \frac{\pi \Delta P r^4}{8\mu l} \tag{18.3}$$

where Q is the volumetric flow rate, v is the blood flow velocity, A is the cross-sectional area of the vessel, ΔP is the pressure gradient between the two points of interest, r is the vessel radius, μ is the viscosity of the blood, l is the distance between the two points, and R is the resistance between the two points [10]. Note that Equation (18.3) is Poiseuille's Law.

In addition to these relationships, the unitless Reynolds number (N_R) is a useful quantity for predicting flow conditions:

$$N_R = \frac{\rho D v}{\mu} \tag{18.4}$$

where ρ is the fluid density and D is the vessel diameter. Laminar flow is typical of healthy vessels and occurs when N_R is below 2000; such flow is characterized by streamlines parallel to the vessel axis. A value of N_R above 3000 is indicative of turbulent flow, with irregular fluid motion and mixing. This type of flow is much less efficient than laminar flow and is

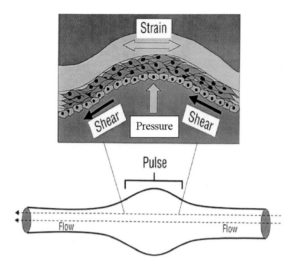

FIGURE 18.2. Schematic diagram of the forces resulting from blood flow which act on an artery wall. (Reprinted and adapted from New Surgery, Vol 1, Araim, Chen, and Sumpio, Hemodynamic forces: effects on atherosclerosis, 92–100, Copyright 2001, with permission from Landes Bioscience)

characteristic of diseased vessels. N_R values between 2000 and 3000 describe transitional flow.

As a result of blood flow, vessel walls are continuously subjected to a range of mechanical stimuli. The most predominant forces experienced by the arteries are shear stress, pressure, and cyclic strain or stretch due to the nature of blood flow [7, 40, 83]. These forces (shown schematically in Figure 18.2) affect the vessel physically, while also altering the function and signaling pathways of arterial cells. Therefore, the involvement of each of these mechanical forces has been investigated with regard to the maintenance of vessel homeostasis and the development of vascular diseases.

Perhaps the most widely studied force in the vascular system is shear stress, which results from the friction created as blood flows parallel to the vessel wall. Shear stress has been shown to modulate vessel homeostasis and cellular functions in many ways, including altered endothelial cell morphology and function; Figure 18.3 is a representation of the various endothelial cell responses to both laminar and atherogenic flows. The expression for wall shear stress, τ (measured in dynes/cm^2), can be obtained based on Poiseuille's Law:

$$\tau = \frac{4\mu Q}{\pi r^3} \tag{18.5}$$

As stated above, μ is the viscosity of the blood, Q is the volumetric flow rate, and r is the vessel radius [10]. Normal shear stress values in the venous system range from 1 to 6 dynes/cm^2, and in the arterial system from 10 to 70 dynes/cm^2 [47]. More importantly, abnormal levels of shear stress are directly correlated with the development of cardiovascular diseases. Though not as widely studied, vessel pressure and strain also contribute to cellular responses and overall vessel health. Hydrostatic pressure within a

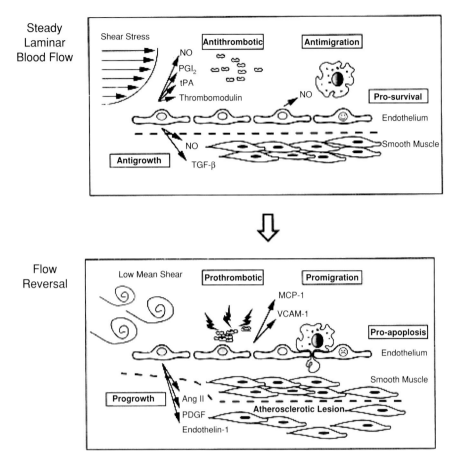

FIGURE 18.3. Endothelial cell responses to shear stress. Steady laminar shear stress promotes a healthy endothelial monolayer, while a low shear stress with flow reversal contributes to the development of atherosclerosis. (Reprinted from Arteriosclerosis, Thrombosis, and Vascular Biology, Vol 18, Traub and Berk, Laminar shear stress: mechanisms by which endothelial cells transduce an atheroprotective force, 677–685, Copyright 1998, with permission from Lippincott, Williams, and Wilkins)

vessel is proportional to the systemic arterial blood pressure. Therefore, any change in blood pressure (due to hypertension or vessel location in the body) results in altered hydrostatic pressure forces on the vessel wall and its constitutive cells. Such changes accompany atherosclerosis development, therefore understanding pressure effects on cells is also necessary if disease mechanisms are to be elucidated. Examples of sustained hydrostatic pressure effects on vascular cells include increased proliferation rates and integrin activation [70]. Cyclic strain arises from the rhythmic nature of blood flow as the vessel wall is stretched and deformed. It is believed that physiological levels of strain are beneficial to vascular endothelial and smooth muscle cells, while elevated levels are harmful. Responses of endothelial cells to cyclic strain include increased production of nitric oxide [8], prostacyclin (PGI$_2$), and endothelin (ET-1) [16] when compared to static controls.

18.3. TECHNIQUES FOR STUDYING THE EFFECTS OF SHEAR STRESS ON CELL CULTURES

Based on its relevance to cardiovascular disease development, many researchers have investigated the effects of shear stress on vascular cells. The two most common devices used for this purpose are the cone and plate viscometer and the parallel plate flow chamber.

18.3.1. Cone and Plate Viscometer

The cone and plate viscometer consists of a cone at a shallow angle, which rotates at a chosen angular velocity to generate fluid shear stress over a confluent layer of cells located on a culture dish (Figure 18.4; [14, 57] etc.). The shear stress generated in this system is given by the following equation:

$$\tau = \frac{\mu w}{\alpha} \quad (18.6)$$

where τ(dynes/cm^2) is the uniform shear stress generated over the cell monolayer, μ(dynes·s/cm^2) is the fluid viscosity, w(radian/s) is the angular velocity of the cone, and α(radians) is the cone angle [14].

This type of apparatus allows for application of both laminar and turbulent flow, within the same experiment, thus the placement of the cell specimen is critical [14]. This design also allows for a wide range of shear stresses while only requiring a small fluid volume. Furthermore, a large number of experimental observations can be performed since the plate can house several specimens [14]. Though this system provides the necessary uniform shear stress over the cell layer (at moderate to low speeds), limitations include a small cell-to-volume ratio, the fact that it does not allow for continuous fluid sampling, the requirement of continuous injections of fresh medium due to evaporation, the fact that it does not allow for real-time microscopic visualization, and the fact that it cannot be modified to accurately mimic the *in vivo* flow behavior [14, 25].

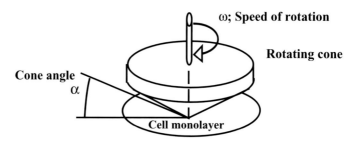

FIGURE 18.4. Schematic diagram of a cone and plate apparatus.

18.3.2. Parallel Plate Flow Chamber

The original parallel plate flow chamber consisted of a carefully milled polycarbonate plate, a rectangular Silastic gasket (which creates the channel depth), and a glass slide to which cells are attached [25]. A vacuum at the periphery of the slide held the system together, while also ensuring a uniform channel depth. The chamber was located between two fluid reservoirs, and the flow rate was controlled by a fluid pressure differential. Namely, a constant fluid flow was maintained by supplying fluid to the top reservoir at a higher rate than fluid flowed over the cells. Any excess fluid drained down the overflow manifold into the bottom reservoir. The polycarbonate plate had two ports: an entrance and exit for fluid flow, thereby allowing for continuous flow. The dimensions of the original chamber were 0.022 cm channel height and 2.5 cm channel width. In this chamber geometry, the shear stress over the cells can be easily calculated for a given flow rate, Q (cm³/s), using the following equation for plane Pouiselle flow:

$$\tau = \frac{6Q\mu}{bh^2} \tag{18.7}$$

where μ(dynes·s/cm²) is the fluid viscosity, h(cm) is the channel height, b(cm) is the slit width, and τ(dynes/cm²) is the wall shear stress [25]. This corresponded to a shear stress of roughly 10 dynes/cm² for a flow rate of 0.2 cm³/s.

The parallel plate flow chamber is the most widely used technique for studying shear stress effects on cells; therefore this system has undergone several modifications. For instance, one version consists of a machine milled polycarbonate plate with two ports for fluid flow, a glass slide to which the cells were attached, and a machined top (Figure 18.5; [48]). The polycarbonate plate possessed two milled recesses: one created the flow channel and the other served as a shelf where the glass cover slide rested. The height difference between the two recesses determined the channel height. Screws located on the chamber periphery held the entire chamber together. The specific dimensions of the flow chamber used in these studies were as follows: 3.429 cm width (b), 0.03302 cm height (h), and 7.4 cm length (l). As in the original parallel plate flow chamber, fluid was delivered to the top reservoir at a faster rate than the fluid flow over the cells, thereby providing continuous flow. Additionally, a valve located prior to the plate inlet was used to control the flow rate. In another parallel plate flow chamber design used by Nauman et al. [55], an adjustable flowmeter was introduced at the chamber outlet to more precisely control flow (as opposed to changing the height difference between the upper and lower reservoirs).

In all parallel plate studies, flow is assumed to be steady, laminar, and uniform over the length of the plate. Therefore, all cells in the chamber are thought to be exposed to the same flow conditions regardless of their location. However, when dealing with modified chambers and dimensions on the order of microns to hundreds of microns, standard machining tolerances (which occur in even the most carefully machined and assembled parallel plate flow chambers) can cause non-negligible variations in the chamber height and local shear stresses. Sensitive cell responses, like those involving gene expression, may not be consistent over the chamber area in the environment of these non-uniform flows. Therefore, flow characterization studies are often used to determine whether flow is uniform in such modified parallel plate flow chambers and, in turn, whether these changes in flow uniformity influence cell responses.

Note: Drawing not to scale

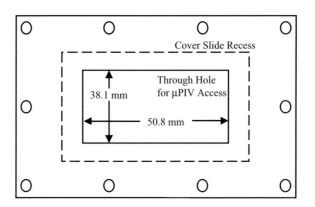

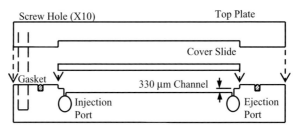

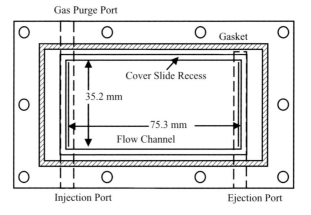

FIGURE 18.5. Schematic diagram of a parallel plate flow chamber. (Reprinted from Annals of Biomedical Engineering, Vol 3, McCann, Peterson, Plesniak, Webster, and Haberstroh, Non-uniform flow behavior in a parallel plate flow chamber alters endothelial cell responses, 327–335, Copyright 2005, with permission from Springer Science)

For example, such flow properties have been studied *in vitro* by Nauman et al. [55], Chung et al. [18], and [48]. Nauman et al. [55] used a variant of μ-PIV to measure the velocity field in a parallel plate flow chamber. Their technique was not a true PIV technique, however, in that the particles did not faithfully follow the local flow field. The mean flow streamlines

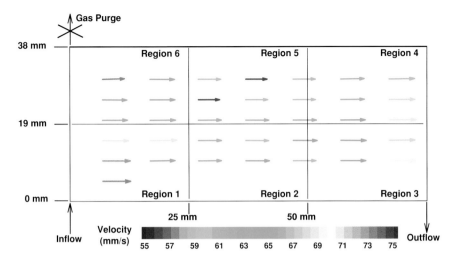

FIGURE 18.6. Velocity vector variations across the modified parallel plate flow chamber. The velocity was not uniform over the plate area, as determined using μ-PIV and *DaVis* software. The color scheme is such that the lowest velocities are represented in blue tones and the highest velocities are represented in red tones. Thus, the regions which experienced the overall lowest velocities were regions 5 and 2 (the two regions in the center of the slide), while region 1 (nearest the inlet) experienced the overall highest velocity. (Reprinted from Annals of Biomedical Engineering, Vol 3, McCann, Peterson, Plesniak, Webster, and Haberstroh, Non-uniform flow behavior in a parallel plate flow chamber alters endothelial cell responses, 327–335, Copyright 2005, with permission from Springer Science)

were determined with particles which had diameters of roughly 1/3 the channel height; these particles, while likely following the general flow path (assuming changes in flow velocity were very slow), can not accurately determine the local flow velocity due to their very large size [23, 63, 69b]. Therefore, McCann et al. [48] combined true micro-Particle Image Velocimetry (μ-PIV) flow measurements and gene expression studies to elucidate flow irregularities and corresponding cell response variations in their modified parallel plate flow chamber. Results from this study indicated that flow was laminar and parabolic, though not uniform over the chamber area (Figure 18.6) as could be assumed in perfectly smooth parallel plates. Specifically, channel height variations resulting from standard machining tolerances caused altered velocities across the parallel plate. The resulting shear stress distribution across the plate is presented in Table 18.1; these values demonstrated that the fluid shear stress in each of the six regions was not equal, as would be expected in parallel plate flow chamber models. The regional shear stress deviated from the average shear stress on the plate by as much as 11%. This change in shear stress was significant, as indicated by the non-uniformity of the endothelial cell gene expression across the parallel plate. In particular, results indicated non uniform up-regulation of COX-2 and ecNOS across the parallel plate chamber area [48]. These variations in gene expression corresponded with velocity values in the same region. For example, the regions with the overall lowest velocities (and shear stresses) also demonstrated the lowest level of gene up-regulation.

Regardless of machining accuracy, entrance, exit, and wall effects need to be considered when determining cell placement and responses [18]. Specifically, the center of the slide is

TABLE 18.1. Normalized shear stress values varied by parallel plate location. Shear stress values in each region were normalized by the measured average shear stress (7.6 dynes/cm^2) over the entire plate. The shear stress values deviated from the average shear stress by as much as 11% across the channel area, ranging from 6.76 dynes/cm^2 in region 5 to 8.43 dynes/cm^2 in region 1. The uncertainty associated with the shear stress data is approximately $+/-$ 0.01 dynes/cm^2. (Reprinted from Annals of Biomedical Engineering, Vol 3, McCann, Peterson, Plesniak, Webster, and Haberstroh, Non-uniform flow behavior in a parallel plate flow chamber alters endothelial cell responses, 327–335, Copyright 2005, with permission from Springer Science.)

Region	Normalized Shear Stress (dynes/cm^2)
1	1.11
2	0.92
3	1.05
4	1.05
5	0.89
6	1.07

typically the best place from which to obtain data; however, chamber inconsistencies are indiscriminant and can be a factor at any chamber location. Therefore, the most trusted cell responses are those averaged over the entire plate.

18.4. MODIFICATIONS TO TRADITIONAL FLOW CHAMBERS

Altered flow patterns occur in diseased vessels and can be simulated *in vitro* by modifying the traditional cone and plate viscometer and parallel plate flow chambers. For example, Freyberg et al. [26] studied laminar flow in a typical cone and plate viscometer and also inserted a triangular sheet of silicone into a parallel plate flow chamber to create a flow disturbance pattern. In addition, Holme et al. [33] placed a cosine shaped eccentric stenosis into a parallel plate perfusion chamber to mimic a coronary artery with advanced single eccentric stenosis. The protrusion of the stenosis into the flow channel determined the wall shear rate at the apex.

In addition to such geometrical changes, alterations to the flow profile can be created. For instance, Bao et al. [9] used a parallel plate flow chamber to study the effects of impulse, step, and ramp flow profiles (generated by a computer-controlled syringe pump) to study endothelial cell responses to steady shear stress as well as to temporal gradients, which are often present in areas of disturbed flow.

18.5. NONTRADITIONAL FLOW DEVICES

In addition to the more traditional cone and plate viscometer and parallel plate flow chambers, several other novel devices have been used to study shear stress effects on cells. In one example, plastic wells were mounted onto a horizontal rotator which moved the samples in a circular pattern (circumference = 3.5 cm); as the height of media on the cells was changed, cells located on the bottom of the wells were exposed to pulsatile turbulent flow [71]. Similar to this approach, Kraiss et al. [37] applied shear stress to cell cultures via an orbital shaker, though this system did not subject cells to a well-defined and uniform laminar flow pattern. Another type of apparatus involves the use of a tube as opposed to a plate, which more accurately mimics the natural geometry of an artery or vein. For example, cells have been seeded onto polystyrene tubes and then exposed to laminar flow; the shear stress experienced by the cells was changed by controlling the flow velocity through the system [77]. Similarly, Ziegler et al. [93, 94] designed a unique system, which places elastic silicone tubes into a perfusion loop. The advantage of this strategy is that cells may be simultaneously exposed to both shear stress and mechanical strain.

18.6. LAMINAR SHEAR STRESS EFFECTS ON ENDOTHELIAL CELLS

Shear stress is studied with respect to endothelial cells because of their anatomical location in the vasculature (i.e., they are directly exposed to blood flow). Furthermore, endothelial cells are transducers of hemodynamic forces: they "feel" these forces via cell-surface receptors and respond appropriately to modulate vessel homeostasis. For example, it has been well-documented that following exposure to laminar shear stress, endothelial cells elongate and align in a direction parallel to the flow [21, 82]. Associated with this cell elongation is cytoskeletal (e.g., F-actin) realignment [19] in the direction of flow.

Exposure to steady laminar shear stress also results in altered endothelial cell proliferation. Akimoto et al. [1] exposed confluent layers of bovine aortic endothelial cells and human umbilical vein endothelial cells to laminar shear stresses of 5 and 30 dynes/cm^2 and observed significantly decreased cell proliferation rates (as measured by DNA synthesis) after just 4 hours. However, at a shear stress of 1 dyne/cm^2, only the human cell line showed decreased cell proliferation [1]. These results suggest that each cell type has its own threshold value at which proliferation is inhibited.

Flow-exposed endothelial cells also produce many bioactive agents which participate in the maintenance of vessel homeostasis; the expression of such mediators is determined by the cell's gene expression. For example, laminar shear stress induces an atheroprotective phenotype in endothelial cells, leading to the expression of many vasodilating agents as well as mitogenic factors; three example molecules will be discussed in detail and Table 18.2 demonstrates a wide range of molecules whose expression are controlled at least in part by shear stress. First, a study by Noris et al. [57] demonstrated significantly increased levels of nitric oxide (NO) in human umbilical vein endothelial cells exposed to a laminar shear stress of 8 dynes/cm^2 for 6 hours. This is important because the production of nitric oxide by endothelial cells is mediated by endothelial cell derived nitric oxide synthase (ecNOS), whose expression is also shear stress dependent. ecNOS is important for its role in catalyzing the reaction of arginine to citrulline, with NO as a by-product. This synthase is activated by exposure to shear stress, acetylcholine, and thrombin [44]. In support of this pathway, ecNOS gene expression was also significantly up regulated after 6 hours of shear exposure; this finding has been confirmed by numerous investigators [29, 81]. Moreover, the addition of an ecNOS inhibitor (N^{ω}-nitro-L-arginine, L-NNA) to the culture abolished the shear induced nitric oxide production. Therefore, the response of ecNOS to various stimuli may well determine a cascade of atheroprotective events within the vessel environment. NO itself also possesses significant biological functions, acting as a vasodilator, inhibiting

TABLE 18.2. Laminar shear stress effects on endothelial cells.

Molecule	Function	Effect	Reference
ecNOS	Synthesis of NO	Sustained up regulation	[81]
COX-2	Synthesis of PGI_2	Sustained up regulation	[81]
TGF-β	Regulation of cell growth	Sustained up regulation	[58]
MnSOD	Antioxidant	Sustained up regulation	[81]
PDGF-β	Growth factor	Sustained up regulation	[65]
c-jun, c-fos	Growth response genes	Sustained up regulation	[35]
ICAM-1	Intracellular adhesion molecule	Transient up regulation	[54]
MCP-1	Chemokine	Transient up regulation	[74]
Egr-1	Transcrition factor	Transient up regulation	[36]
LOX-1	Endothelial receptor for Ox-LDL	Transient up regulation	[53]
VCAM-1	Vascular cell adhesion molecule	Down regulation	[5]
Endothelin-1	Vasoconstrictor	Down regulation	[46, 49]
ACE	Angiotensin converting enzyme	Down regulation	[67]
MCP-1	Chemokine	Down regulation	[49]

platelet and leukocyte adhesion to the endothelial cell layer, and inhibiting smooth muscle cell proliferation [44]. Each of these activities is directly related to atherosclerosis; for instance, without biologically active amounts of NO, smooth muscle cell proliferation would significantly increase, resulting in a situation where thrombus formation is likely.

Another atheroprotective molecule is prostacyclin (PGI_2); prostacyclin is a vasodilator, an inhibitor of platelet and leukocyte aggregation at the vessel wall, and an inhibitor of smooth muscle cell contraction, migration, as well as cholesterol accumulation [81]. Research has demonstrated that atherosclerotic vessels possess less PGI_2 than normal vessels, further suggesting its role as an atheroprotective molecule [88]. Moreover, laminar shear stress (over a range from 0.016 to 24 dynes/cm^2) resulted in increased PGI_2 production [25]. The gene responsible for the production of prostacyclin by endothelial cells is cyclooxygenase (COX)-2, which regulates the synthesis of prostanoids from arachidonic acid, with PGI_2 as the main product. This molecule is also responsive to laminar shear stress [81]; COX-2 mRNA expression was significantly up regulated after 1 and 6 hours of exposure to a shear stress of 10 dynes/cm^2.

Finally, the platelet derived growth factor family represents mitogenic factors which are released by endothelial cells; particularly, platelet derived growth factor (PDGF)-α and -β are involved in the progression of atherosclerosis. PDGF-α is a potent vasoconstrictor and smooth muscle cell mitogen. A report by Khachigian et al. [36] demonstrated significant induction of PDGF-α gene expression following exposure to a physiological level of steady laminar shear stress (10 dynes/cm^2); this response was correlated with the expression of and interaction with the immediate-early gene (erg-1). However, results of Bao et al. [9] contradict this finding; their study suggested that NO release by endothelial cells in response to steady laminar shear stress inhibited the expression of PDGF-α. Similar to PDGF-α, PDGF-β is a potent mitogen for vascular smooth muscle cells, inducing smooth muscle cell growth, proliferation, and migration. The effect of shear stress on endothelial cell expression of PDGF-β has been studied by Resnick et al. [65]. Their study exposed bovine aortic endothelial cells to laminar shear stresses of 10 dynes/cm^2 for four hours and demonstrated elevated levels of PDGF-β mRNA. Moreover, they reported that a shear stress responsive element located in the PDGF-β gene was responsible for such induction. These results were important clinically since PDGF-β expression by vascular endothelial cells may constitute a pathogenic risk factor.

Shear induced up regulation of each of these genes suggests that endothelial cells constantly adjust to flow patterns *in vivo* to maintain homeostasis. Furthermore, these studies demonstrate that the overall cellular response to shear stress is a combination of both proatherogenic and atheroprotective mediators.

18.7. ENDOTHELIAL CELL RESPONSE TO ALTERED FLOWS

Healthy blood vessels generally experience laminar fluid flow; however, flow alterations (including lower than physiological shear stresses, turbulent flows, oscillatory flow patterns, temporal and spatial gradients in flow, etc.) exist in diseased vessels and often compound further disease development.

For instance, shear stresses at lower than physiological values mimic *in vivo* stagnation points, where inflammatory cells may attach to and act on the endothelial layer. This is supported by the finding that exposure to steady shear stresses of 2 dynes/cm^2 resulted in increased endothelial cell expression of ET-1 (endothelin-1, a vasoconstricting agent and smooth muscle cell mitogen), VCAM-1, MCP-1 (monocyte chemoattractant protein-1, a potent chemotactic agent for monocytes), and the translation factor NF-κB [38, 50, 75]. In addition, turbulent flow profiles are found in diseased vessels; in contrast to laminar flow, such patterns have a range of shear stress frequencies and flow directions. Davies et al. [20] therefore characterized endothelial cell responses to both laminar and turbulent flow. Results indicated that following exposure to laminar flow ($\tau = 8$ dynes/cm^2) for 24 hours, cells became ellipsoidal shaped and aligned in the direction of flow. However, applying low shear stresses ($\tau = 1.5$ dynes/cm^2) in a turbulent flow profile for 16 hours resulted in randomly oriented cells; in addition, many of these cells were rounded yet still attached. After exposure to higher levels ($\tau = 14$ dynes/cm^2) of turbulent flow for 24 hours, cell retraction and cell loss became evident, as holes appeared in the monolayer. Furthermore, when the authors investigated cell cycle activity (via [^{3}H]thymidine-labeled nuclei) following 24 hours of exposure to each flow pattern, it became evident that cells maintained under either low or high turbulent shear stresses exhibited higher turnover rates compared to cells maintained under laminar flow profiles.

Besides the adverse effects of stagnation regions, oscillatory flow patterns, described by a low mean component and flow reversals, have also been implicated in regions of atherosclerosis initiation. Ziegler et al. [93, 94] therefore studied the effects of this type of flow behavior (mean $\tau = 0.3$ dynes/cm^2, ranging from -3 to $+3$ dynes/cm^2) on endothelial cells and demonstrated decreased ecNOS expression when compared to static cell cultures and cells exposed to unidirectional flow. Furthermore, exposure of endothelial cells to oscillatory flow patterns resulted in up regulation of ET-1 mRNA after only four hours; while ET-1 was also up regulated following exposure to low steady shear stresses ($\tau = 0.08 - 0.3$ dynes/cm^2) and pulsatile shear stress ($\tau = 6$ dynes/cm^2), its maximum expression occurred following oscillatory flow exposure.

In addition to these patterns, both temporal (an increase or decrease of shear stress over a certain time at a given location) and spatial (the difference in shear stress between two points of a vessel or individual cell at the same time) gradients have also been implicated in atherosclerosis development. To better understand this theory, White et al. [92] studied temporal gradients (attained with a 0.5 second flow impulse at 10 dynes/cm^2 followed by 20 minutes of no flow) in shear stress on endothelial cell responses and found increased endothelial cell proliferation. In contrast, this same group found that four hours of exposure to spatial gradients (attained with a smooth ramped flow model with flow beginning at 0 mL/s and reaching 3.5 mL/s within 15 seconds; the shear stress in this system was 10 dynes/cm^2) had the same effect as steady laminar shear stress. Furthermore, specific genes were more responsive to temporal and spatial shear stresses. For example, while steady shear stress induced expression of MCP-1 and PDGF-α in endothelial cells, ramp flow caused a slightly greater increase and step and impulse flow models resulted in significant increases in these same genes [9]. Specifially, for MCP-1, a 2 fold increase over ramp flow occurred due to step flow while a 7 fold increase occurred for impulse flow. Similarly, for PDGF-α, a 3 fold increase occurred with step flow and a 6 fold increase occurred in response to impulse flow when compared to the results obtained for ramp flow.

Finally, to mimic flow patterns found in stenotic vessels, an object is often introduced into the flow models discussed previously. For instance, Freybery et al. [26] introduced flow disturbances in a flow chamber with the addition of a triangular piece of silicon to create proatherogenic flow profiles; the mean shear stress created in this system was 5 dynes/cm^2. This study demonstrated increased endothelial cell apoptosis in response to such flow conditions. The authors used this finding to predict an atherosclerosis initiation mechanism. Specifically, their model suggests that endothelial cell injury and apoptosis is hemodynamically induced, causing increased endothelial cell proliferation and ultimately yielding a dysfunctional endothelial cell layer.

18.8. LAMINAR SHEAR STRESS EFFECTS ON VASCULAR SMOOTH MUSCLE CELLS

Shear stress over smooth muscle cells occurs *in vivo* via transmural flow [78, 79, 90]. A 2-D analysis (assuming that smooth muscle cells reside as cylindrical vessels in a uniform matrix) of the medial layer of the arterial wall suggested that, under normal transmural interstitial flow conditions, smooth muscle cells can be subjected to shear stresses on the order of 1 dyne/cm^2 [90]. This model was later extended by Tada and Tarbell [78] to include the more complex flow conditions that are present at the entrance of fenestral pores. Their study modeled the medial layer of the vessel wall as a heterogeneous medium with square arrays of cylindrical smooth muscle cells, which were assumed to be impermeable to flow. The interstitial fluid (assumed to be Newtonian) was modeled as a continuous phase with a uniform matrix, and the internal elastic lamina was taken to be impermeable except for the openings in the fenestral pores. Further assumptions used in their study included the following: steady and uniform flow perpendicular to the smooth muscle cell axis, fenestral pores distributed in a periodic array, no-slip at the smooth muscle cells and the internal elastic lamina, and a uniform velocity profile. Results of this 2-D study suggested that smooth muscle cells immediately below the internal elastic lamina experienced shear stresses 10–100 times higher than those located further away from the internal elastic lamina; the exact shear level depended on the parameters used to describe the size of the fenestral pores [78].

These findings imply that smooth muscle cells nearest the internal elastic lamina may be the most active due to the elevated shear stresses that they experience.

To more fully characterize the flow on smooth muscle cells located directly under the internal elastic lamina, their 2-D model was further developed into a 3-D model. Results from the 3-D model suggested that smooth muscle cells adjacent to the internal elastic lamina may experience shear stresses 3–14 times higher than smooth muscle cells located further away, a value much lower than predicted by the previous 2-D model [79]. However, the average shear stress experienced by smooth muscle cells remained at 1 dyne/cm^2, suggesting that spatial gradients exist throughout the medial wall.

In addition to transmural flow, the effect of shear stress on vascular smooth muscle cells is also important in pathological situations when endothelial cells are removed from the luminal surface. In this case, smooth muscle cells experience direct shear exposure. As with endothelial cells, smooth muscle cells have been shown to respond to shear stress with altered cell morphology, function, and gene expression.

For example, in one study Lee et al. [39] exposed canine vascular smooth muscle cells to laminar shear stress ($\tau = 20$ dynes/cm^2) for 48 hours. This group found that the cells and their F-actin cytoskeleton aligned in a direction perpendicular to the flow; this realignment occurred in a magnitude and shear dependent manner. In contrast, a study by Sterpetti et al. [76] found that bovine aortic smooth muscle cells aligned slightly in the flow direction at a laminar shear stress of 6 dynes/cm^2, while Papadaki et al. [60] observed no alignment of human abdominal aortic smooth muscle cells exposed to shear stresses up to 25 dynes/cm^2. These contrasting results suggest that cell alignment may be dependent on factors including cell origin (species and vessel wall layer), cell conditions at the time of shear exposure (state of confluency), shear level, and duration of flow exposure [39, 60, 76].

In addition to morphology changes, smooth muscle cell proliferation was also inhibited in response to laminar shear stress exposure [39, 76, 85]. For example, in a study by Ueba et al. [85], human vascular smooth muscle cell numbers were significantly decreased after exposure to shear stresses of 14 and 28 dynes/cm^2 for 24 hours. The higher shear level resulted in even further suppression of cell proliferation rates. This implicates shear stress as an inhibitor of smooth muscle cell proliferation at lesion sites, thus higher shear levels may play a role in preventing atherogenesis.

As with endothelial cells, shear stress also elicits a wide range of biochemical changes in vascular smooth muscle cells. Though it is likely that numerous molecules are affected by shear stress, only a few examples are provided. Specifically, an important atheroprotective mediator is PGI_2, which is an inhibitor of smooth muscle cell proliferation and is also shear responsive in smooth muscle cells. When rat aortic smooth muscle cells were exposed to laminar shear levels of 1 and 20 dynes/cm^2, enhanced production of PGI_2 resulted [3]. The level of 1 dyne/cm^2 was chosen to represent transmural flow; these findings therefore suggest that normal transmural flow may play a role in vessel remodeling.

In addition to the production of vasodilators such as PGI_2, the release of mitogenic factors by smooth muscle cells is also controlled by fluid shear stress. For example, platelet derived growth factor (PDGF) and basic fibroblast growth factor (bFGF) are potent mitogens for smooth muscle cells [66, 77] and are likely involved in the proliferative response of smooth muscle cells to vascular injuries. Sterpetti et al. [77] applied laminar shear stresses of 3, 6, and 9 dynes/cm^2 to bovine thoracic smooth muscle cells and found that with increasing shear levels, the release of both PDGF and bFGF into the circulating medium was significantly increased. Rhoads et al. [66] exposed human aortic smooth muscle cells to laminar shear stresses of 1, 5, and 25 dynes/cm^2 and found significant increases in fibroblast growth factor (FGF)-2 release at each shear level. However, no specific relationship was found between shear level and FGF-2 release, suggesting that its expression was flow rather than shear stress mediated. Interestingly, the FGF-2 release was not sustained over the course of the experiment and returned to near control levels after 24 hours of flow exposure. This finding was attributed to the 4 hour half-life of FGF-2, where the decline was due to FGF-2 absorption onto the surfaces within the flow loop. In combination, these studies provided evidence that smooth muscle cells are capable of secreting growth factors to sustain their own growth. In pathological conditions, the release of such mitogens could also affect the proliferation rates of cells in areas of plaque formation.

18.9. MECHANOTRANSDUCTION

18.9.1. Shear Stress Receptors

The precise mechanisms by which cells recognize and respond to shear stress (and other mechanical forces) are unclear. Potential mechanoreceptors include but are not limited to ion channels, G-proteins, and integrins.

For example, ion channels are responsible for transferring specific ions across the cell membrane; these channels play a critical role in maintaining intracellular solute concentrations. Several channels have shown shear stress responsiveness, including the K$^+$ channel. Endothelial cell exposure to physiological shear stresses resulted in activated K$^+$ channels after only seconds of exposure [2]; the magnitude of activation was dependent on the shear magnitude. Once the channel was activated, K$^+$ influx was linked to hyperpolarization, which participated in increased Ca$^+$ and altered signaling mechanisms [59]. Moreover, blockage of K$^+$ channels has been shown to inhibit shear-stimulated gene expression (e.g., TGF-β1 when $\tau = 20$ dynes/cm^2 [58] and ecNOS when $\tau = 15$ dynes/cm^2 [86]).

A second class of mechanoreceptors is the membrane-bound G protein family, which includes heterotrimeric G proteins, Ras and its homolog, and translation elongation factors. Heterotrimeric G proteins (composed of α, β, and γ subunits) are located on the phospholipid bilayer membrane of cells and are activated by cell-surface receptor interactions. When G proteins are associated with an activated receptor, the α subunit separates from the $\beta\gamma$ unit (Figure 18.7) and each acts separately to affect signal transduction. These G-protein linked receptors are activated directly by fluid shear stress; Gudi et al. [30] demonstrated that this activation can occur just seconds after the onset of shear exposure. Moreover, this activation was followed by intracellular calcium release, which plays a role in many transduction pathways.

Finally, integrins are heterodimeric transmembrane receptors which connect the extracellular matrix and cell cytoskeleton; these molecules can also transmit signals across the cell membrane. Different combinations of α and β integrin subunits have unique binding characteristics and may therefore result in distinct intracellular effects. For example, Urbich et al. [87] demonstrated α_5 and β_1 mRNA and protein up regulation in endothelial cells following exposure to laminar shear stress. Such integrin activation can be linked to downstream effects; for instance, fluid shear stress experiments have shown that $\alpha_v\beta_3$ is

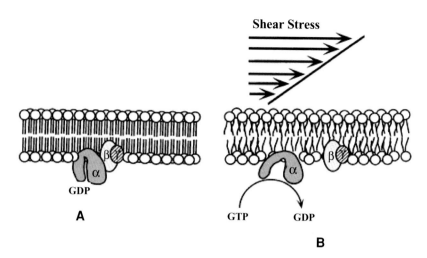

FIGURE 18.7. Schematic diagram depicting the mechanism by which G proteins can be directly activated by shear stress. (Reprinted from Proceedings of the National Academy of Sciences, Vol 95, Gudi, Nolan, and Frangos, Modulation of GTPase activity of G proteins by fluid shear stress and phospholipid composition, 2515–2519, Copyright 1998, with permission from the National Academy of Sciences, U.S.A.)

involved in the activation of IκB [11, 17], while β_2 activation is involved in ERK1/2 activation [81]. In addition to affecting protein kinases, integrin activation also plays a role in gene expression (e.g. $\alpha_5\beta_1$ integrin is linked to ecNOS expression in endothelial cells [22]) and other cell functions like adhesion and migration [6, 52].

Finally, evidence has suggested that there may be significant interaction among these membrane receptors. For instance, G_α activation has been linked to both Ca^{2+} and K^+ ion channel activation [31].

18.10. TRANSDUCTION PATHWAYS

Following activation of shear stress receptors, nonlinear, complex signaling cascades begin and ultimately result in the transduction of an external mechanical force into a chemical signal received by the nucleus. As such, signal transduction cascades are an area of significant scientific effort. For example, a wide range of molecules participating in these complex cascades have been identified, including G proteins (e.g., Rho and Ras), tyrosine kinases (Src, FAK, MAPKs), and many others. In general, these pathways involve a complex set of reactions which occur in the cell cytoplasm, followed by activation of a transcription factor which can move into the nucleus to affect cell functions. Two potential shear-induced signaling mechanisms are outlined below.

18.10.1. Ras-MAPK Pathways

p21^{ras}, a small G protein, is activated by laminar shear stress and triggers the MAPK-JNK/ERK pathways (Figure 18.8; [41]). Upstream of p21^{ras}, several potential mechanisms exist. For instance, integrins or G proteins can activate p60src [34] via a Shc-Grb2-Sos

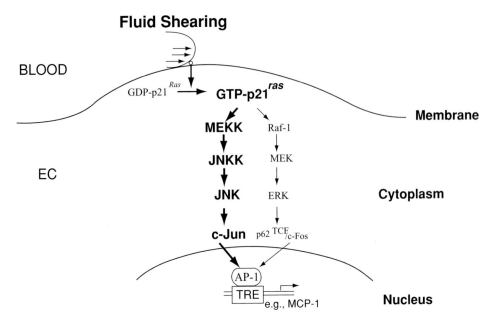

FIGURE 18.8. Schematic diagram of the Ras-MAPK signal transduction pathways in response to fluid shear stress. (Reprinted from Molecular and Cellular Biology, Vol 16, Li, Shyy, Li, Lee, Su, Karin, and Chien, The Ras-JNK pathway is involved in shear-induced gene expression, 5947-5954, Copyright 1996, with permission from the American Society for Microbiology)

pathway, which allows for $p21^{ras}$ activation. Another potential upstream mediator involves increased phosphorylation and activity of focal adhesion kinase (FAK). This activation leads to recruitment of Src and continuation of FAK phosphorylation, which allows FAK to associate with the complex growth factor receptor-binding protein 2 (Grb2) and guanine nucleotide exchange factor, son of sevenless (Sos) [17]. It is this complex which further activates Ras (Figure 18.9).

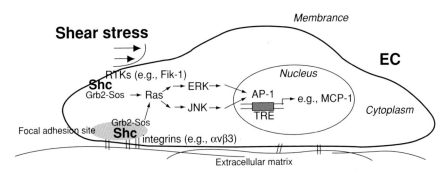

FIGURE 18.9. A proposed transduction pathway in endothelial cells in response to shear stress. (Reprinted from the Journal of Biological Chemistry, Vol 274, Chen, Li, Kim, Li, Yuan, Chien, and Shyy, Mechanotransduction in response to shear stress, 18393-18400, Copyright 1999, with permission from the American Society for Biochemistry and Molecular Biology)

Once activated, p21^{ras} triggers Raf-1 and MEKK (MAPK kinase kinase); in turn these stimulate ERK (extracellular signal regulated kinases) and JNK (c-Jun NH$_2$-terminal kinases), respectively (Figure 18.8). ERK and JNK are MAPKs; ERK mediates cell growth while JNK mediates programmed cell death. When shear stress alone is the extracellular stimulus, JNK is activated to a much greater extent than ERK. In the case of JNK activation, c-Jun is expressed and crosses into the nucleus where it interacts with the transcription factor AP-1. AP-1 then interacts with the 12-O-tetradecanoyl-phorbol-13-acetate-responsive element (TRE) which mediates gene expression [42]; for example, MCP-1 expression is mediated through TRE [72]. Activation of ERK1 and ERK2 result in up regulation of Elk-1, [28] which crosses into the nucleus to interact with SRE. In turn, SRE affects gene expression of such molecules as c-fos [85].

18.10.2. IKK-NF·κB Pathway

The IKK-NK-κB pathway is likely initiated by the $\alpha_v\beta_3$ integrin, which is stimulated by shear stress as was previously discussed. NF·κB is a transcription factor which mediates gene expression [12]; laminar shear stress up regulates its transcriptional activity [73]. NF·κB activation requires phosphorylation and then degradation of the IκB proteins, which inhibit NF·κB activity. This process is mediated by the IKKs (IkB kinases), which are in part mediated by $\alpha_v\beta_3$ integrin expression in response to flow (Bhullar, 1998). The IKKs phosphorylate IκB, which subsequently degrades, allowing NFκB to translocate into the nucleus where it affects gene expression. It has also been suggested that MEKK is capable of stimulating this pathway, demonstrating the intricate and complex network involved in mechanotransduction [42].

18.11. APPLICATIONS TO CLINICAL TREATMENT

Understanding these critical signaling mechanisms involved in disease progression may allow for the design of pharmacological treatments with increased efficacy. In addition, knowledge of the relationship between gene expression and cardiovascular disease development may lead to gene manipulation strategies to restore or prevent specific gene expression (Heiko, 2001). As an example, it may be possible to restore the effects of NO via ecNOS gene transfer. Whatever the strategy, ultimately, results obtained from the fluid shear stress studies outlined in this chapter will aid physicians in diagnosing and treating the clinical presentation of atherosclerosis.

18.12. SUMMARY

Vascular cell responses to hemodynamic forces and biochemical stimulation dictate vessel behavior. For this reason, endothelial cell responses to fluid flow have been widely studied using flow models (e.g., the parallel plate flow chamber) which allow scientists and engineers to investigate cell functions under well defined flows. In such models, cells sense the flow via surface receptors, which activate one to many intracellular signaling cascades, thereby affecting the cell nucleus. In turn, cell functions including cell growth, mRNA expression, and protein production are altered. These responses are directly related to the flow environment; laminar fluid flow yields healthy/physiological cell functions, while disturbed flow patterns favor cardiovascular disease development. Understanding the relationships between such intracellular signaling pathways and cell functions in response to these flow types (through studies such as those outlined in this chapter) will aid in understanding disease initiation and progression, as well as in the development of clinical treatment strategies.

REFERENCES

[1] S. Akimoto, M. Mitsumata, T. Sasaguri, and Y. Yoshida. *Circ. Res.*, 86:185, 2000.
[2] B.R. Alevriadou, S.G. Eskin, L.V. McIntire, and W.P. Schilling. *Ann. Biomed. Eng.*, 21:1, 1993.
[3] S.N. Alshihabi, Y.S. Chang, J.A. Frangos, and J.M. Tarbell. *Biochem. Biophys. Res. Commun.*, 224:808, 1996.
[4] American Heart Association. (2002) *Heart and Stroke Statistical Update*. Dallas, Texas, American Heart Association, 2001.
[5] J. Ando, H. Tsuboi, R. Korenga, Y. Takada, N. Toyama-Sorimachi, M. Miyasaka, and A. Kamiya. *Am. J. Physiol. Cell Physiol.*, 267:679, 1994.
[6] A.E. Aplin, S.M. Short, and R.L. Juliano. *J. Biol. Chem.*, 274:31223, 1999.
[7] O. Araim, A.H. Chen, and B. Sumpio. *New Surgery*, 1:92, 2001.
[8] M.A. Awolesi, W.C. Sessa, and B.E. Sumpio. *J. Clin. Invest.*, 96:1449, 1995.
[9] X. Bao, C. Lu, and J.A. Frangos. *Arterioscler. Thromb. Vascul. Biol.*, 19:996, 1999.
[10] R.M. Berne and M.N. Levy. *Physiology*, (4th Ed.), Chapters 21 and 25, Mosby Inc, St. Louis, 1998.
[11] I.S. Bhullar, Y.S. Li, H. Miao, E. Zandi, M. Kim, J.Y. Shyy, and S. Chien. *J. Biol. Chem.*, 273:30544, 1999.
[12] D.R. Blake, P.G. Winyard, and R. Marok. *Ann. NY Acad. Sci.*, 723:308, 1994.
[13] G.J. Blake and P.M. Ridker. *Circ. Res.*, 89:763, 2001.
[14] S.R. Bussolari, C.F. Dewey, and M.A. Gimbrone. *Rev. Sci. Instrum.*, 53:1851, 1982.
[15] C.G. Caro, J.M. Fitz-Gerald, and R.C. Schroter. *Proc. Roy. Soc. Lond. Ser. B: Biol. Sci.*, 177:109, 1971.
[16] J.A. Carosi, S.G. Eskin, and L.V. McIntire. *J. Cell Physiol.*, 151:29, 1992.
[17] K.D. Chen, Y.S. Li, M. Kim, S. Li, S.Yuan, S. Chien, and J.Y.J. Shyy. *J. Biol. Chem.*, 274:18393, 1999.
[18] B.J. Chung, A.M. Robertson, and D.G. Peters. *Comp. Struct.*, 81:535, 2003.
[19] A. Cucina, A.V. Sterpetti, G. Pupelis, A. Fragale, S. Lepid, A. Cavallaro, Q. Giustiniani, and L.S. D'Angelo. *Eur J. Vasc. Sur.*, 9:86, 1995.
[20] P.F. Davies, A. Remuzzi, E.J. Gordon, C.F. Dewey, and M.A. Gimbrone. *Proc. Natl. Acad. Sci.*, 83:2114, 1986.
[21] C.F. Dewey, S.R. Bussolari, M.A. Gimbrone, and P.F. Davies. *J. Biomech. Eng.*, 103:177, 1981.
[22] S. Dimmeler, B. Fisslthaler, I. Fleming, C. Hermann, R. Busse, and A.M. Zeiher. *Nature*, 399:601, 1999.
[23] R.P. Dring. *J. Fluids Eng.*, 104:15, 1982.
[24] A. Farzaneh-Far, J. Rudd, and P.L. Weissberg. *Brit. Med. Bull.*, 59:55, 2001.
[25] J.A. Frangos, L.V. McIntire, and S.G. Eskin. *Biotechnol. Bioeng.*, 32:1053, 1988.
[26] M.A. Freyberg, D. Kaiser, R. Graf, J. Buttenbender, and P. Friedl. *Biochem. Biophys. Res. Commun.*, 286:141, 2001.
[27] M.G. Frid, E.C. Dempsey, A.G. Durmowicz, and K.R. Stenmark. *Arterioscler. Thromb. Vascul. Biol.*, 17:1203, 1997.
[28] H. Gille, M. Kortenjann, O. Thomas, C. Moomaw, C. Slaughter, M.H. Cobb, and P.E. Shaw. *Euro. Mol. Biol. Organ. J.*, 14:951, 1995.
[29] T. Gloe, S. Riedmayr, H.Y. Sohn, and U. Pohl. *J. Biolog. Chem.*, 274:1599, 1999.
[30] S. Gudi, C.B. Clark, and J.A. Frangos. *Circ. Res.*, 79:834, 1996.
[31] S. Gudi, J.P. Nolan, and J.A. Frangos. *Proc. Natl. Acad. Sci.*, 95:2515, 1998.
[32] G.K. Hansson. *Arterioscler. Thromb. Vascul. Biol.*, 21:1876, 2001.
[33] P.A. Holme, U. Orvim, M. Hamers, N.O. Solum, F.R. Brosstad, R.M. Barstad, and K.S. Sakariassen. *Arterioscler. Thromb. Vascul. Biol.*, 17:646, 1997.

[34] S. Jalali, Y.S. Li, M. Sotoudeh, S.Yuan, S. Li, S. Chien, and J.Y.J. Shyy. *Arterioscler. Thromb. and Vascul. Biol.*, 18:227, 1998.
[35] L.M. Khachigian, N. Resnick, M.A. Gimbrone, and T. Collins. *J. Clin. Invest.*, 96:1169, 1995.
[36] L.M. Khachigian, K.R. Anderson, N.J. Halnon, M.A. Gimbrone Jr., N. Resnick, and T. Collins. *Arteriosler. Thromb. Vascul. Biol.*, 17:2280, 1997.
[37] L.W. Kraiss, A.S. Weyrich, N.M. Alto, D.A. Dixon, T.M. Ennis, V. Modur, T.M. McIntyre, S.M. Prescott, and G.A. Zimmerman. *Am. J. Physiol. Heart Circulat. Physiol.*, 278:1537, 2000.
[38] M.J. Kuchan and J.A. Frangos. *Am. J. Physiol. Heart Circulat. Physiol.*, 264:150, 1993.
[39] A.A. Lee, D.A. Graham, S.D. Cruz, A. Ratcliffe, and W.J. Karlon. *J. Biomech. Eng.*, 124:37, 2002.
[40] S. Lehoux and A. Tedgui. *Hypertension*, 32:338, 1998.
[41] S. Li, M. Kim, Y.L. Hu, S. Jalali, D.D. Schlaepfer, T. Hunter, S. Chien, and J.Y.J. Shyy. *J. Biolog. Chem.*, 272:30455, 1997.
[42] Y.S. Li, J.Y.J. Shyy, S. Li, J. Lee, B. Su, M. Karin, and S. Chien. *Mol. Cell. Biol.*, 16:5947, 1996.
[43] J.K. Liao. *Clin. Chem.*, 44:1799, 1998.
[44] D.M. Lloyd-Jones and K.D. Bloch. *Ann. Rev. Med.*, 47:365 1996.
[45] A.J. Lusis. *Nature*, 407:233, 2000.
[46] A.M. Malek, A.L. Greene, and S. Izumo. *Proc. Natl. Acad. Sci. USA*, 90:5999, 1993.
[47] A.M. Malek, S.L. Apler, and S.I. Izumo. *J. Am. Med. Assoc.*, 282:2035, 1999.
[48] J.A. McCann, S.D. Peterson, T.J.Webster, M.W. Plesniak, and K.M. Haberstroh. *Ann. Biomed. Eng.*, 33:327, 2005.
[49] S.M. McCormick, S.G. Eskin, L.V. McIntire, C.L. Teng, C.M. Lu, C.G. Russell, and K.K. Chittur. *Proc. Natl. Acad. Sci.*, 98:8995, 2001.
[50] S. Mohan, N. Mohan, and E.A. Sprague. *Am. J. Physiol. Cell Physiol.*, 273:572, 1997.
[51] J.E. Moore and D.N. Ku. *J. Biomech. Eng.*, 116:337, 1994.
[52] L. Moro, M. Venturino, C. Bozzo, L. Silengo, F. Altruda, L. Beguinot, G. Tarone, and P. Defilippi. *Euro. Mol. Biol. Organ. J.*, 17:6622, 1998.
[53] T. Murase, N. Kume, R. Korenga, J. Ando, T. Sawamura, T. Masaki, and T. Kita. *Circul. Res.*, 83:328, 1998.
[54] T. Nagel, N. Resnick, W.J. Atkinson, C.F. Dewey, and M.A. Gimbrone Jr. *J. Clini. Invest.*, 94:885, 1994.
[55] E.A. Nauman, K.J. Risic, T.M. Keaveny, and R.L. Satcher. *Ann. Biomed. Eng.*, 27:194, 1999.
[56] R.M. Nerem, P.R. Girard, G. Helmlinger, O. Thoumine, T.F. Wiesner, and T. Ziegler. Cell mechanics and cellular engineering. *Cell Mechanics and Cellular Engineering*, Chapter 5. Springer-Verlag, New York, 1994.
[57] M. Noris, M. Morigi, R. Donadelli, S. Aiello, M. Foppolo, M.Todeschini, G. Remuzzi, and A. Remuzzi. *Circul. Res.*, 76:536–543, 1995.
[58] M. Ohno, J.P. Cooke, V.J. Dzau, and G.H. Gibbons. *J. Clin. Invest.*, 95:1363, 1995.
[59] S.P. Olesen, D.E. Clapham, and P.F. Davies. *Nature*, 331:168, 1988.
[60] M. Papadaki, L.V. McIntire, and S.G. Eskin. *Biotechnol. Bioeng.*, 50:555, 1996.
[61] E.M. Pederson, O. S. Agerbaek, and I.B. Kristensen. *Euro. J. Vascul. Endovascul. Surg.*, 18:238, 1999.
[62] W.K. Purves, G.H. Orians, and H.C. Raig Heller. In Sinauer Associates and WH Freeman, *Life: The Science of Biology*, 4th Ed. New York, 1995.
[63] M. Raffel, C.Willert, and J.Kompenhans. *Particle Image Velocimetry: A Practical Guide*. Springer Press, Germany, 1998.
[64] G.E. Rainger and G.B. Nash. *Circul. Res.*, 88:615, 2001.
[65] N.T. Resnick, T. Collins, W. Atkinson, D.T. Bonthron, C.F. Dewey Jr., and M.A. Gimbrone. *Proc. Natl. Acad. Sci. USA*, 90:4591, 1993.
[66] D.N. Rhoads, S.G. Eskin, and L.V. McIntire. *Arteriosler. Thromb. Vascul. Biol.*, 20:416, 2000.
[67] M.J. Rieder, R. Carmona, J.E. Krieger, K.A. Pritchard, Jr, and A.S. Greene. *Circul. Res.*, 80:312, 1997.
[68] H. Robenek and N.J. Severs. *Cell Interactions in Atherosclerosis*. CRC Press, Boca Raton, 1992.
[69a] R. Ross. *Endothelial Dysfunction and Atherosclerosis*. Plenum Press, New York, 1992.
[69b] M. Samimy and S.K. Lele. *Physics of Fluids A*, 3:1915, 1991.
[70] E.A. Schwartz, R. Bizios, M.S. Medow, and M.E. Gerritsen. *Circul. Res.*, 84:315, 1999.
[71] K. Shigematsu, H. Yasuhara, H. Shigematsu, and T. Muto. *Internat. Angiol.*, 19:39, 2000.
[72] J.Y.J. Shyy, H.J. Hsieh, S. Usami, and S. Chien. *Proc. Natl. Acad. Sci.*, 91:4678, 1994.
[73] J.Y.J. Shyy, M. Lin, J. Han, Y. Lu, M. Petrime, and S. Chien. *Proc. Natl. Acad. Sci.*, 92:8069, 1995.
[74] J.Y.J. Shyy, Y.S. Li, M.C. Lin, W. Chen, S. Yuan, S. Usami, and S. Chien. *J. Biomech.*, 28:1451, 1995.

[75] E.A. Sprague and J. Luo. *Circulation*, 90:I-84, (Abstract) 1994.
[76] A.V. Sterpetti, A. Cucina, L. Santoro D'Angelo, B. Cardillo, and A. Cavallaro. *J. Cardiovascul. Surg.*, 33:619, 1992.
[77] A.V. Sterpetti, A. Cucina, A. Fragale, S. Lepidi, A. Cavallaro, and L. Santoro-D'Angelo. *Euro. J. Vascul. Surg.*, 8:138, 1994.
[78] S. Tada and J.M. Tarbell. *Am. J. Physiol. Heart Circulat. Physiol.*, 278:1589, 2000.
[79] S. Tada and J.M. Tarbell. *Am. J. Physiol. Heart Circulat. Physiol.*, 282:576, 2001.
[80] M. Takahashi and B.C. Berk. *J. Clini. Invest.*, 98:2623, 1996.
[81] J.N. Topper, J. Cai, D. Falb, and M.A. Gimbrone. *Proc. Natl. Acad. Sci.*, 93:10417, 1996.
[82] J.N. Topper and M.A. Gimbrone. *Mol. Med. Today*, 5:40, 1999.
[83] O. Traub and B.C. Berk. *Arterioscler. Thromb. Vascul. Biol.*, 18:677, 1998.
[84] R. Treisman. *Trends Biochem. Sci.*, 17:423, 1992.
[85] H. Ueba, M. Kawakami, and T. Yaginuma. *Arteriosler. Thromb. Vascul. Biol.*, 17:1512, 1997.
[86] M. Uematsu, Y. Ohara, J.P. Navas, K. Nishida, T.J. Murphy, R.W. Alexander, R.M. Nerem, and D.G. Harrison. *Am. J. Cell Physiol.*, 269:1371, 1995.
[87] C. Urbich, D.H. Walter, A.M. Zeiher, and S. Dimmeler. *Circul. Res.*, 87:683, 2000.
[88] J.R. Vane and R.M. Botting. *J. Lipid Med.*, 6:395, 1993.
[89] H.E. von der Leyen and V.J. Dzau. *Circulation*, 103:2760, 2001.
[90] D.M. Wang and J.M. Tarbell. *J. Biomech. Eng.*, 117:358, 1995.
[91] R.A. White. *Atherosclerosis and Arteriosclerosis*. CRC Press, Boca Raton, 1989.
[92] C.R. White, M. Haidekker, X. Bao, and J.A. Frangos. *Circulation*, 103:2508, 2001.
[93] T. Ziegler, P. Silacci, V.J. Harrison, and D. Hayoz. *Hypertension*, 32:351, 1998.
[94] T. Ziegler, K. Bouzourene, V.J. Harrison, and D. Hayoz. *Arterioscler. Thromb. Vascul. Biol.*, 18:686, 1998.

About the Editors

Professor Mauro Ferrari is a pioneer in the fields of bioMEMS and biomedical nanotechnology. As a leading academic, a dedicated entrepreneur, and a vision setter for the Nation's premier Federal programs in nanomedicine, he brings a three-fold vantage perspective to his roles as Editor-in-Chief for this work. Dr. Ferrari has authored or co-authored over 150 scientific publications, 6 books, and over 20 US and International patents. Dr. Ferrari is also Editor-in-Chief of Biomedical Microdevices and series editor of the new Springer series on Emerging Biomedical Technologies.

Several private sector companies originated from his laboratories at the Ohio State University and the University of California at Berkeley over the years. On a Federal assignment as Special Expert in Nanotechnology and Eminent Scholar, he has provided the scientific leadership for the development of the Alliance for Cancer Nanotechnology of the National Cancer Institute, the world-largest medical nanotechnology operation to date. Dr. Ferrari trained in mathematical physics in Italy, obtained his Master's and Ph.D. in Mechanical Engineering at Berkeley, attended medical school at The Ohio State University, and served in faculty positions in Materials Science and Engineering, and Civil and Environmental Engineering in Berkeley, where he was first tenured. At Ohio State he currently serves as Professor of Internal Medicine, Division of Hematology and Oncology, as Edgar Hendrickson Professor of Biomedical Engineering, and as Professor of Mechanical Engineering. He is Associate Director of the Dorothy M. Davis Heart and Lung Research Institute, and the University's Associate Vice President for Health Science, Technology and Commercialization.

Rashid Bashir completed his Ph.D. in 1992. From Oct 1992 to Oct 1998, he worked at National Semiconductor in the Process Technology Development Group as Sr. Engineering Manager. He is currently a Professor of Electrical and Computer Engineering and Courtesy Professor of Biomedical Engineering at Purdue University. He has authored or coauthored over 100 journal and conference papers, has over 25 patents, and has given over 30 invited talks. His research interests include biomedical microelectromechanical systems, applications of semiconductor fabrication to biomedical engineering, advanced semiconductor fabrication techniques, and nano-biotechnology. In 2000, he received the NSF Career Award for his work in Biosensors and BioMEMS. He also received the Joel and Spira Outstanding Teaching award from School of ECE at Purdue University, and the Technology Translation Award from the 2001 BioMEMS and Nanobiotechnology World Congress Meeting in

Columbus, OH. He was also selected by National Academy of Engineering to attend the Frontiers in Engineering Workshop in Fall 2003. https://engineering.purdue.edu/LIBNA

Professor Steve Wereley completed his masters and doctoral research at Northwestern University and joined the Purdue University faculty in August of 1999 after a two-year postdoctoral appointment at the University of California Santa Barbara in the Department of Mechanical and Environmental Engineering. At UCSB he focused exclusively on developing diagnostic techniques for microscale systems, work which ultimately led to developing, patenting, and licensing to TSI, Inc., the micro-Particle Image Velocimetry technique. His current research interests include designing and testing microfluidic MEMS devices, investigating biological flows at the cellular level, improving micro-scale laminar mixing, and developing new micro/nano flow diagnostic techniques. Professor Wereley has co-authored Fundamentals and Applications of Microfluidics, Artech House, 2002.

Index

Page numbers with *t* and *f* indicate table and figure respectively.

1-bromonaphthalene (1-BrNp), 333
2-[Methoxy(polyethylenoxy)propyl] trimethoxysilane (PEG-silane), 31
3,3-Dithiobis [sulfo-succinimidylpropionate] (DTSSP), 31
3T3 fibroblast cells, 174
5-carboxymethoxy-2-nitrobenzyl (CMNB)-caged fluorescein, 333

Absorption measurements of a pH sensitive dye, 12
AC electrokinetic phenomena, 244
  AC electroosmosis, 244
  dielectrophoresis, 244
  electrophoresis, 244
AC electroosmosis. *See* AC electrokinetic phenomena
AC MHD micropump, 141–142
  measurements in, 152,187
Acrylic. *See* PMMA, 98
Active mixers, 233
Active valves, 228
Adhesive bonding, 111
Adhesive stamping technique, 28
Airy function, 266
Amperometric detection, 57
Amplitude modulation, 76
Analytical instruments, 95
Anionic networks, 122
Antibodies, characteristics of, 59
Antibody binding sites, free streptavidin blocking of, 48
Antibody-antigen bindings, 28
Antibody-antigen interactions. *See* Protein-protein bindings, 30
Antigen antibody interaction, 10
Anti-lysozyme F-ab fragment, 7
Array of signaling systems characteristic of cell-based sensors, 87
Arthrobacter nicotiane microorganism, 10

Artificial membranes, use of, 11
Atherosclerosis initiation, 371
Atherosclerotic vessels,
  pathology, 374
  physiology, 372
Atomic force microscope, 196
Avidin oxidase, 7
AZ 4620, 95

Bacteria, 171
Big I/O, 28
Biochip capability, 89
Biochip technologies, application of, 18
BioChip, definition of, 190
Biochips
  microarray systems, 14
  integrated biochip systems, 16
Biocompatibility of materials & processing,
  biocompatibility tests, 113
  material response, 113
  tissue & cellular response, 113
Biocompatibility, 63–64
Biological flurescence. *See* Bioluminescence
Bioluminescence, 57, 195
Bioluminescent catabolic reporter bacterium, 57
Biomedical micro devices (BMMD), 95, 113
  advantages of, 110
  drawbacks of, 95
  fabrication of, 94
  size of, 96
Biomedical micro devices, materials for,
  polymers, 95–96
  silicon & glass, 93–94
Biomedical micro devices, packaging of,
  adhesive bonding, 110
  interconnects, 109
  thermal direct bonding, 109

BioMEMS for cellular detection
  electrical detection, 197
  mechanical detection, 196
  optical detection, 194–195
BioMEMS for cellular manipulation and separation
  dielectrophoresis, 191–196
  electrophoresis, 189
BioMEMS sensors, use of, 194
Biomimetic ligand, 10
Biomimetic receptors, construction methods of
  recombinant techniques, 11
Biomolecular reaction arrays
  deflection of DNA, 29–31
  deflection of PSA, 28–30
Biomolecular separators, 304
Biomolecules, immobilization of, 25
Bioreceptors, bases of
  cellular systems, 9
  enzymes, 9
  non-enzymatic proteins, 9
Bioreceptors, forms of
  antigen interactions, 3
  cellular interactions, 5
  enzymatic interactions, 5
  interactions using biomimetic materials, 5
  nucleic acid interactions, 5
Bioreceptors, schematic diagram of, 6f
Bioreceptors, types of
  antibody bioreceptors, 5
  biomimetic bioreceptors, 10
  cellular bioreceptors, 9
  enzyme bioreceptors, 8
  nucleic acid bioreceptors, 14
Bioreceptors, types of, 14
  antibody, 5–8
  biomimetic, 10
  cellular, 9–10
  enzyme, 8
  nucleic acid, 14
Biosensing process, principle of, 4f
Biosensing system
  chip assembly, 67–68
  environmental chamber, 68
Biosensor for protein and amino acid estimation, 13
Biosensor, definition of, 55
Biosensors, application of, 66,80,82
  biological monitoring, 5
  environmental sensing applications, 5
Biotin, 11
Biotinylated antibody, 7
Bisensors
  types of bioreceptors. See Bioreceptors, types of
  types of transducers. See Transducers, forms of
Blood flow, hemodynamics of, 375
  cyclic stain forces on, 377
  shear stress force on, 380

Bnzo[a]pyrene tetrol (BPT), 7–8
Bovine Serum Albumen (BSA), 45
BRCA1 gene. See cancer gene, screening of
Breast cancer gene BRCA1, 18
Bromine production, 13
Brownian diffusion effects in nPIV. See Nano-particle
  image velocimetry
Brownian motion of particles, 162, 248, 275, 338
Bulk micromachining, 94
Bulk silicon micromachining, 105

Calcium Green-1, 58
Calibration of the sensor, 14
Caliper life science microfluidic chip, 365
Cancer gene, screening of, 15
Cantilever bending, 21, 23
Cantilever deflection, 30
Cantilever response to DNA hybridization plotted as a
  function of target DNA concentration, 30f
Cantilever sensor, noise of, 24
Cantilever, fabrication of, 36
Cantilever, surface-stress change, measurement of,
  27
Cantilever's deflection, 23
Cantilevers curvature, use of, 26
Cantilevers, signal-to-noise ratio of, 27
Capacitance cytometry, 152
Cardiac pacemakers, 205
Cardiovascular diseases, 376
Cascaded sensing, 79
Catch-and-tell operation, 58
Causative agent, 10
CCD screen, 26
CCD spot, 26
Cell based biosensors designs & methods of
  biosensing system, 67
  cell culture, 68–69
  cell manipulation techniques, 64
  cell manipulation using dielectrophoresis
    (DEP), 66
  cell types and parameters for dielectrophoretic
    patterning, 67–68
  experimental measurement system, 67
  principles of dielectrophoresis (DEP), 89
  requirements for cell based sensors, 63
Cell based biosensors, measurements of
  influence of geometry and environmental factors on
    the noise spectrum, 74
  interpretation of bioelectric noise, 76
  long-term signal recording in vivo, 69
  selection of chemical agent, 76
  signal processing, 76
Cell based biosensors, types of
  cellular microorganism based biosensors, 57–58
  extra cellular potential based biosensors, 61–63
  fluorescence based cellular biosensors, 58–59

impedance based cellular biosensors, 59
intracellular potential based biosensors, 59–61
Cell based sensing, 56
Cell biosensor for environmental monitoring, 200
Cell biosensor, specific for, formaldehyde, 57
Cell culture
  neuron culture, 68
  primary osteoblast culture, 69–70
Cell monitoring-system, 198
Cell organelles, 10
Cell patterning methods
  biocompatible silane elastomers, use of, 64
  micro-contact printing (μCP), 64
  topographical method, 64
Cell positioning over the electrodes, technique for isolating and positioning, 69
Cell signal transducting, approaches or, 57
Cell-based biosensors, applications of, 87
Cell-gel sensors. See Microfluidic tectonics, sensors in
Cells, use of, in environmental applications, 57
Central difference interrogation (CDI), 271
Charged couple device (CCD) camera, 26–27
Charged couple device (CCD) detection, 26
Chemical agent sensing
  cascaded sensing of chemical agents using single osteoblast, 87
  comparison of detection limits and response times, 82
  EDTA sensing using single osteoblast, 85
  effect of verifying concentration of chemical agents, 85
  ethanol sensing using single osteoblast, 79
  hydrogen peroxide servicing using single osteoblast, 79–81
  pyrethroid servicing using single osteoblast, 79–80
Chemiluminescence, 194
CHF3 RIE, 36
Chip assembly, integration with silicone chamber, 67
Chip-base capillary electrophoresis devices, 223
Chip-based fabrication, advantages of, 45
Cholera toxins, binding of, 11
Claussius-Mossotti factor 65, 164–168,190,261
CMOS biochip coupled to multiplex capillary electrophoresis (CE) system, 17
CMOS technology, 17, 211
Cochlear prosthetics, 205
Codeine, 11
Coenzyme, 8
Colloid concentration, 46, 50
Colloids, binding capacity of, 46
Colloids, binding of, to antibodies, 46
Compartmentalization, 231–232

Complementary metal oxide silicon technology. See CMOS technology
Computer generated holograms (CGHS), 350–351
Cone and plate viscometer, 382
Conformal mapping, 176
Continuous window shifting (CWS), 269
Continuum model of EOF, 324
  experimental data of, 320–324
Coulomb body forces in fluid, 244
Coulombic interactions, 9
Coulter counter principle, 199
Coulter counter technique, 35
Coulter counter
  application of, 36
  sensitivity of, 37
Covalently binding antibodies, self assembled method for, 14
COX-2 mRNA expression, 384
Carcinogen benzo[a]pyrene, 7
Crossover frequency, 65
Cy5-labeled antibody, 17
Cyclooxygenase (COX) 2, 388
Cyclopentanone (CP), 97
Cytosensor microphysiometer, 198
Cytotoxicity, 114
Czarnik's compound, 58

DC MHD micropump, 144
de Silva's compound, 58
Debye-Huckel approximation, 306
Deconvolution method, 262–263
  drawbacks of, 263
  hypothesis of, 263
Deep reactive etching (DRIE), 95
Deionized (DI), 99
DEP. See Dielectrophoresis
Detection of E. coli, 18
Detect-to-treat. See Cell based sensors
Detect-to-warn. See Cell based sensors
Diagnostic wireless microsystems, 210–213
Dielctrophoretic force, 245
Dielectrophoretic device, performance quantification experiments of, 266–268
Dielectrophoresis (DEP), 64, 66, 79, 160, 162–165, 190–196. See also AC electrokinetic phenomena
  applications of, 192–193
  Claussius-Mossotti factor, 164–168
  electrothermal forces, effect of, 171–172
  interaction with gravitational force, 169–171
  interaction with hydrodynamic drag forces, 170–171
  multipolar effects, 167–169
  negative, 190
  positive, 190
  scaling, 168–169

Dielectrophoretic force, 260–262
  application of Maxwell's stress tensor, 167
  computation of, at various electric fields, 246–247
Dielectrophoretic microdevice, 259
Dielectrophoretic-field flow fractionation (DEP-FFF), 190
Diffractive optical elements, 349–352
Diffuse layer capacitance, 148
Diverging fringe doppler sensor, 360–363
DNA analysis, novel approach to, 18
DNA biosensors, 9
DNA chips, 16
DNA hybridization, 29–31
DNA microarray detection system, 15
DNA microarrays, case of, 21
DNA multiarray device, use of, 16
DNA pairing, 9
DNA sequencing, 153
DNA–fragments, 9
DNA-ligand interactions, 9
Donnan equilibrium, 123
Donnan exclusion, 123
Doppler pedestal, 354
Doppler shift, 353
DRIE, 111
Drilling molecular-scaled holes, 35
Drug delivery microship. *See* Therapeutic Microsystems
DS19 murine erythroleukemia cells, 174
Dual velocity sensor, 358–360
Dye-immobilized-gel, use of, 235–236

E. coli detection, 10, 17
EBL-defined pore, 38
Edmund Scientific mini electromagnet, 139, 141
EDTA. *See* Chemical agent sensing
Elastomeric micromolding, 225
Electrical double layer (EDL), 146, 148
Electrical double layer field (EDL), 279
Electrical double layer
  structure of, 306–308
Electrical impedance spectroscopy, 152
Electrochemical detection methods, 12
Electrochemical detection, 13
Electrochemical mmunoassay, 7
Electrofusion of cells, 174
Electrokinetic flow, governing equations for, 308–313
Electron-beam lithography (EBL), 36
Electroosmotic channel flow, 304–306
  asymptotic and numerical solutions for electric double layer thickness, 314–317
  equilibrium considerations, 318–320
  molecular dynamic simulations, 324–328
  nano-PIV results in, 342–346

Electroosmotic flow, case studies of
  in a rectangular channel, 290–293
  pressure driven flow over heterogeneous surfaces, 293–298
Electrophoresis, 302. *See also* AC electrokinetic phenomena
  application of, 189
  definition of, 189
Electroporation of cells, 174
Electro-readout technique, advantages of, 27
Electrothermal stirring, 249–251
  use of, 253
Electrothermally driven flow (ETF), 244, 252
ELISA (enzyme linked immunoassay), 239
Ellipsometry, 9
Endoradiosonde, 205
Endoscopic wireless camera-pill, 210, 212–213
Endothelial cells, 374, 385
  laminar shear stress effect on, 383, 384t, 393
  response to altered flows, 385–386
Engineered bacteria, 57
Engineering transmembrane protein, 35
Enhanced species mixing in electroosmotic flow, case study of
  flow simulation, 288–289
  mixing simulation, 290
Enzyme immunoassays, 7
Enzyme-linked immunosorbent assay (ELISA), 18
Epoxy group, 97
Ethanol sensing. *See* Chemical agent sensing
Ethylene diamene tetra acetic acid (EDTA), 77–79, 82
Evanescent wave fibre optic biosensor, 10
Evanescent wave illumination, 342
Evanescent wave immunosensor, 12
Evanescent waves, theory of, 334–337
Excitation light, 7
Extracellular based biosensors
  applications of, 61
Extracellular recording, 55
Extracellular signal spectrum, 55

Fast Fourier transformation (FFT), 73, 354–356, 362
Femlab, 249
Fetal Bovine Serum (FBS), 69–70
FFT analysis, 76
Field effect transistors (FET), 64, 127
Field-flow fractionation (FFF), 155
Flatworm checking valve, 229
Flow devices, non-traditional, 382–383
Fluidic interconnects, 109
  general requirements of, 110
  types of, 111–112
Fluorescence base cellular biosensors
  application of, 58
  intracellular ionic signals developed by, 58

INDEX
401

Fluorescence based sensors, applications of, 58
Fluorescence detection techniques, 194
Fluorescent analyte, 7
Fluorophor-labeled antigen, 7
Food and drug administration (FDA), 94
Fourier series, 176
Fourier transform, iterative, 350
Free antigen, 48
Free energy change, advantage of, 22
Free energy reduction, 22, 30
Free ligand, 49
Free-flow electrophoresis (FFE), 156
Frequency modulation, 76

Gamma-butyrolacetone (GBL), 97
Gaussian-shaped microchannels, 107
Gene chips, 223
General purpose interface bus (GPIB), 69
Genosensor, 9
Gerchberg-Saxton algorithm, 350
Gibbs free energy, 119
Glass transition point, 109
Glass, usage, reason for, 95
  polymers, types of, 95–96
  PDMS, 96
  PMMA, 96–97
  SU-8, 97
Glucode optodes, 8
Glucose microtransponder. See Therapeutic
  Microsystems
Glued interconnects, 111
Gold-thiol (Au-S) bonds, 25, 29
Golgi apparatus, 373
Gouy-Chapman model of EDL, 308
Gradient refractive index (GRIN), 15
Green fluorescent protein (GFP), 58
Green's function, 176

H19-7 cell line, 67
H19-7/IGF-IR cells, 69
Hard bake. See SU-8, process, steps in
Hard-baked, 101
Helmholtz layer, 146
Helmholtz plane capacitance, 148
Heterogeneous reactions, 251–253
Hetrotrimeric G proteins, 388
Hexahstidine modified extension, 7
H-filter^{TM} design, 154
High performance liquid chromatography (HPLC), 111
High-density oligonucleotide arrays, 16
Histidine-tagged antibodies, 7
Horizontal tubing interconnects. See process fit
  interconnects, 113
Horsweradish peroxidase, 7

Hot embossing techniques, 225
Hot Embossing, 96-97, 105–106, 225–226
Human serum albumin (HSA), 31
Hydrodynamic off-axis effect, 43, 41
Hydrogel sensors. See Microfluidic tectonics, sensors
  in
Hydrogel valves, 228
Hydrogels, 119-121, 231, 234, 238
  classification of, 119
  water uptake, importance of, 121
Hydrogen peroxide sensing, 79 –80
Hydrogen peroxide, 80
Hydrophilic polymeric networks. See Hydrogels
Hydrophobic surface patterning, 229

IC microchip-CE system, advantages of, 18
IGF-IR protein, 69
ILS250TLL, 259
iMEDD device, 304
Immobilized ligand, 7
Immunoassay-based sensors, 243
Immunoassays, types of, 44
Immunosensors, 7
Impedance techniques
  extracellular signal strengths, 59
  reliability of, 58–59
  use of, 59
Implantable wireless microsystems
  battery operated, 208
  coding scheme, 208
  data transmission, 208
  degree of protection, 209
  design considerations, 207
  factors influencing choice of power source,
    208–209
  hybrid approach designing, 207
  sealing techniques, 209–210
  transmitter efficiency, 209
  wireless communication, 207–208
Infrared spectroscopy, 7, 10
Inhibition assay, 44
Inhibition immunoassay, 50
Injection Molding, 106
Injection molding. See Micromolding
In-plane Brownian diffusion, 338, 345
  errors due to, 345
Integrated artificial pore on chip, 35
Integrated biosensor arrays, 14
Intelligent polymer networks
  advantages of, 119
  applications of, 126-129
  different polymer systems, 119
  major classes of, 119f
Intercellular adhesion molecule(ICAM)-1, 374
Interdigitated electrodes. See n-DEP trap geometries

Interspike interval histogram (ISIH), 75
Intracellular based bio sensors
  advantages, 60
  disadvantages, 60
Intracellular calcium release, 388
Intracellular ionic signals developed by fluorescence based biosensors, 58
Intraocular pressure (IOP) measurement microsystem, 210
Ion sensitive field-effect transistors (ISFET), 198
Isoelectric focusing, 155
Isopropyl alcohol (IPA), 99

Knock down effect, 77
Knudesn number, 302
Kovar, 111–112

L929 fibroblast cells, 174
L-phenylalanine via NADH, indirect detection of, 13
Labeled analyte, 7
Lab-on-a-chip, 187, 193, 228, 243–244, 301, 304
  use of, 187
Laminar shear stress effects
  application to clinical treatment, 391
  on endothelial cells, 383-385
  on vascular smooth muscle cells, 386–388
Langmuir-Blodgett layers, 11
Laser doppler velocimetry, 332, 352
  application of doppler effect, 353
Laser machining, 96, 106
Laser miniaturization, 5
Leaching, 113
Leviation measurements, 181
LIGA, 105
Ligand molecule, 7
Light emitting diodes (LEDs), 17
Light-addressable potentiometric sensors (LAPS), 198
  principle of, 198–199
Limit of detection (LOD), 18
Linear PSD, 24
Lipopeptide, 10
Lipophilic carboxylated polyvinyl chloride, 12
Liposome sensors. See Microfluidic tectonics, sensors in
Liquid adhesive, 110
Liquid phase photopolymerization, 235
Liquid phase photopolymerization. See Microfluidic tectonics
Listeria innocua cells, 190
Listeria monocytogenes, detection of, 13
Listeria, 13, 66
Lithography
  of thick resists, 98
  steps in, 98
  SU-8 on PMMA techniques, 98–99
Lithography, 97
Living cells, use of, as sensing elements, 55
Lower critical miscibility or solution temperature (LCST), 121
Low-pressure chemical vapor deposited (LPCVD), 25
LU factorization, 287
Lucite. See PMMA, 98

Mammalian cell death, 172
Mammalian cells, 168
Mammalian cells, use of, 56
Maskless fabrication method, 16
Mass detection methods, 11
Mass-sensitive Techniques, 13–14
Matlab script, 26
MatLab, 161, 343
Maxwell-Wagner interfacial polarization, 165
Medical diagnostics, 118
MEMS based biosensors
  advantages of, 207
  disadvantages of, 207
MEMS, 224, 233, 302, 332, 348, 358, 360, 365
MEMS-technology, 97
Metabolism cell stress, 57
METGLAS, 141
MFB biochip device, 17
MHD electrode pairs, 142–143
MHD microfluidic pumps, 138–139
MHD micropump for sample transport using microchannel parallel electrodes
  AC MHD micropump, measurements in, 141
  electromagnet field strength, 140–141
  fabrication of silicon MHD microfluidic pumps, 138–139
  Lorentz force acting on the pump, 138, 144
  measurement setup and results, 139–140
  MHD microfluidic switch, 142–144
  other MHD micropumps and future work, 144–145
  principle of operation, 136–138
Micro Coulter particle counter principle, 152
Micro electromechanical systems (MEMS), 93
Micro IEF technique, 156
Micro particle image velocimetry, 261–262
Micro sensor development
  micro sensor for flow shear stress measurement, 358–363
  micro velosimeter, 356–358
Micro sensor for flow shear stress measurement
  diverging fringe doppler sensor, 360–363
  dual velocity sensor, 358–360
Micro total analysis systems (μTAS), 135, 187
  integrated fluids for, 135–136
Microarray of electrochemical biosensors
  for the detection of glucose on line, 14
  for the detection of lactate on line, 14

INDEX

Microcantilever, 21
Microcantilever, use of, 29
Microcantilevers, 21
Microchannel parallel electrodes for sensing biological fluids
    MHD based flow sensing, 145–153
    MHD based viscosity meter, 146
    impedance sensors with microchannel parallel electrodes, 146–153
Microchannel parallel electrodes, impedance sensors with
    electrical double layer, 146–147
    fabrication of channel electrodes and microfluidic channel, 147–148
    flow sensing, 148–149
    measurement of particles in solution, 151–152
    measurement of solution properties, 150–151
    particle cytometry using capacitance and impedance measurements, 152–153
Micro-elctro mechanical systems, 224–225
Microelectrode array technology, 62
Microelectromechanical systems (MEMS), 28, 128
Microfabrication, utilization of, 44
Microfluidic chambers, integration with cantilevers, advantages of, 28
Microfluidic devices, 156, 224
Microfluidic devices, polymeric fabrication methods
    components used, 228
    design considerations, 227–228
    laser ablation of polymer surface method, 226
    liquid phase photopolymerization. *See* Microfluidic tectonics
    soft lithography, 225–226
    system integration, 223–224
Microfluidic devices, traditional fabricating methods
    micromachining, 224
    micromolding, 224–225
Microfluidic electrostatic DNA extractor
    demonstration, 155
    design and experiment, 154–155
    principle, 154–156
Microfluidic technology, 93
Microfluidic tectonics, 226, 227–229
    advantages of, 229
Microfluidic tectonics, components used in
    filters, 230–231
    mixers, 232–233
    pumps, 229–230
    valves, 228–229
Microfluidic tectonics, sensors in
    cell gel sensors, 236
    E-gel, 237–238
    hydrogels, 235
    liposome sensors, 237

    sensors that change color, 235–236
    sensors that change shape, 234–235
Microfluidic transverse isoelectric focusing device, 156
Microfluidics system, 17
Microfluidics, 27–28
Microfluidics, diagnostic techniques in
    challenges associated with, 349
    hot-wire anemometry, 332
    laser-doppler velocimetry, 332
    micro-particle image velocimetry, 332–334
    molecular tagging velocimetry, 333
    nano-particle image velocimetry, 334
    particle-image velocimetry, 332
Microgripper. *See* polymeric microgripper
Micromachining, 225
Micro-molding techniques, utilization of, 44
Micromolding, 225–226
Micro-PIV, 100
MicroS3^{TM}, 364
Microscale flow and transport simulations, fundamentals of
    applied electric field, 279
    electrical double layer (EDL), 280–282
    inlet/outlet boundary conditions, 247, 249, 279
    microscale flow analysis, 278–280
    microtransport analysis, 286–287
    numeric modeling of, 287, 294
Microscale impedance-based detection systems, 199
Microsystem components
    interface components, 207
    packaging and encapsulation, 209–210
    power source, 208–209
    transducers, 208–209
    wireless communication, 219, 213
MicroV^{TM}, 331
Mini LDV^{TM}. *See* Miniature laser doppler velocimeter
Miniature laser doppler velocimeter
    features of, 351
    principle of measurement, 353-355
    siganal processing, 355–358
Miniature laser doppler velocimeter, 352
Miniaturized total analysis systems (μ-TAS), 118
Modeling of AC electroosmotic flow in a rectangular channel, case study of, 290–293
Molecular adsorption, 21
Molecular beacon (MB) detection, 18
Molecular dynamics simulations in electroosmotic flow, 326–328
    fluid dynamics, 326–328
    fluid statics, 325–326
Molecular light emission, 58
Molecular sensing, 35
Monitoring of phosphorylcholine, 11
Monocyte-derived macrophages, 374

Monolayers, activation of
 1-ethyl-3-[3-(dimethylamino)propyl] carbodiimide hydrochloride, 14
 N-hydroxysulfosuccinimide, 14
Monolithic horizontal integration, 109
Monte Carlo simulations, 341
Morphine binding, 11
Mouse anti-human antibody (MAH-PSA), 31
Multi-chip module (MCM) concept, 109
Multifunctional biochip (MFB), development of, 16
Multiplexing, 27
Multi-shelled particulate models, 165

N,N′-Cystamine-bisacrylamide, 234
Nanochannel systems, 304
Nano-particle image velocimetry
 Brownian diffusion effects in, 338–342
 generation of evanescent waves, 337–338
Nanopores for molecular sensing, 35
Natural receptors, limitations of, 124
Navier-Stokes equations, 301, 312, 327
n-DEP trap geometries
 interdigitated electrodes, 180–181
 octopole electrodes, 178
 other electrode structures, 178
 quadrupole electrodes, 176–177
 strip electrodes, 178
n-DEP trapping, 247–248
Negative temperature-sensitive systems, 121
Negatively-charged (carboxyl-coated) latex colloids, 37
Nernst potential, 319
Neuronal networks for biosensor applications, 61
Non-covalent recognition process, 125
Non-enzymatic protein biorecognition, 10
Non-equilibrium molecular dynamics (NEMD), 324
Non-spherical cells, 167
Nucleic acids, use in array chips, 15

Octopole electrodes. See n-DEP trap geometries
Oligo amide, 9
On-chip artificial pore for molecular sensing
 applications, 44–51
 fabrication of the pore, 36–37
 features of the chip device, 37
 pore-measurement, 37–40
 PDMS-based pore, 40–44
Optical biosensors, properties of,
 amplitude, 12
 decay time and/or phase, 11
 energy, 12
 polarization, 12
Optical detection methods, 12
Optical MEMS sensor feasibility study, 365–369

Optical transduction, 11
O-rings, 111
Oroglass. See PMMA, 96
Oscillating ferromagnetic micropump, 230
Oscillatory flow patters, 385
Osteoblast cells, 77
Out-of-plane Brownian diffusion, 339

Packaging concepts for biomedical micro devices
 monolithic horizontal integration, 109
 multi-chip module (MCM), concept, 109
 stacked modular system, 109
Parallel microchannel electrodes for sample preparation
 channel electrodes for isoelectric focusing with field flow fractionation, 155–156
 microfluidic electrostatic DNA extractor, 153–155
Parallel plate flow chamber, 378–382
Passive mixers, 232
Passive valves, 229
Patch clamp technique, drawback of, 59–60
PCR amplification, 15
p-DEP trap geometries. See Trap geometries
p-DEP trapping, 163
PDMS
 application of, 82
 properties of, 98–99
PDMS, 195, 209, 225
PDMS-based pore, 40–44
 hydrodynamic off-axis, effect of, 41
 PDMS mold, 40
 PDMS slab, 40
 PDMS sealing, 40
 pressure usage in, 44
PDMS-based pore, 51
PECVD silicon dioxide, 259
Peptide nucleic acid. See oligo amide
Perspex. See PMMA, 96
PGDF-β mRNA, 388
PGMEA developer
Phase-resolved fiberoptic fluoroimmunosensor (PR-FIS), use of, 12
Phosphate buffered saline (PBS), 302, 306,
Phosphorylcholine analog, 11
Photodiodes, role of, 17
Photo-induced electron transfer (PET), 58
Photometrics CoolSNAP HQ interline transfer monochrome camera, 261
Photo-polymerization, 227
Physiocontrol-microsystem, 198
Picornavirus, 10
Piezoelectric bimorph disc. See pump actuator, 101
Piezoelectric crystals, use of, 13
Piezo-electric transducer, 196

INDEX

Piezoresistance cantilever array, 27
PIV velocity measurements, 261–262
Pixel binning, 261
Planar microelectrode arrays, 61
Planar quadrupoles, 170, 174, 178, 180
Plane Couette flow, 170
Platelet derived growth factor (PGDF), 374
PL-defined pore, 38
Plexiglas. *See* PMMA, 94
Pluronics F127 surfactant, 45
PMMA
  structure of, 98
  surface properties of, 96
PMMA, 94
PNIPAAm, 122
Poisson-Bolzmann theory, 328
Poly (bis-GMA), 228
Poly (iso-bornyl acrylate), 228
Poly dimethylsiloxane. *See* PDMS
Polycyclic aromatic hydrocarbons (PAHs), 7
Polydimethylsiloxane (PDMS), 148
Polyetheretherketone (PEEK), 111
Polymer complexes, 121
Polymer matrix, 13
Polymer systems, types of
  biohybrid hydrogels, 123–124
  biomolecular imprinted polymers, 124–125
  environmentally responsible hydrogels, 124
  hydrogels, 123–124
  pH-responsive hydrogels, 128–129
  star polymer hydrogels, 125–126
  temperature sensitive hydrogels, 121–122
Polymerase chain reaction (PCR), 193
Polymeric microgripper, 104
Polymeric micromachining technologies
  laser machining, 106–108
  lithography, 98–100
  polymeric surface micro machinery, 101, 103
  replication of technologies, 104–105
Polymeric micropump, 100
Polymeric surface micro machining techniques
  polymeric micro gripper, 101–104
  polymeric micro pump, 101-103
Polymyxin B, 10
Polystyrene latex (PSL) microspheres, 261
Pore-measurement of on chip device, 37
  prediction of resistance changes, 37
Position sensitive detector (PSD), 24
Positive temperature-sensitive systems, 121
Pre-fabricated double-sided adhesive, 110
Press-fit interconnects, 111
Primary rat osteoblast, 69
Printed circuit board (PCB), 109
Prism method, 337–338
Prismless method, 337–338

Propylene glycol methyl ether (PGMEA), 103
Prostacy (PCI$_2$), 373
Prostate-specific antigen (PSA), 31
Protein engineering techniques, use of , 11
Protein microarrays, disadvantages of, 22
Protein-protein bindings, 30
Protoplasts, 166
PSD signal, 24
Pseudomonas fluorescens HK44, 57
Pump actuator, 101
Pyrethroid sensing, 80
Pyrethroids, 77

Quadrupole electrodes. *See* n-DEP trap geometries
Quartz crystal microbalance biosensor, 13

Radioimmunoassay (RIA), 5, 44
Rapid prototyping techniques, 277
Reactive ion etching (RIE), 36
Read out techniques
  optical beam deflection of 1D cantilever array, 24–25
  optical beam deflection of 2D cantilever array, 24–25
Readout techniques
  optical beam deflection of ID cantilever array, 24–25
  optical beam deflection of 2D array, 25–27
  piezoresistance cantilever array, 27
Recycling of catechol, 13
Redox hydrogel, 7
Rehabilitative Microsystems, 216–219
Renal arteries, physiology of, 372
Replication technologies
  advantages of, 110
  drawbacks of, 105
  hot embossing, 106
  injection molding, 106
  soft lithography, 225–226

Sacrificial layer lithography method, 304
Sandwich assay, 44, 45, 49
Scaling in DEP-based systems, 168
Scatterers, use of, 350
Second version of the device, 40
Self-assembled monolayer (SAM), 196
Sensitivity of device, factors affecting, 149
Sensor applications of intelligent polymer networks, 126–129
Sepis, 10
Shear stress effects on cell cultures, techniques to measure,
  cone and plate viscometer, 382
  parallel plate flow chamber, 382–383
Shear stress receptors, 388–389

Signals from microelectrodes
  sealing of cells, 70–71, 73
  signal-to-noise ratios (SNR), 70–71
Signature pattern, 55, 62, 89
Silicon micromachining. *See* Micromachining
Silicon technology, 93, 95
Single channel neuromuscular microstimulator.
  See Rehabilitative Microsystems
Single linear PSD, 24
Single-stranded DNA (ssDNA), 29–30
Small I/O, 27
Smart pill. *See* Therapeutic Microsystems
Smooth muscle cells, 371
Soft bake. *See* SU-8, process, steps in,
Soft lithography, 225–226
Soft-operator IFT algorithm, 351
Spectrochemical instrumentation, 5
Spin coating. *See* SU-8, process, steps in
SPV (Signature Pattern Vector), 76
Stacked modular system, 11
Star polymer hydrogels, 125–126
Stokes flow, 170
Stokes' drag, 170–172, 181–182
Stoney's formula, 23
Streaming current, 148
Streptavidin monolayer, 11
Streptavidin-coated latex colloids, 46
Strip electrodes. *See* n-DEP trap geometries
SU-8 photoresist, components of, 97
SU-8 process, 98, 101
SU-8, 116
SU-8, fabrication of, 105
SU-8, process, steps in, 99
SU-8, use of, 102
Subtler effects, 175
Surface acoustic wave biosensor, inductively coupling of, 14
Surface enhanced Raman scattering (SERS), 9
Surface micromachining, 94
Surface plasmon resonace measurement, 7, 10
Surface plasmon resonance (SPR), 9, 22, 44
Surface roughness effects, 59
Surfactant Tween 20, 37
SV40 T antigen, 69
Swelling, 113
Symptoms of inflammation, 113
Synthetic image method. *See* Velocity profile extraction process, methods for
Synthetic thioalkane chelator, 7

Therapeutic Microsystems, 213–216
Thermal bimorph effect 26
Thermal direct bonding, 109
Thermal expansion coefficients (TEC)
  of PMMA, 100
  of SU-8, 100
Thermoplastic polymers, 96
Thin-film batteries, use of, 209
Total internal reflection fluorescence, 11
Total internal reflection microscopy (TIRM), 336
Total internal reflection, application of, 334
Toxic heavy metal effects, methods to evaluate,
  bioluminescence, 9
  cell metabolism, 9
  cell respiration, 9
Toxins, different types of,
  ricin, 15
  staphylococcal enterotoxin B (SEB), 15
  yersinia pestis, 15
Tranducers, definition of, 206
Transducer classification, techniques of
  electrochemical, 7
  mass-sensitive measurements, 5
  optical measurements, 5
Transducer, 3
Transducers, forms of
  electrochemical detection methods, 11–12
  mass sensitive detection methods, 13
  optical detection methods, 11–12
Transduction pathways,
  IKK-NF.k B pathway, 391
  Ras-MAPK pathways, 389–391
Transduction pathways, 389
Trap design for cells, factors influencing
  current-induced heating, 172
  direct electric-field interactions, 173–175
Trap design
  conditions for stability, 160
  considerations for, 182–183
  finite-element modeling, 161
  trapping point, 161
  with cell manipulation, 172–175
Trap geometries
  comparison of p-DEP and n-DEP approaches, 176t
  n-DEP trap geometries, 175–178
  p-DEP trap geometries, 179
Trap parameters, quantification of, 180–183

Ultrasonic bath treatment
Universal M-300 Laser Platform of Universal Laser Systems Inc., 106
Unlabeled antibody-antigen pairs, 44
Urease, 12
UV-lasers, use of, 106

Vascular cell adhesion molecule(VCAM)-1, 374
Velocity measurement in polymer micro channels
  using micro velocitymetry, 365
  using mini LDV, 361–363

Velocity profile extraction process, methods for
    comparison of two methods, 264–265
    deconvolution method, 262–263
    synthetic image method, 263–264
Vertical cavity surface-emitting laser (VCSEL), 24
Vertical tubing interconnects. *See* process fit
    interconnects, 111
Video pill. *See* Endoscopic wireless camera-pill
Virtual walls. *See* Compartmentalization
Virus, 167

Visual prosthetic device. *See* Rehabilitative
    Microsystems

Wheatstone bridge, 27
WT (Wavelet Transformation), 76
WT analysis, 79, 80, 82, 85

X-ray lithography, 99

Yeast, 166

Abbreviated Table of Contents

List of Contributors xv
Foreword xix
Preface xxi

I. **Micro and Nanoscale Biosensors and Materials** 1

1. **Biosensors and Biochips** 3
   Tuan Vo-Dinh

2. **Cantilever Arrays** 21
   Min Yue, Arun Majumdar, and Thomas Thundat

3. **An On-Chip Artificial Pore for Molecular Sensing** 35
   O. A. Saleh and L. L. Sohn

4. **Cell Based Sensing Technologies** 55
   Cengiz S. Ozkan, Mihri Ozkan, Mo Yang, Xuan Zhang, Shalini Prasad, and Andre Morgan

5. **Fabrication Issues of Biomedical Micro Devices** 93
   Nam-Trung Nguyen

6. **Intelligent Polymeric Networks in Biomolecular Sensing** 117
   Nicholas A. Peppas and J. Zachary Hilt

II. **Processing and Integrated Systems** 133

7. **A Multi-Functional Micro Total Analysis System (μTAS) Platform** ... 135
   Abraham P. Lee, John Collins, and Asuncion V. Lemoff

8. **Dielectrophoretic Traps for Cell Manipulation** 159
   Joel Voldman

9. **BioMEMS for Cellular Manipulation and Analysis** 187
   Haibo Li, Rafael Gómez-Sjöberg, and Rashid Bashir

10. **Implantable Wireless Microsystems** 205
    Babak Ziaie

11. **Microfluidic Tectonics** 223
    J. Aura Gimm and David J. Beebe

12. **AC Electrokinetic Stirring and Focusing of Nanoparticles** 243
    Marin Sigurdson, Dong-Eui Chang, Idan Tuval, Igor Mezic, and Carl Meinhart

III. Micro-fluidics and Characterization 257

13. **Particle Dynamics in a Dielectrophoretic Microdevice** 259
    S.T. Wereley and I. Whitacre

14. **Microscale Flow and Transport Simulation for Electrokinetic
    and Lab-on-Chip Applications** 277
    David Erickson and Dongqing Li

15. **Modeling Electroosmotic Flow in Nanochannels** 301
    A. T. Conlisk and Sherwin Singer

16. **Nano-Particle Image Velocimetry** 331
    Minami Yoda

17. **Optical MEMS-Based Sensor Development with Applications
    to Microfluidics** 349
    D. Fourguette, E. Arik, and D. Wilson

18. **Vascular Cell Responses to Fluid Shear Stress** 371
    Jennifer A. McCann, Thomas J. Webster, and Karen M. Haberstroh

About the Editors 395
Index 397

Printed in Singapore